PHYLOGEOGRAPHY OF SOUTHERN EUROPEAN REFUGIA

Phylogeography of Southern European Refugia

Evolutionary perspectives on the origins and conservation of European biodiversity

Edited by

Steven Weiss
Karl-Franzens Universität Graz, Austria

and

Nuno Ferrand
CIBIO, University of Porto, Portugal

Springer

A C.I.P. Catalogue record for this book is available from the Library of Congress.

ISBN-10 1-4020-4903-X (HB)
ISBN-13 978-1-4020-4903-3 (HB)
ISBN-10 1-4020-4904-8 (e-book)
ISBN-13 978-1-4020-4904-0 (e-book)

Published by Springer,
P.O. Box 17, 3300 AA Dordrecht, The Netherlands.

www.springer.com

Graphical assistance was provided by Gabi Kriks, and the layout was done by Gabi Kriks and Estella Weiss-Krejci. Financial support was provided by CIBIO, Portugal and Karl-Franzens University of Graz, Austria

Cover photos:
From left to right, photos 1-4 (Quercus leaf, Green frog, Wall lizard and Viper) are from Jorge Coutinho, and photo 5 (European rabbit) is from Paulo Magalhães

Printed on acid-free paper

Contents

Contributors

Agúndez, Dolores Department of Forest Systems and Resources, CIFOR-INIA, Carretera de la Coruña km 7.5, 28040 Madrid, Spain, agundez@inia.es

Alexandrino, João Departamento de Zoologia, Instituto de Biociências, Universidade Estadual Paulista, 13506-900 Rio Claro, São Paulo, Brazil, jalex@rc.unesp.br

Alía, Ricardo Department of Forest Systems and Resources, CIFOR-INIA, Carretera de la Coruña km 7.5, 28040 Madrid, Spain, alia@inia.es

Arntzen, Jan W National Museum of Natural History, P. O. Box 9517, 2300 RA Leiden, The Netherlands, arntzen@nnm.nl

Atkinson, Rachel J National Parks and Conservation Service, Ministry of Agro-Industry and Fisheries, Réduit, Mauritius

Avise, John C Department of Ecology and Evolutionary Biology, University of California, Irvine, CA 92697, USA, javise@uci.edu

Berrebi, Patrick Ecosystemes Lagunaires, UMR 5119, Equipe EFEP, Université Montpellier 2 Place E. Bataillon, 34095 Montpellier Cedex, France, berrebi@univ-montp2.fr

Branco Madalena CIBIO, Centro de Investigação e Biodiversidade e Recursos Genéticos, ICETA, Campus Agrário de Vairão, Rua do Monte, Crasto, Vairão, 4480 Vila do Conde, Portugal, msbranco@fc.up.pt

Campos, Rita CIBIO, Centro de Investigação e Biodiversidade e Recursos Genéticos, ICETA, Campus Agrário de Vairão, Rua do Monte, Crasto, Vairão, 4480 Vila do Conde, Portugal, ritacampos@fc.up.pt

Carrión, José S Department of Plant Biology, Faculty of Biology, University of Murcia, 30100 Murcia, Spain, carrion@um.es

Casula, Paolo Dipartimento di Biologia Sperimentale, Sezione di Genetica, Università di Cagliari, Viale Poetto 1, 09126 Cagliari, Italy

Crnobrnja-Isailovic, Jelka Department of Evolutionary Biology, Institute for Biological Research, Belgrade 11060, Serbia, jelka@ibiss.bg.ac.yu

Ferrand, Nuno CIBIO, Centro de Investigação e Biodiversidade e Recursos Genéticos, ICETA, Campus Agrário de Vairão, Rua do Monte, Crasto, Vairão, 4480 Vila do Conde, Portugal, nferrand@mail.icav.up.pt

Gil, Luis Unit of Anatomy, Physiology and Genetics, ETSIM, Ciudad Universitaria s/n., 28040 Madrid, Spain

Gómez, Africa Department of Biological Sciences, University of Hull, Hull HU6 7RX, UK, a.gomez@hull.ac.uk

Gómez, Aranzazu Biosafety Unit, IMIDRA, 'El Encín' A-2 km 38.200, 28800 Alcalá de Henares (Madrid), Spain

González-Martínez, Santiago C. Department of Forest Systems and Resources, CIFOR-INIA, Carretera de la Coruña km 7.5, 28040 Madrid, Spain, santiago@inia.es

Grill, Andrea Institute for Biodiversity and Ecosystem Dynamics, Faculty of Science, University of Amsterdam, P. O. Box 94766, 1090 GT Amsterdam, The Netherlands, grill@science.uva.nl

Kotlík, Petr Department of Vertebrate Evolutionary Biology and Genetics, Institute of Animal Physiology and Genetics, Academy of Sciences of the Czech Republic, 277 21 Libichov, Czech Republic, kotlik@iapg.cas.cz

Lecis, Roberta Institut de Recerca i Tecnologia Agroalimentàries (IRTA), Genètica de la Conservació, Centre de Cabrils, Carretera de Cabrils s/n, 08348 Cabrils (Barcelona), Spain, roberta.lecis@irta.es

Lunt, David H Department of Biological Sciences, University of Hull, Hull HU6 7RX, UK, d.h.lunt@hull.ac.uk

Menken, Steph Institute for Biodiversity and Ecosystem Dynamics, Faculty of Science, University of Amsterdam, P. O. Box 94766, 1090 GT Amsterdam, The Netherlands, menken@science.uva.nl

Petit, Rémy J Directeur de Recherche, INRA, UMR Biodiversité, Gènes & Ecosystèmes, Equipe de Génétique, 69 Route d'Arcachon, 33612 Cestas Cedex, France, remy@pierroton.inra.fr

Randi, Ettore Istituto Nazionale per la Fauna Selvatica (INFS), Via cá Fornacetta 9, 40064 Ozzano dell' Emilia (BO), Italy, met0217@iperbole.bo.it

Rokas, Antonis The Broad Institute of MIT & Harvard, 320 Charles Street, Cambridge, MA 02141, USA, arokas@mit.edu

Stone, Graham N Institute of Evolutionary Biology, Edinburgh University, The Kings Buildings, West Mains Road, Edinburgh EH9 3JT, UK, graham.stone@ed.ac.uk

Teixeira, José CIBIO, Centro de Investigação e Biodiversidade e Recursos Genéticos, ICETA, Campus Agrário de Vairão, Rua do Monte, Crasto, Vairão, 4480 Vila do Conde, Portugal, jteixeira@mail.icav.up.pt

Vargas, Pablo Real Jardín Botánico de Madrid (CSIC), Plaza de Murillo 2, 28014 Madrid, Spain, vargas@ma-rjb.csic.es

Vendramin, Giovanni G Consiglio Nazionale delle Ricerche, Istituto di Genetica Vegetale, Via Madonna del Piano 10, 50019 Sesto Fiorentino, Italy, giovanni.vendramin@igv.cnr.it

Weiss, Steven Institute of Zoology, Karl-Franzens University of Graz, 8010 Graz, Austria, steven.weiss@uni-graz.at

Introduction

Nuno Ferrand* and Steven Weiss**
E-mail: *nferrand@mail.icav.up.pt; **steven.weiss@uni-graz.at

This volume contains 14 contributions, many of which have been developed from presentations made at the first international symposium devoted exclusively to phylogeography, held in Vairão, Portugal, in March 2002, at what has now become the Research Center in Biodiversity and Genetic Resources (CIBIO). Approximately 150 participants from over 20 countries attended, all sharing their enthusiasm for the growing number of phylogeographic oriented studies in Europe with a particular emphasis on the circum-Mediterranean region, nominated as one of the world's 25 biodiversity hotspots. The conservation relevance of the Mediterranean basin is exemplified by the fact that it holds approximately 20% of the world's floristic diversity (see Chapter 10).

The symposium witnessed over 100 oral and poster communications including the plenary talks of John Avise (The history and development of phylogeography); Ettore Randi (Mammalian phylogeography of South Europe); Rémy Petit (Phylogeography of temperate trees and shrubs in Europe and the importance of southern refugia); and Pierre Taberlet (Comparative phylogeography of Alpine plants). Taxonomic coverage of the presentations was broad including mammals (26), plants (20), invertebrates (17), amphibians and reptiles (17), fish (15), and birds (8).

Section one opens with a chapter by John Avise, outlining 25 evolutionary insights that have arisen from what he coins a phylogeographic revolution. Key to this perspective is the notion that the multidisciplinary field of phylogeography is serving as an epistemological bridge between the formerly distinct fields of population genetics and phylogenetics. More succinctly, phylogeography has reoriented and extended the field of population genetics incorporating the gene-tree perspective into assessment of intraspecific variation and historical demography. Many of these insights stem from his early work screening variation in the mtDNA molecule and its unique properties that have served as the working horse for the development of this nascent discipline. The next chapter, authored by another invited guest Rémy Petit

S. Weiss and N. Ferrand (eds.), Phylogeography of Southern European Refugia, 1–4.

and his colleague Giovanni Vendramin is a studious and thoroughly comprehensive review of the history and development of plant organelle genetics (both mtDNA and cpDNA) and their application in population studies. Insights are provided on the structure, levels of variation, modes of inheritance, vegetative segregation, and well documented recombination of organelle genomes in plants with particular emphasis on the unique opportunities that these characteristics provide evolutionary research.

Section two contains two chapters characterized by their broad-scale coverage. Chapter 3, also authored by an invited guest, Ettore Randi, reviews the phylogeography of mammals in southern Europe in the context of our knowledge on the paleoecological conditions that have shaped current patterns of lineage distribution. Emphasis is given on the taxon-specific patterns and growing appreciation for the complexity of glacial refugia beyond the simple three Peninsula model, while highlighting the implications for taxonomy and conservation. And finally, Chapter 4 presents a comparative study assessing species richness and genetic diversity in a coevolutionary system of oaks and their obligate parasitic gallwasps. The study is additionally unique in that it assesses and compares diversity along a longitudinal rather than latitudinal axis crossing four major glacial refugia, extending over from Iberia to Asia Minor. While requiring more organisms to test, the authors postulate that the lower species richness of the Iberian Peninsula compared to other major refugia is a result of stronger demographic fluctuations stemming from a more arid climate.

The next section (Chapters 5-9) reports on a series of review perspectives and case studies on the Iberian Peninsula, the best studied refugial region in Europe. The section begins with a comparative phylogeographic review of the Iberian Peninsula outlining not only the concordant patterns but moreover an emerging and important concept of 'refugia-within-refugia'. The message is clear and impressive in its taxonomic coverage as multiple refugia are verified for species or species complexes in plants (8), mammals (6), reptiles and amphibians (6), fish (7), and invertebrates (5). This perspective underlines the emerging research focus on complexity and dynamics of contact zones, hybridization, introgression and population diversification of Iberian biota but also warns against making potentially misleading conclusions concerning so-called northern (postglacial) refugia, before we understand the cryptic and underappreciated lineage diversity existing in southern Europe. Chapter 6 is noted for its integration of ecological, phenotypic and phylogeographic data in characterizing the historical biogeography of an endemic Iberian salamander, *Chioglossa lusitanica*, a system that is demonstrating its amenability to the current development of landscape genetics and GIS-based approaches to evolutionary and conservation oriented research. The study provides a framework for comparative phylogeographic research that can be used to designate key areas for multi-species conservation. Chapter 7 unveils a data set on protein polymorphisms (20 loci) collected over a 10-year period that provides the basis of a long-term comprehensive research program on the European rabbit, *Oryctolagus cuniculus*. These data serve as a multi-locus baseline for the study of introgression, hybridization, and selection within the explicitly defined phylogeographic context of two divergent lineages and a temporally dynamic contact zone, existing within a Peninsula refuge. Chapter 8, reports on hemoglobin polymorphisms (HBA and HBB) across the contact zone of the European rabbit in Iberia. The starkly contrasting spatial patterns of the HBA six allele system suggest that strong selective forces are operative over a large spatial scale and further reveal the presence of a hybrid allele, two observations that underlie the value of studying contact zones in a refuge that has persisted throughout the Pleistocene. Chapter 9 presents a case study of a commercially important species, the maritime pine *Pinus pinaster*, investigating spatial structure and the effect of specific landscape features on gene flow using coalescent theory. Inferences are drawn on the role of mountain ranges within refugia in serving to both allow altitudinal migration and to isolate specific populations. The biological inferences also provide the basis for specific conservation and management recommendations.

Section four (Chapters 10-13) includes studies that survey organisms or review phylogeographic patterns in non-Iberian refugia. Chapter 10 is the first multi-taxa review of the remarkable endemic floral and faunal diversity of Sardinia. The dating of the fundamental biogeographical phenomena associated with the island's formation, compared with genetic divergences suggests that the present diversity has arisen subsequent to the marine transgressions five million years ago. Characterization of the state of the island's present system of nature preserves, emphasizes the threats to these unique biota. Among the endemics covered are 14 species of butterflies, nine plants, six cave beetles, four salamanders, two lizards, two frogs, and one mammal. Chapter 11 takes an interesting phylogenetic approach to characterizing the rich floral diversity of the five volcanic archipelagos (Azores, Madeira, Selvagens, Cape Verde and Canaries) that comprise Macaronesia. Boasting some 3100 plant species, the role of Macaronesia as a lineage refuge is assessed by applying a comparative phylogenetic analysis. The analysis distinguishes various classes of relictualism and further proposes that endozoochory has played a major role in promoting multiple colonizations of Macaronesia, aiding its status as a floral refuge. Chapter 12 employs a nested clade phylogeographic analysis (NCPA) on an mtDNA data set of a common cyprinid fish, *Barbus barbus*. A number of the historical biogeographic inferences drawn *ad hoc* from a previous study were supported by the NCPA, but several others were not, exemplifying the conservative, speculation hindering tendency of the analysis. Chapter 13 reviews the diversity, levels of endemism and available genetic data on reptiles and amphibians of the Balkan Peninsula, clearly the least studied of Europe's three major refugial Peninsulas. Emphasis is placed on the region's historical complexity and how it serves as a mountainous cross-road for range expansions and contractions during postglacial and glacial episodes.

The book closes with an up-to-date review on the current perspectives, caveats and prospects for phylogeography as a discipline, with particular emphasis on its importance in both understanding and conserving European biodiversity. Current controversies and criticisms concerning phylogeographic data analysis and inference are discussed with an optimistic view of new methodologies in development, recognizing that the stochastic variance of the coalescent process must be more seriously taken into consideration. As in any scientific endeavor, theory and methodology will only become broadly accepted through the rigors of repeated observation and hypothesis testing within legitimate statistical frameworks. Nonetheless, as phylogeography deals with the uncertainty of history combined with the tremendous complexity of evolutionary pathways, a plea is also made for maintaining an open-minded and pluralistic approach to study design and data analysis.

Part I

Historical foundations and perspectives

Chapter 1

Twenty-five key evolutionary insights from the phylogeographic revolution in population genetics

John C. Avise
E-mail: javise@uci.edu

Abstract
An overview is provided of 25 novel perspectives that the field of phylogeography has brought to scientific studies of population genetics and speciation. A unifying theme is that microevolution can be described as an extended genealogical process played out in space and time, and reflecting the oft-idiosyncratic biological and environmental factors that have impinged on historical population demography. Most of the empirical and conceptual methods of phylogeography depart considerably from conventional equilibrium approaches, and they are helping to reorient and extend traditional population genetics in realistic directions that emphasize historical demography and genealogy.

Keywords: phylogeography, genealogy, gene trees, demography, speciation

Introduction

Phylogeography is a relatively young discipline concerned with the principles and processes governing the geographic distributions of gene lineages, especially within and among closely related species (Figure 1). The phylogeographic revolution, inspired by mitochondrial (mt) DNA analyses that were introduced nearly three decades ago, has transformed the study of population genetics and speciation in several ways. In particular, this ongoing reformation has drawn closer empirical and conceptual connections between microevolutionary genetics and phylogenetic biology.

Here I substantiate these claims by compiling more than two dozen salient insights about microevolution that seldom (sometimes never) were an explicit part of the fabric of population genetics in the pre-phylogeography era. The entries in this list appear in a sequence generally consistent with the underlying train of logic, rather than necessarily in order of importance or date of development. Many concepts in the list are nested or partially overlapping, yet each qualifies for inclusion by virtue of having been quite unorthodox when

S. Weiss and N. Ferrand (eds.), Phylogeography of Southern European Refugia, 7–21.

introduced. More importantly, most of these insights from the field of phylogeography will (I suspect) prove to be enduring truths about microevolutionary processes in nature.

This list provides a brief overview and historical backdrop, and is intended to encapsulate phylogeography's principal contributions. Thus, for each of the 25 entries, I have cited just one key reference that either was seminal in the history of ideas or is particularly informative as a more recent review. For much fuller treatments of all topics considered, readers should consult Avise (2000) and the extensive primary literature that it summarizes.

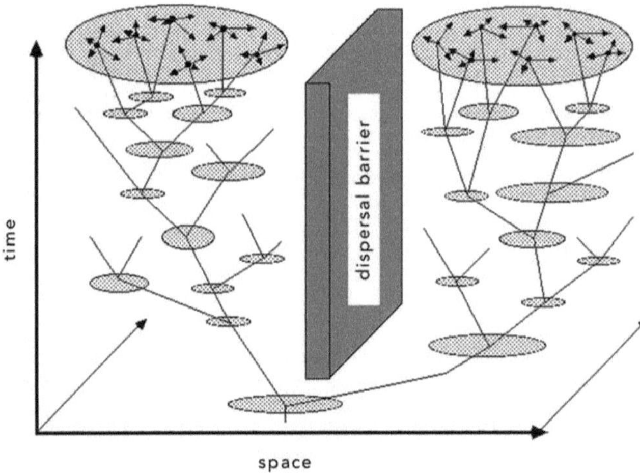

Figure 1. *The axes of phylogeography are space and time, across which gene genealogies are scrutinized (modified from Avise 2000).*

Twenty-five primary revelations

1. Cytoplasmic genomes add a new hierarchical level to population genetics

In diploid organisms, nuclear genomes typically exist as two copies per somatic cell and are transmitted across generations in a single-copy (haploid) molecular fashion. By contrast, mitochondrial and other cytoplasmic genomes [such as chloroplast (cp) DNA (see also Petit & Vendramin, this volume)] exist as populations of molecules within each somatic and germline cell, and usually are transmitted from parent to offspring as multiple copies. This realization led to a novel research arena directed toward the intra-individual population dynamics of cytoplasmic genomes in somatic cells and germlines – a newly recognized level in the population genetic hierarchy (Birky *et al.* 1983).

2. Germline bottlenecks in mtDNA numbers attend mitochondrial inheritance

Notwithstanding concept #1 (above), the vast majority of mtDNA variation is apportioned among (not within) individual animals, even in local populations. Thus, an individual typically displays a single predominant cytoplasmic genotype that often differs clearly in DNA sequence from other conspecifics. Coupled with experimental findings from pedigree analyses (Hauswirth & Laipis 1982), this observation indicates that relatively small effective population sizes often characterize the intracellular pool of mtDNA molecules that transmits from one generation to the next through animal germlines. Thus, significant heteroplasmy (the joint appearance of two or more mtDNA genotypes within an individual) is normally a transient condition lasting only a small number (i.e. tens or perhaps hundreds) of organismal generations. This discovery carries a huge pragmatic benefit: within-individual sequence heterogeneity seldom seriously compromises mtDNA's utility for genealogical assessments at proximately higher levels in the biological hierarchy (e.g. local demes and geographic populations).

3. DNA repair mechanisms can influence molecular evolutionary rates

A traditional paradigm of molecular evolution is that genes with conserved function evolve slowly. The mitochondrial genome, with its central role in cellular energy metabolism, was thought to be a paradigm if not the epitome of functional conservatism. Thus, early reports that animal mtDNA evolves rapidly (about 5-10x faster than single-copy nuclear DNA) came as a great surprise (Brown *et al.* 1979). Subsequent studies showed that mtDNA sequence evolution is concentrated at synonymous sites and non-coding regions of the molecule, as might be expected. However, a totally unexpected factor contributing to mtDNA's rapid evolution was also intimated and later confirmed: a severe deficiency of DNA repair mechanisms within the mitochondrion.

4. Some DNA sequences in sexual species show asexual inheritance

This notion was not entirely novel because mammalian Y-chromosomes were long known to be paternally inherited, and cytoplasmic genomes such as mtDNA were assumed to be transmitted maternally. Nonetheless, explicit analyses of cytoplasmic gene sequences and their transmission genetics through organismal pedigrees soon confirmed mtDNA's maternal inheritance at the molecular level. This validation compelled the field of population genetics to reconsider the ramifications of asexual inheritance in sexually reproducing organisms. In particular, it prompted the field to more fully embrace the notion that asexual genomes exist and are transmitted across successive generations without the normal complications of intermolecular recombination otherwise attendant with sex (Hutchinson *et al.* 1974).

5. Matrilineal histories within species can be recovered

The rapid pace of mtDNA sequence evolution in animals, coupled with the molecule's maternal inheritance, meant that species display a wealth of non-recombining markers suitable for deciphering the matriarchal component of an extended organismal pedigree (Wilson *et al.* 1985). Population geneticists thereby were afforded unprecedented access to genealogical information at the intraspecific level.

6. A gene tree is a recognizable component of a population pedigree

Considerations of matrilineal ancestry pioneered the gene-tree concept. Animal mtDNA consists of about 37 functional genes, but the entire non-recombining mitochondrial genome can be considered a single locus from a genealogical perspective. Like the traditional phylogenies of higher-level systematics, an intraspecific gene tree for mtDNA is hierarchically branched and non-reticulate. The gene-tree notion can also be extended to particular sequences in the nuclear genome, at least in principle (Tajima 1983). Thus, the term gene tree gained a generalized definition: the genealogical history of any defined segment of DNA. Such gene trees are real and discrete components of population pedigrees (Figure 2).

7. Gene trees in sexual species are multitudinous and non-isomorphic

For sexually reproducing organisms, a matrilineal gene tree represents only a minuscule fraction of a species' hereditary history. More than 99% of that total history resides instead in nuclear genes whose alleles have been transmitted along multi-generation genealogical pathways involving both genders. Due to Mendelian segregation and independent assortment, the realized transmission histories of unlinked DNA sequences inevitably differ from locus to locus. Thus, gene trees for unlinked loci are highly unlikely to be strictly isomorphic (identical in branching structure). This insight led to the notion that any pictured cladogram summarizing historical relationships of populations or species is actually a much-simplified representation of an

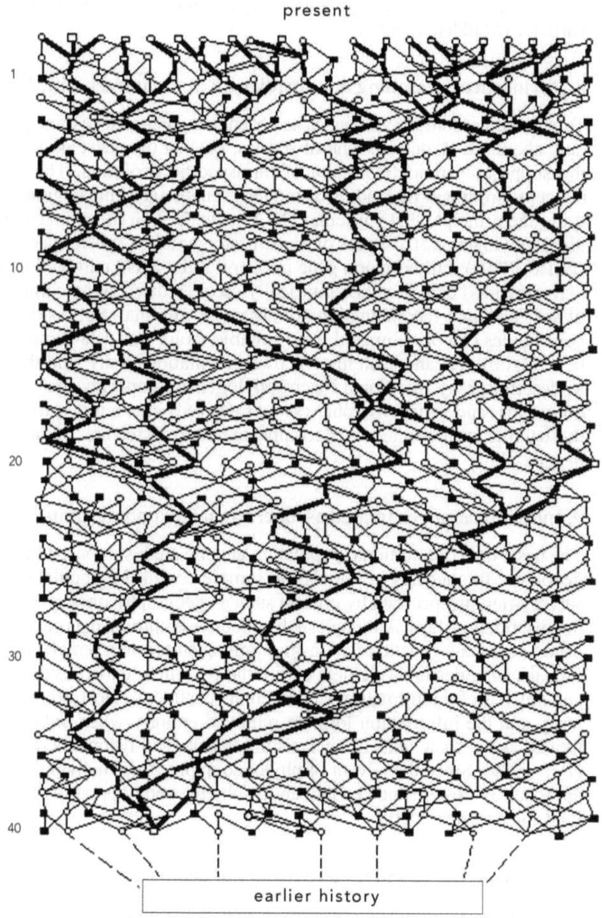

Figure 2. *Highlighted in heavy lines is one (among multitudinous) gene trees within an organismal pedigree across 40 generations. Note how the lineages coalesce to shared ancestors (modified from Avise 2000).*

underlying statistical 'cloudogram' of quasi-independent gene trees with a variance (Maddison 1997).

8. Phylogenetic reasoning is quite relevant at the intraspecific level

Traditional wisdom in systematics had been that phylogenetic principles apply only above the taxonomic rank of species, i.e. that they have no meaning in the supposedly reticulate realm of intraspecific evolution. The gene-tree concept, prompted by studies of mtDNA, challenged this dogma by clarifying the principle that particular DNA sequences do in fact have genealogical (phylogenetic) histories of transmission within a species pedigree also (Avise 1989a). Furthermore, gene trees (or at the least, unrooted genealogical networks) can often be empirically recovered when the complications of sex-mediated recombination are absent or minimal. So, historical gene genealogies can be estimated from suitable molecular data using phylogenetic algorithms, and such historical representations aptly lend themselves to description by traditional macro-phylogenetic concepts such as clades, outgroups, and synapomorphic (shared-derived) characters (provided that these terms are now interpreted to apply explicitly to features of the gene tree *per se*).

9. Individuals can be treated as 'operational taxonomic units' (OTUs)

The basic data of traditional population genetic analysis consist of allelic or genotypic frequencies in population samples, with the populations themselves often prespecified by criteria such as geography or suspected reproductive relationships. Although such collective empirical data can always be used to estimate genetic relationships among suites or assemblages of individuals, an undesirable element of circular reasoning underlies the exercise, and much useful information on particular specimens is lost. Both the circularity and the information loss are removed entirely when *individuals* are treated as the basic units of analysis in genealogical reconstructions (Avise *et al.* 1979). Since the advent of phylogeography, this 'individual as OTU' approach is now included routinely in phylogenetic appraisals of mtDNA (and some nuclear loci).

10. Intraspecific genealogy and historical demography are intertwined

At the intraspecific level, concepts of genealogy and historical population demography are inextricably associated. Precise mating relationships of individuals, coupled with generation-by-generation means and variances in individual reproductive success, describe the extended pedigree of a population, thereby defining the genealogical pathways that were available for allelic transmission. Any gene tree is one realized subset, or historical sample, from this constellation of pathways (Figure 2). As such, each gene tree is an ineluctable reflection of historical population demography. This realization gave rise to 'coalescent theory' (Hudson 1990), a burgeoning discipline in mathematical population genetics that seeks to uncover and formalize the relationships between historical population demography and the structure of intraspecific gene genealogies.

11. Evolutionary effective population sizes of most animal species are relatively small

In most surveyed animal species (especially those that are relatively abundant today, and whose populations are characterized by high historical levels of gene flow), estimated evolutionary effective population sizes (N_e values over the long term) have proved to be orders-of-magnitude smaller than contemporary census numbers (N). This conclusion stems from coalescent theory as applied to the surprisingly shallow intraspecific gene trees for such species, as evidenced in empirical mtDNA data sets (Avise *et al.* 1988). Two explanations are likely, the first probably being of greater importance: a) population-demographic histories *per se*, such as occasional bottlenecks in population size, or large variances among females in reproductive success; and b) rare 'selective sweeps' that purge existing variation as selectively advantageous mutations course through a species to fixation. Either way, gene lineages that survived for current observation have been historically squeezed through many fewer ancestors than otherwise might have been supposed, thereby constraining what would otherwise be greater temporal depths in intraspecific mtDNA gene trees.

12. Cytonuclear associations matter

The joint availability of molecular data from nuclear and cytoplasmic (e.g. mitochondrial or chloroplast) loci prompted an important new research area dealing with 'cytonuclear' patterns. Of special interest is how natural selection and other biological factors interact to produce the non-random associations (cytonuclear disequilibria) often observed between uni-parentally and bi-parentally inherited alleles in particular populations or species (Asmussen *et al.* 1987).

13. Key behavioral and demographic parameters can differ between the genders

Several parameters relevant to intraspecific gene genealogies often show fundamental asymmetries between the genders. For example, males are the primary dispersers in many animal species, females so in others. In many plant species, seed propagules to which cytoplasmic genomes are confined may be far less dispersive than pollen granules that typically carry nuclear genes only. In both animal and plant taxa, variances in individual reproductive success often differ between the sexes. In general, matrilineal genetic markers in conjunction with those from nuclear loci have opened many novel opportunities to empirically assess the population genetic consequences of such gender-associated biological asymmetries (Melnick & Hoelzer 1992).

14. Conspecific populations are genealogically allied yet often highly distinctive from one another

Molecular data from mtDNA are especially useful in revealing phylogeographic structure of populations within a species. An important realization is that these historical population structures can range along a continuum from evolutionarily (temporally) shallow to deep (the latter being especially true for species that have had severe restrictions on historical gene flow). Within a species, the most distinctive deeper units (the major matrilineal branches) sometimes are referred to as intraspecific 'phylogroups' (Avise & Walker 1998). Such phylogroups often, but not invariably, are also apparent in appropriate assays of nuclear genes, in which case they may warrant potential recognition as evolutionarily significant units for purposes of taxonomy or conservation efforts (see concept # 20 below).

15. Principles of genealogical concordance assess the depth and strength of phylogeographic structure

Not all phylogeographic population structures are equal in magnitude. To distinguish the historically deep from the shallow population separations, four distinct aspects of phylogeographic concordance are employed (Avise & Ball 1990). Each examines the level of agreement or consensus among multiple classes of information: across multiple sequence characters within a single gene tree (aspect 1 of genealogical concordance); across multiple gene trees within a species (aspect 2); across multiple species within a regional biota (aspect 3); and across multiple categories of data, such as molecular genetics and historical geography (aspect 4). By hard criteria, only when concordance has been demonstrated in at least some (preferably several) of these various aspects is it proper to conclude that the available data register salient evolutionary separations among the conspecific populations examined.

16. The number of phylogroups per species usually is small

In most vertebrate species and many invertebrate and plant species surveyed to date, the number of highly distinctive intraspecific phylogroups is small or modest – typically only about 1-5 per taxonomic species (Avise & Walker 1999). This observation, coupled with the finding that cytoplasmic gene sequences even in closely related biological species usually tend to be readily distinguishable, suggests that historically distinctive units identified in molecular-genetic analyses often conform quite well both in composition and number (at least within an order-of-magnitude) to the arrays of taxonomic species recognized in more traditional biological classifications. Thus, when judged from the newer vantages and criteria of molecular phylogeography, traditional non-molecular systematists generally seem to have done an excellent job in identifying and classifying salient historical discontinuities in the biological world.

17. Intraspecific phylogroups are nearly always allopatric

Because individuals can be considered OTUs in gene-genealogical analyses (concept #9), there is no logic demanding that major branches in gene trees must be allopatric. Empirically, however, most such intraspecific phylogroups have proved to be non-overlapping or nearly so in geographic distribution (and when this is not the case, secondary overlap often seems to be the most plausible explanation). Furthermore, these phylogroups are often spatially arrayed in coherent regional patterns such that they can be thought of as corresponding roughly to what was implied under the traditional concepts of subspecies, incipient species, or (in more recent literature) 'evolutionarily significant units' (Moritz 1994).

18. The geographic distributions of intraspecific phylogroups usually make biogeographic sense

In specific instances, the spatial arrangements of major branches in intraspecific gene trees usually orient well with known or suspected biogeographic agents, such as obvious environmental barriers to historical gene flow, or the locations of Pleistocene refugia. Indeed, the primary aim of most phylogeographic studies has been to employ gene-tree data to help recover and interpret the genealogical history of conspecific populations and closely related species in the context of historical geography and other relevant factors. In recent years, this approach often has been extended to multiple codistributed species, thereby revealing the composite histories of regional biotas. This type of endeavor has blossomed into a new subdiscipline in its own right that can be termed comparative phylogeography (Avise 1992).

19. Species' natural histories also impact phylogeographic patterns

In addition to vicariant historical factors associated with changes in the physical environment, endogenous biological factors – species' ecologies, behaviors, and natural histories – play key roles in shaping phylogeographic patterns (Avise 2000). To mention just one example, highly dispersive marine fishes, as a rule, have proved to show far less phylogeographic population structure than most of their freshwater counterparts when sampled across ranges of comparable size.

20. Phylogeographic units have primary importance for taxonomy and conservation

Biodiversity (which in the final analysis is genetic diversity) is what taxonomy seeks to name and conservation biology seeks to preserve. By describing the spatial distributions of genealogical variety within and among related species, the data of phylogeography can help tremendously in recognizing historical biotic partitions that should be of central relevance both to microevolutionary systematics and to biodiversity preservation (Avise 1989b).

21. Gene trees can differ in topology from population trees and species trees

This statement applies to sexually reproducing species, but not to strictly asexual taxa (where, in principle, one-and-the-same historical transmission pathway characterizes all loci). The fundamental distinction between a gene tree and a species tree in sexual species was unappreciated until fairly recently (Hey 1994). For example, an earlier paradigm in systematics stated that even one synapomorph (shared-derived character) is enough to define a clade. This is patently false (unless 'clade' refers solely to a branch in the particular gene tree in question). It is for such reasons that principles of genealogical concordance (concept #15) are important in deciding whether or not deep historical partitions in particular gene trees accurately register genome-wide partitions that should distinguish long-separated populations, intraspecific phylogroups, or species.

22. The phylogenetic status of sister populations or species can itself be evolutionarily dynamic

With respect to gene genealogies, it is no longer adequate to consider recently separated populations or species as having a fixed phylogenetic relationship to one another (Neigel & Avise 1986). Due to lineage sorting across the generations mediated by demographic turnover (organismal reproduction and death), extant populations at any point in time carry only a subset of the lineage diversity of their ancestors, plus newly arisen lineage diversity postdating the vicariant separations. Thus, a common phylogenetic progression for gene trees in recently separated sister taxa is initial polyphyly or paraphyly, only eventually followed by reciprocal monophyly (Figure 3). The rate at which this genealogical transition proceeds (under neutrality) is a function of population demographic events immediately preceding, during, and following the vicariant split. The transition also takes longer in principle, all else being equal, for autosomal than for cytoplasmic genes, due to a four-fold larger effective population size for genes that are diploid (as opposed to haploid) and inherited bi-parentally (as opposed to uni-parentally).

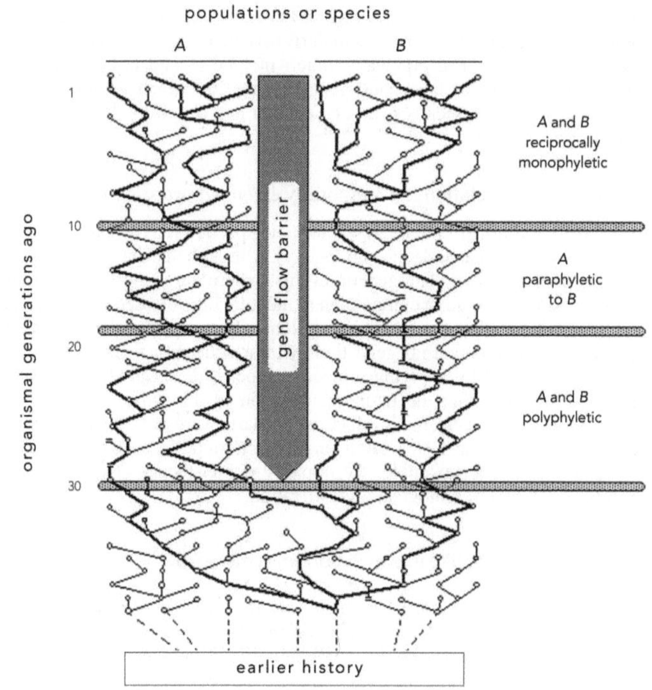

Figure 3. *Illustration of the shifting phylogenetic status of a gene tree (heavy lines) through time in two recently separated populations or species.*

23. Discordant gene trees and species trees can also characterize ancient taxa

Suppose that two or more successive cladogenetic events occurred long ago but close together in evolutionary time. If evolutionary effective population sizes of the species traversing the relevant nodes in the phylogenetic tree were larger than the internodal times as measured in organismal generations, then lineage sorting may not have proceeded to reciprocal monophyly in all gene trees of the immediate descendents (Tateno *et al.* 1982). Then, a topological discordance between a gene tree and a species tree will become evolutionarily 'locked in' as subsequent lineage sorting results in the eventual fixations of ancestral lineages in derivative taxa. Such idiosyncratic lineage sorting is one of several ways (others include secondary hybridization and various means of horizontal gene transfer) by which gene trees can come to differ topologically from one another and also from the composite species tree.

24. Allopatric speciation is a temporally extended process

A habit in traditional systematics is to view speciation as a point event in time (i.e. as a discrete node in a phylogeny). Although this may generally be acceptable for ancient cladogenetic events (where any temporal durations for speciation are small in comparison to the total time elapsed since), it can be grossly inadequate for recent speciations. By comparing the branching structures of mitochondrial gene trees within and among extant pairs of sister species, and by applying molecular clocks, recent phylogeographic appraisals suggest that the temporal duration of allopatric speciation in many vertebrate taxa averages (albeit with a large variance) about two million years (Avise *et al.* 1998). Such lengthy timeframes cannot be neglected when appraising, for example, the impacts of Pleistocene or more recent events on patterns of biological diversification.

25. Microevolution, like macroevolution, is historical

This catch-all truism sums up many of the phylogeographic insights described above. It is a basic realization that too often was overlooked in conventional population genetics, probably due in large part to that discipline's underlying formal theoretical framework. For reasons of mathematical tractability, many derivations and formulations in traditional population genetics dealt with equilibrium expectations (e.g. between mutation and selection, or genetic drift and migration) in unrealistically simplified contemporary settings (e.g. an 'island model' in which equal-sized populations are all assumed to exchange genes at equal rates). These were always presented as simplifying assumptions, but the net result nonetheless was a discipline too seldom focused on historical idiosyncrasies and non-equilibrium outcomes that are a *sine qua non* of real-life intraspecific evolution. Phylogeographic perspectives have enriched population genetics by adding an explicit focus on historical genealogy, and thereby drawing the field much closer to allied disciplines such as population demography, biogeography, and phylogenetic biology (Avise *et al.* 1987).

Synopsis

Historical reasoning and phylogenetic analysis have long been central themes of macroevolutionary biology and higher-level systematics, but until recently they had not permeated studies of intraspecific evolution to nearly the same extent. Thus, throughout most of the 20th century, there was a major gulf between the fields of phylogenetic biology and population genetics, to the detriment of both. Phylogeography is helping to bridge this gulf.

Microevolution too is a historical-genealogical process. Indeed, all limbs, branches, and twigs in any phylogenetic tree summarizing species' relationships ultimately consist of generation-to-generation organismal pedigrees through which genes were transmitted. The tools of molecular biology can now provide explicit historical information about genealogical tracings through such extended pedigrees, within as well as among living species. The net result has been the birth and growth of phylogeographic perspectives that promise to forge a useful new synthesis of micro- and macroevolutionary thought.

Acknowledgements

The author's recent work has been supported by a Fellowship from the Pew Foundation.

References

Asmussen MA, Arnold J, Avise JC (1987) Definition and properties of disequilibrium statistics for associations between nuclear and cytoplasmic genotypes. ***Genetics* 115**, 755-768.

Avise JC (1989a) Gene trees and organismal histories: a phylogenetic approach to population biology. ***Evolution* 43**, 1192-1208.

Avise JC (1989b) A role for molecular genetics in the recognition and conservation of endangered species. ***Trends in Ecology and Evolution* 4**, 279-281.

Avise JC (1992) Molecular population structure and the biogeographic history of a regional fauna: a case history with lessons for conservation biology. ***Oikos* 63**, 62-76.

Avise JC (2000) ***Phylogeography: The History and Formation of Species.*** Harvard University Press, Cambridge.

Avise JC, Ball RM Jr (1990) Principles of genealogical concordance in species concepts and biological taxonomy. ***Oxford Surveys in Evolutionary Biology*** 7, 45-67.

Avise JC, Walker D (1998) Pleistocene phylogeographic effects on avian populations and the speciation process. ***Proceedings of the Royal Society of London, B* 265**, 457-463.

Avise JC, Walker D (1999) Species realities and numbers in sexual vertebrates: perspectives from an asexually transmitted genome. ***Proceedings of the National Academy of Sciences, USA* 96**, 992-995.

Avise JC, Giblin-Davidson C, Laerm J, Patton JC, Lansman RA (1979) Mitochondrial DNA clones and matriarchal phylogeny within and among geographic populations of the pocket gopher, *Geomys pinetis*. ***Proceedings of the National Academy of Sciences, USA* 76**, 6694-6698.

Avise JC, Arnold J, Ball RM, Bermingham E, Lamb T, Neigel JE, Reeb CA, Saunders NC (1987) Intraspecific phylogeography: the mitochondrial DNA bridge between population genetics and systematics. ***Annual Review of Ecology and Systematics* 18**, 489-522.

Avise JC, Ball RM Jr, Arnold J (1988) Current versus historical population sizes in vertebrate species with high gene flow: a comparison based on mitochondrial DNA lineages and inbreeding theory for neutral mutations. ***Molecular Biology and Evolution* 5**, 331-344.

Avise JC, Walker D, Johns GC (1998) Speciation durations and Pleistocene effects on vertebrate phylogeography. ***Proceedings of the Royal Society of London, B* 265**, 1707-1712.

Birky CW Jr, Maruyama T, Fuerst PA (1983) An approach to population and evolutionary genetic theory for genes in mitochondria and chloroplasts and some results. ***Genetics* 103**, 513-527.

Brown WM, George, M Jr, Wilson AC (1979) Rapid evolution of animal mitochondrial DNA. ***Proceedings of the National Academy of Sciences, USA* 76**, 1967-1971.

Hauswirth WW, Laipis PJ (1982) Mitochondrial DNA polymorphism in a maternal lineage of Holstein cows. ***Proceedings of the National Academy of Sciences, USA* 79**, 4686-4690.

Hey J (1994) Bridging phylogenetics and population genetics with gene tree models. In: ***Molecular Ecology and Evolution: Approaches and Applications*** (eds. Schierwater B, Streit B, Wagner GP, DeSalle R), pp. 435-449. Birkhäuser Verlag, Basel.

Hudson RR (1990) Gene genealogies and the coalescent process. ***Oxford Surveys in Evolutionary Biology*** 7, 1-44.

Hutchinson CA III, Newbold JE, Potter SS, Edgell MH (1974) Maternal inheritance of mammalian mitochondrial DNA. ***Nature* 251**, 536-538.

Maddison WP (1997) Gene trees in species trees. ***Systematic Biology* 46**, 523-536.

Melnick DJ, Hoelzer GA (1992) Differences in male and female macaque dispersal lead to contrasting distributions of nuclear and mitochondrial DNA variation. ***International Journal of Primatology* 13**, 379-393.

Moritz CC (1994) Defining 'evolutionarily significant units' for conservation. ***Trends in Ecology and Evolution*** 9, 373-375.

Neigel JE, Avise JC (1986) Phylogenetic relationships of mitochondrial DNA under various demographic models of speciation. In: ***Evolutionary Processes and Theory*** (eds. Nevo, Karlin, S), pp. 515-534. Academic Press, New York,

Tajima F (1983) Evolutionary relationships of DNA sequences in finite populations. ***Genetical Research*** **123**, 437-460.

Tateno Y, Nei M, Tajima F (1982) Accuracy of estimated phylogenetic trees from molecular data. I. Distantly related species. ***Journal of Molecular Evolution*** **18**, 387-404.

Wilson AC, Cann RL, Carr SM, George M, Gyllenstein UB, Helm-Bychovski KM, Higuchi RG, Palumbi SR, Prager EM, Sage RS, Stoneking, M (1985) Mitochondrial DNA and two perspectives on evolutionary genetics. ***Biological Journal of the Linnean Society*** **26**, 375-400.

Chapter 2

Plant phylogeography based on organelle genes: an introduction

Rémy J. Petit* and Giovanni G. Vendramin
*E-mail: petit@pierroton.inra.fr

Abstract
Plants have unique biological attributes of great interest to researchers investigating population dynamics. Yet, until recently, organelle DNA had been seldom utilized for phylogeographic studies in plants. While cpDNA variation has been used extensively to construct interspecific phylogenies, many researchers have considered that the relatively low levels of intraspecific variation revealed by early studies of plant organelle DNA render these genomes of little use for intraspecific studies. In this review we introduce the field of plant phylogeography based on organelle polymorphisms by providing a detailed discussion of the processes underlying this variation. Progress in molecular organelle genetics has provided insights into the structure, variation, inheritance, vegetative segregation and recombination of organelle genomes. While some of these features (e.g. low substitution rates) may complicate phylogeographic studies, others (e.g. presence of two genomes and frequency of atypical modes of transmission) offer unique opportunities, many of which are virtually unexplored.

Keywords: cpDNA, mtDNA, population, history, structure, variation, inheritance, recombination, gene flow, dispersal, selection

Introduction

Plants have many unique features with great appeal for researchers trying to unravel spatio-temporal dynamics of populations and their consequences for evolution, the objects of phylogeography (see Avise, this volume). In particular, they stand still during most of their life. This explains the central importance of space in their study (Silvertown & Charlesworth 2001) and considerably facilitates their sampling. Despite this immobility, seed plants can move their genes in two specialized vehicles during short but critical phases of their life cycle: before fertilization, in the male gametophyte (pollen), and later in the young sporophyte (the seed). Hence, in stark contrast to many animals, the new diploid embryo is normally mobile while juveniles and adults are sedentary. Even the closest relatives of seed plants, the

S. Weiss and N. Ferrand (eds.), Phylogeography of Southern European Refugia, 23–97.

ferns, have a distinct life cycle and live in a different world with respect to migration processes (Sauer 1988). In angiosperms and gymnosperms, which will be the target of this review, pollen plays a major role in connecting extant populations with gene flow, but seeds (or other plant parts) are necessary to establish new populations of plants. Consequently, maternally inherited genes (which are not transmitted to the next generation by pollen) should be of special value for clarifying the spatio-temporal dynamics of plant populations.

Phototrophic plants are central components of ecosystems, responsible for the primary production of biomass, providing food and shelter to most animals. At the same time, plants often depend on animals for reproduction and/or dispersal. Zoologists are starting to realize that the timing and pattern of colonization by plants are of particular importance in understanding animal phylogeography, whereas the phylogeographic patterns studied by botanists have actually been shaped by the movements and behavior of animals as well as by the distribution of pathogens or symbionts.

Given their central place in the ecology of life, the vital importance of their genetic resources for sustainable agriculture and forestry, and their unique biological attributes, one would expect plants to be at the forefront of phylogeographic research. Surprisingly, however, there were until recently few explicit phylogeographic studies of plants (Schaal *et al.* 1998), in contrast with the situation for animals (Avise 2000). Fortunately, this situation is changing rapidly, and plant phylogeography is attracting increasing interest.

In this review, we provide an introduction into the field of population genetics and phylogeography using genetic variation in plant organelles. Typically, one of two approaches is used by geneticists: either they focus on the frequencies of variants (haplotypes) within and among populations, but do not attempt to take haplotype similarities into account, or they focus instead on intraspecific phylogenies but do not consider within population variation. Ideally, population sampling should be combined with intraspecific phylogenies of the variants, bridging the gap between these two approaches. Besides reviewing information useful to interpret population or phylogeographic surveys (or a combination of both), we identify some of the opportunities – many of them unexplored – that this field might provide in the future.

Part 1 summarizes some key steps in the history of plant organelle genetics, and the changing perception of the usefulness of organelle DNA as a source of markers for population and phylogeographic surveys. Part 2 examines in some detail the characteristics of plant organelle genomes that are relevant for population studies. Plants are unique among eukaryotes in possessing two DNA-containing organelles, the ubiquitous mitochondrion and the distinctive plastid. We consider in turn, genome structure and variation organelle inheritance and vegetative segregation, as well as recombination and the association between chloroplast (cp) DNA and mitochondrial (mt) DNA. Part 3 considers important parameters that can influence the geographic structure of organelle genes: intraspecific gene flow, as seen from a genetic and from an ecological perspective, history, including past and ongoing human influences, selection, interspecific gene flow and horizontal transfers.

Although the development of plant phylogeographic studies using organelle DNA (oDNA) is discussed in Part 1, a synthesis of phylogeographic patterns found in these studies is not the focus of this review. Rather we discuss in detail the underlying processes that determine plant organelle phylogeography. We focus on those features that are unique or at least more typical of plants and, where deemed important, contrast the situation found in plants with that in animals, a long tradition in comparative biology (e.g. Bradshaw 1972). Phylogeographic studies of plants also benefit from more general progress in molecular techniques, population genetics, and paleoecology, but this is also true of studies on animals and is therefore beyond the scope of this review.

Some of their features make the study of organelle genomes of plants somewhat more complicated, at least from a technical perspective. However, other features, such as the presence of two organelle genomes (compared to a single one – the mitochondrial genome – in most other eukaryotes), the relatively high frequency of atypical modes of transmission (maternal, paternal or bi-parental inheritance, often distinct for mtDNA and cpDNA), and the existence of mitochondrial plasmids, offer unique opportunities for phylogeographic studies. Besides the characteristics of organelle genomes themselves, we consider some life historic attributes of plants that can affect their population dynamics and evolution (Harper 1977; Silvertown & Charlesworth 2001) and ultimately the spatial distribution of genealogical lineages. These traits include indeterminate and modular growth (hence the elusive individuality of plants and potentially extreme longevity), sessileness, the evolution of a variety of seed dispersal mechanisms, and the retention of a significant if much reduced gametophytic phase. Finally, plants often have very weak interspecific barriers (Levin 1979), which may have considerable impact on their phylogeography: this will certainly complicate interpretations but we will argue that it should rather be viewed as an advantage as it may bridge further the gap between micro- and macroevolutionary studies.

1. Historical development of population and phylogeographic studies based on organelle variation in plants

The study of plant organelles has played a major role in the development of genetics, and cytoplasmic genomes were extensively used as plant geneticists applied developing molecular techniques to investigate plant diversity and evolution. However, early indications of a relatively low level of variation in plant organelle genomes meant that (despite some promising evidence) much of this work focused on interspecific variation, whereas the use of cytoplasmic markers in plant population and phylogeographic studies lagged behind similar studies in animals. More recently, there has been a rapid rise in the number of plant oDNA population studies. We provide an overview of the key steps in the history of plant organelle genetics that illustrates these changing perceptions of oDNA's utility for population and phylogeographic studies.

Plants play a major role in the early history of organelle genetics

Mendel's (1866) study on peas and beans is the best-known example of the role of plants in the early history of genetics (for a recent botanical account see Fairbanks & Rytting 2001). Similarly, all three rediscoverers of Mendel's laws in 1900 were botanists. One of them (Correns) discovered the first case of cytoplasmic inheritance. In gynodioecious species (i.e. made up of females and hermaphrodites), females crossed with hermaphrodites produce only female plants, indicating uni-parental inheritance of sex (Correns 1906, 1908). Other reports of non-Mendelian inheritance quickly followed using crosses between variegated (white and green) and normal (green) plants (Baur 1909; Correns 1909). Correns found strict maternal inheritance of variegation in *Mirabilis* but was reluctant to attribute this to particulate inheritance of the plastids, whereas Baur reported the first case of bi-parental inheritance and vegetative segregation (sorting out of heteroplasmic oDNA during mitosis; see 2.3) in *Pelargonium* and correctly attributed this to genetic material present in the plastids (Hagemann 2000). This latter interpretation was long considered as doubtful (e.g. Wright 1968-1978).

Following the pioneer work of Baur and Correns, cytoplasmic genetics developed during the first half of the 20th century, principally in Europe (Sapp 1987). It was dominated by work on plants, thanks to the widespread availability of phenotypic markers, especially chlorophyll defects and male sterility. Patterns of inheritance attracted much attention (reviewed in Kirk & Tilney-Bassett 1978). Nucleo-cytoplasmic interactions were also studied, by Renner and by Stubbe on *Oenothera*, and by Michaelis and his coworkers on *Epilobium* (see references in Hagemann 2000 and in Grun 1976). These studies form *"an extraordinary continued line of research that has stretched over more than 50 years"* (Grun 1976). In particular, they show that selection may be involved in shaping geographic variation at organelle genes, a possibility that has been largely ignored in recent studies.

The first report of DNA in plant organelles (Chiba 1951) was based on Feulgen-staining of chloroplasts in *Selaginella* as well as two flowering plants, but another decade passed before reports of DNA in both chloroplast and mitochondria became common, bringing cytoplasmic genetics back into mainstream biological research (Sapp 1987).

The very homogeneous appearance and size of the plastids had been described since the beginning of the 20th century (e.g. Möbius 1920). Similarly, as plant molecular biology developed in the early 1970s, physico-chemical investigations of plant oDNA revealed a fairly homogeneous genome. Kirk (1971) showed that GC content of cpDNA varies little across species. Electronic microscopy revealed that the circular cpDNA was comparatively large (10 times larger than animal mtDNA) and did not vary much in size across species (Kolodner & Tewari 1972). Chan & Wildman (1972) studied the most abundant chloroplast protein, ribulose bisphosphate carboxylase/oxygenase (called Fraction I protein at the time), pointing out the importance of both chloroplast and nuclear genomes for making chloroplast proteins. They also provided the first insight into the evolutionary rate of cpDNA: *"...only one mutation of the cistron in chloroplast DNA coding for the more than 200 amino acids in the large subunit of fraction I protein has survived during 150 million years of evolution separating the Australian from the Western Hemisphere* Nicotiana *species."*

Perception of low variation discourages the use of organelle DNA in intraspecific studies

Plant cytoplasmic genomes were the subject of many studies, from 1975 to 1980 (Bogorad 2001). The first restriction enzymes (Hedgpeth *et al.* 1972) were rapidly applied to the analysis of cpDNA (Atchison *et al.* 1976; Vedel *et al.* 1976) and of plant mtDNA (Levings & Pring 1976). The first restriction map of cpDNA was for maize (Bedbrook & Bogorad 1976). The advantages of the use of restriction patterns to measure similarity among species were immediately apparent, compared to physico-chemical measures that indicated great homogeneity across taxa.

Early application of restriction enzymes to study intraspecific polymorphism of organelle DNA in plants might have indicated the potential of oDNA for such studies. The first survey of cpDNA variability in natural populations of a plant species, *Nicotiana debneyi* (Scowcroft 1979) [following studies looking for male sterility markers in mtDNA of maize (Levings & Pring 1976) and wheat (Quétier & Vedel 1977)] was contemporary with the first studies based on animal mtDNA (Avise *et al.* 1979). However, whereas the study by Avise *et al.* (1979) on a small rodent identified many (61) mitotypes and allowed the reconstruction of an intraspecific maternal phylogeny, which fitted with their geographic distribution, only one restriction site polymorphism was found in *N. debneyi*. Yet, as stressed by Scowcroft, the *Nicotiana* study revealed a high degree of geographic structure between populations. An *Eco*RI site was fixed in seven of nine populations studied but absent in the remaining two. Scowcroft reasoned that, as the restriction sites screened represented only a small percentage of the entire chloroplast sequence, many substitutions should exist throughout the genome.

Despite this optimistic conclusion, until at least the late 1980s many researchers believed that cpDNA lacked sufficient diversity to make it a useful tool in microevolutionary studies (e.g. Clegg 1987; Palmer 1987). Informative intraspecific phylogenies, which are the basis for phylogeographic surveys, did not appear until very recently, lagging significantly behind comparable animal studies (e.g. Avise *et al.* 1979, 1983; see Figure 1). A study by Clegg *et al.* (1984) on *Hordeum vulgare* cultivars revealed some potential for cpDNA markers, since only the wild relatives of this species had cpDNA diversity, pointing to a bottleneck that had not been identified using isozymes. However, the first detailed population study published was that of Banks & Birky (1985) who entitled their paper: *"Chloroplast DNA diversity is low in a wild plant,* Lupinus texensis*"*. These authors had sampled 100 individuals from 21 populations across the North American range of this wild legume, and applied seven restriction enzymes to purified chloroplast DNA. Only three variants were detected, with one present in 88% of the individuals, in stark contrast with the high levels of isozyme diversity that had been revealed in the same species. This widely cited study had a lasting effect on the scientific community. The low cpDNA diversity that was revealed, especially considering the efforts made, probably discouraged many investigators. However, a closer look at the original data shows that one population was completely fixed for one of the two rare variants, pointing to a much stronger geographic structure compared to that revealed with isozymes. As in Scowcroft's study, the potential of these cpDNA markers for geographic surveys should have struck population geneticists, but was masked by the low overall level of variation detected.

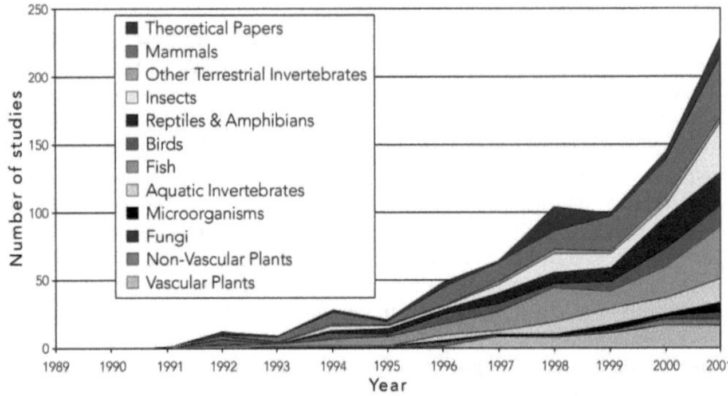

***Figure 1**. Classification of studies explicitly focusing on phylogeography, published since Avise* et al. *(1987) first coined this term. Data were compiled from exhaustive searches in Medline, BIOSIS and CAB databases.*

On the other hand, approaches using cpDNA for investigating interspecific relationships in plants developed considerably during the 1980s (Vedel *et al.* 1978; Palmer 1987), along with cladistic methodology. Although phylogenetic methods are a prerequisite for a development of phylogeography, this focus may have actually played a role in slowing down intraspecific research. Indeed, Soltis *et al.* (1992) pointed out that the perceived invariance of cpDNA within species is a comfortable situation for researchers engaged in interspecific phylogenies: *"Scientists have long sought methodologies that would facilitate phylogenetic reconstructions based on at most a few representatives per species"*. The focus on interspecific phylogenies meant that the majority of studies using cpDNA involved sampling strategies (few populations and few individuals per population) that are inefficient for demonstrating intraspecific variation. The first reviews dealing with intraspecific cpDNA diversity in plants were primarily focused on the consequences of this variation for interspecific phylogeny and insisted that cases of intraspecific variation (among populations, but also within populations and individuals) were actually frequent (Harris & Ingram 1991; Soltis *et al.* 1992). These authors pointed out that in most comprehensive studies some intraspecific variation had been detected: in 1991, 25 species had been studied with samples of at least 100 individuals (or at least 10 populations); in all but one case (*Pennisetum glaucum,* Clegg *et al.* 1984) variation was detected. However, only eight of these 25 studies can be considered as true population surveys, with a hierarchical sampling allowing measurement of variation within and among populations. Indeed, many studies at that time pooled individuals before analysis, assuming little or no variation within populations.

Much work during the 1980s dealt with the molecular organization of plant organelle genomes, confirming their conservative evolutionary rates, especially when compared with the mitochondrial genome of animals [by two orders of magnitude according to the estimate of Zurawski *et al.* (1984)], while also providing new tools for studying population variation. The first seed plant genome to be completely sequenced – a real *tour de force* at that time – was cpDNA from tobacco (Shinozaki *et al.* 1986). This genome is roughly ten times longer than mammal mtDNA and is comprised of 42% of non-coding sequences (including 30% of intergenic spacers and 12% of introns). It was not until ten years later that the first complete sequence of a seed plant mitochondrial genome (*Arabidopsis thaliana*) was achieved (Unseld *et al.* 1997). The complete cloning of cpDNA and partial cloning of plant mtDNA provided probes that could be used for population or inheritance studies, using the technique of Southern (1975), and avoiding the need for purification of chloroplasts or mitochondria. The first studies applying that technique for population surveys were from the laboratory of Allard at the University of California (Davis). These studies revealed significant levels of diversity in large sample sizes. Neale *et al.* (1986) sampled 245 individuals of wild barley in Israel, whereas Wagner *et al.* (1987) studied 371 individuals from two pine species in the USA and Canada. However, diversity and its apportionment within and among populations were not quantified in either study, nor were intraspecific phylogenies obtained.

The late but rapid rise of population and phylogeographic studies of plants

Despite the delay in realizing the potential of plant cytoplasmic markers for use in plant population and phylogeographic research, such studies have now begun to increase rapidly (largely based on cpDNA) (Figure 2A). The first intraspecific plant phylogeographic studies that explicitly cited the approach pioneered by Avise (and used the term phylogeography) are those of Soltis *et al.* (1991) on *Tellima grandiflora*, a Saxifragaceae from the west coast of North America, and of Lavin *et al.* (1991) on *Gliricidia sepium*, a leguminous tree from Mesoamerica. Major range discontinuities were identified in both species and this called for an historical interpretation (involving range contractions and Pleistocene refugia). Subsequently, studies of cytoplasmic gene flow at various geographic scales appeared. They had been motivated in part by the finding of strongly contrasting fixation indices (F_{ST} or G_{ST}) for nuclear *versus* cytoplasmic markers in oaks (Kremer *et al.* 1991; Petit *et al.* 1993a). Models were proposed to explain this large asymmetry. Petit (1992), Petit *et al.* (1993b), and Ennos (1994) adapted the island model proposed by Birky *et al.* (1989) to organelle and nuclear genes in plants, and showed that the most likely explanation was a high pollen flow associated with a very low seed flow. The power to detect geographic structure using markers from maternally inherited genomes became obvious, and appropriate methods were sought to overcome the difficulties inherent with low levels of variation. In particular, the development of plant phylogeography has particularly benefited from PCR-based techniques, which allow the use of larger sample sizes. Such techniques are currently used in the majority of population studies using plant oDNA (Figure 2B). The availability of complete cpDNA sequences made it possible to design consensus primers (see 2.1) that further boosted these studies.

2. Characteristics of plant organelle genomes

The two DNA-containing plant organelles are remnants of two independent invasions of ancestral eukaryotic cells by free-living organisms. The mitochondria are derived from α-Proteobacteria and the plastids from cyanobacteria. With the long ($>10^9$ years) period of cohabitation and dependency on the nucleus (Palmer 1990), it might be expected that these ancient endosymbionts would have converged to a similar genome structure and would be

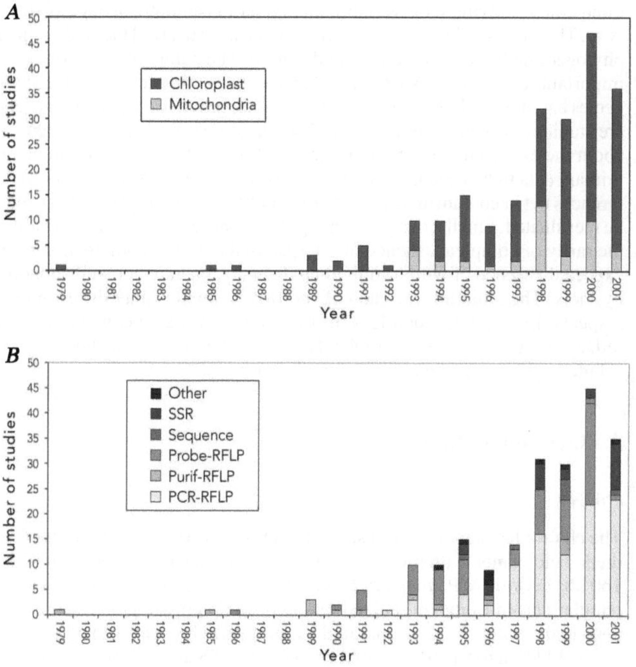

Figure 2. *Number of studies published per year using plant organelle genomes and considered as population studies. Only those studies comprising five or more natural populations and three or more individuals per population were included. The proportion of these studies based on* **A**) *mtDNA or cpDNA, and* **B**) *different methods of detection, is indicated. These methods include: PCR-based analyses of microsatellites (SSR for Simple Sequence Repeats), systematic sequencing of portions of the oDNA, restriction fragment length polymorphism (RFLP) based on hybridization with probes, purification of oDNA, or PCR-amplification of specific fragments.*

evolving in similar ways. However, although both organelles have many genome copies per cell and obey the fundamental rules of organelle genetics (vegetative segregation and predominant uni-parental inheritance), their genomes differ fundamentally in many respects. Furthermore, and quite strikingly, plant mitochondria differ radically from animal mitochondria. We provide an overview of the features of plant organelles that should be kept in mind when studying their variation in population and phylogeographic surveys. The causes of low levels of variation that have delayed the study of plant phylogeography need to be well understood. The nature of variation is also important to consider, especially when reconstructing intraspecific phylogenies based on oDNA variation. Equally fundamental for population genetics are studies of organelle inheritance. The literature is replete with statements about the expected maternal inheritance of organelles. However, in plants, primary data indicate that this expectation is not always justified. Alleged differences between transmission of mitochondria and chloroplasts also need to be reevaluated. Finally, many sweeping statements have been made regarding the rarity or complete absence of interparental oDNA recombination or the expected absence of within-individual diversity, that are not well supported by facts. The situation for each of the two organelles may differ in these respects. In general, much less information is available concerning plant mtDNA (due in part to the complexity of its evolution), which may have contributed to its neglect in phylogeographic surveys.

2.1 Genome structure

cpDNA

The chloroplast genome has a size ranging from 120 to 217 kb in photosynthetic land plants, with most angiosperm species having genomes of 135 to 160 kb (Downie & Palmer 1992). The chloroplast genome exists as a single circular chromosome, organized into two single-copy regions (large: LSC and small: SSC) usually separated by an inverted repeat (IR) of 10-76 kb (20 to 30 kb in most species) that encodes the 18S and 23S ribosomal RNA genes. Due to recombination between repeats (flip-flop recombination), cpDNA actually exists as a 50-50 mixture of two isomeric forms (Palmer 1985). Much of the (limited) variation in overall size is accounted for by variation in the length of the IR, which can spread to encompass regions that are typically unique, or contract to expose new regions of unique sequences, or disappear nearly completely. Such losses have been reported in five separate lineages of land plants, including conifers and some legumes (Palmer 1987). Structure and gene content are generally highly conserved in land plants, although a few large inversions are encountered. Exceptions include a few legume species, some members of the Campanulaceae and *Pelargonium*, which show extensive rearrangements. Species with highly rearranged genomes often present short dispersed repeats that seem to be absent from more typical genomes (Palmer 1987).

mtDNA

The 'elusive' plant mitochondrial genome is larger (at least 200 kb and up to 2500 kb). It contains one or more duplicated or triplicated sequences (1-15 kb), generally directly oriented, called recombination repeats (Hanson & Folkerts 1992). Short dispersed repeats (50-1000 bp) are also common. As a consequence of recombination, a multi-partite structure is generated (Mackenzie *et al.* 1994). This fluid genome exists as a heterogeneous population of DNA molecules of various size (Quétier & Vedel 1977), a constellation of rosettes, subgenomes and higher order multimers, both double-stranded and single-stranded (Backert *et al.* 1996, 1997). Such a complex arrangement requires a multi-scale selection process for the conservation of its informational content during development (Albert *et al.* 1996). Repeated sequences that do not recombine (Hanson & Folkerts 1992), and persistent sublimons (substoichiometric DNA fragments) (Lonsdale 1989) have also been described. In addition, mtDNA frequently integrates foreign sequences (up to 12 kb in length) such as cpDNA sequences (5-10% of the mitochondrial genome) (see 2.2). It often hosts small (1-11 kb) circular or linear plasmids (see 3.6). Most aspects of genome structure (genome size, configuration and gene order) change very rapidly (Sederoff *et al.* 1981; Palmer 1992). An extreme illustration is provided by the comparison of two maize mtDNAs that had diverged about only 100 000 years ago but differ by at least 30 rearrangements whose complexity defies resolution into discrete events (J. Doebley, cited in Palmer 1990).

Differences in structure

Palmer (1990) summarizes the difference between these two genomes: *"the loose, open structure of plant mtDNA, with genes floating singly amidst a sea of large intergenic spacers replete with short dispersed repeats"* is opposed to cpDNA, with its *"compact gene arrangement, paucity of dispersed repeats and hence constrained gene order."* The different constraints on genome size for the two organelle genomes are illustrated by the fact that, in the non-photosynthetic, parasitic plant *Epifagus virginiana,* the photosynthetic genes have disappeared, implying that all unnecessary cpDNA sequences can be readily deleted; simultaneously, a photosynthetic gene had remained trapped (in the form of a pseudogene) in the mitochondrial genome of the same species, witnessing the much more limited selection for size in this genome (dePamphilis & Palmer 1990). Table 1 summarizes the differences in genome structure and variation between plant and animal organelles. The larger genome sizes and higher proportion of non-coding sequences in plant oDNA relative to animal mtDNA suggests that variation useful for population genetic studies may be detectable in plant oDNA despite the lower substitution rates observed (see 2.2).

Table 1. *Comparison of the structure and variation of organelle genomes. (The Arabidopsis Genome Initiative 2000 and other sources cited in the text).*

Feature	cpDNA	Plant mtDNA	Animal mtDNA
Size	120 - 217 kb	200 - 2500 kb	14 - >30 kb
Number of protein genes	79[1]	58[1]	13
Gene order	Conserved	Highly variable	Conserved
Non-coding sequences	30-40%	60% to >90%	5-10%
Density (kb per protein gene)	1.2	6.3	1.0
Genes with introns	18%[1]	12%[1]	0%
Synonymous substitution rate[2]	1.5 (0.3)[3]	0.5	20-50
Non-synonymous substitution rate[2]	0.2 (0.1)[3]	0.1	2-3
Rearrangements	Rare	Frequent	Very rare
Transitions:transversions	<2:1	<2:1	20:1
Edited nucleotide sites	~30	>400	0
Genetic code	universal	universal	specific

[1] Values derived from the completely sequenced organelle genomes of *Arabidopsis*.
[2] Per site per 10^9 years.
[3] In brackets: estimates for the Inverted Repeat (IR) region.

Complete sequences of plant organelle genomes and their utility for intraspecific studies

In plants, as in animals, population and phylogeographic surveys have greatly benefited from PCR approaches. The availability of completely sequenced choloroplast genomes in the 1980s as well as the increase of available oDNA sequences has allowed development of *consensus* (if not *universal*) primers of interest for intraspecific studies. Such primers are anchored in conserved (generally coding) regions separated by more variable regions. The region amplified may be short enough to allow direct sequencing (e.g. Taberlet *et al.* 1991; Chiang *et al.* 1998; Hamilton 1999; Grivet & Petit 2002a), longer for use in combination with restriction enzymes (usually 4-bp cutters) (Demesure *et al.* 1995; Dumolin-Lapègue *et al.* 1997a) or much smaller (<200 bp) but containing potentially variable mononucleotide single strand repeats (cpSSRs) (Vendramin *et al.* 1996; Weising & Gardner 1999).

A set of 38 consensus cpDNA primers spanning the large single copy region from Eudicots (>86 kb) is now available (Grivet *et al.* 2001), as well as 35 mtDNA primers amplifying some 40 kb of plant mtDNA (introns, intergenic spacers and genes) (Duminil *et al.* 2002). A database for chloroplast primers is available online (Heinze 2001).

In the case of mtDNA, a possible solution in coping with the lack of synteny is to use a gene-anchored approach, which could be of great help in detecting polymorphism in the vicinity of genes (Loridon & Saumitou-Laprade 2002). To identify more variable regions within cpDNA (such as cpSSRs), or to anchor primers in the highly fluid plant mtDNA, additional completely sequenced genomes would be of considerable benefit. Unfortunately, there has been only slow progress since 1986, when the cpDNA sequence of tobacco was obtained in Japan. In April 2003, there were only 12 completely sequenced chloroplast genomes of seed plants (one conifer and 11 angiosperms; http://megasun.bch.umontreal.ca/ogmp/projects/other/cp_list.html), representing in total ~2000 kb, which is less than 2% of the 118 700 kb nuclear genome of *Arabidopsis* released in 2000. The situation is even worse for mtDNA: only two seed plant species (both eudicots, *Arabidopsis* and *Beta*) have had their mitochondrial genome completely sequenced. No complete monocot or conifer mtDNA sequence is available. A few projects are under way, and the model species *Oryza sativa* and *Medicago truncatula* should have both of their organelle genomes sequenced soon. Clearly, a more ambitious oDNA sequencing project involving some representatives from all major seed plant families would be very beneficial (this has been initiated in the case of cpDNA: see http://www.jgi.doe.gov/programs/comparative/cover-page.htm).

2.2 Level and nature of variation

Phylogeographic studies require accurate intraspecific phylogenies. This implies that 1) sufficient variation can be identified, and 2) that the mutations are phylogenetically ordered. Here, we examine oDNA substitution rates, as determined from interspecific comparisons, and levels of intraspecific variation. We discuss some of the mechanisms that have been proposed to account for the comparatively low levels of diversity in these organelles. We also provide information on the nature of oDNA variation and its significance for reconstruction of intraspecific phylogenies and mention the special problems created by the presence of oDNA copies in the nuclear genome.

Substitution rates in plant oDNA

In plants, very low substitution rates have been estimated for both organelle genomes (see Table 1 and Wolfe *et al.* 1987). The difference with animal (vertebrate) mtDNA is striking. Estimated non-synonymous substitution rates for plant mtDNA are as low as one event per site every 10 billion years, in stark contrast with animal mtDNA. These rates differ between genomes, within a genome and across species (Table 2). The deceleration seems quite general as even cpDNA microsatellites have reduced mutation rates (Provan *et al.* 1999).

Levels of intraspecific variation in plant oDNA

As seen in Part 1, levels of intraspecific sequence variation in oDNA have been considered to be generally very low in plants, in line with estimates of

Table 2. *Evidence of differences between substitution rates in plant oDNA at different levels.*

Contrast	Evidence	Reference
Between genomes	cpDNA sequences integrated into plant mtDNA evolve at a reduced rate	Schuster & Brennicke 1987
Within genomes	Sequences included in the IR cpDNA region evolve at a reduced rate (at least as low as that of plant mtDNA, see Table 1). May be due to a bias in copy correction between repeats in favor of the wild-type sequence	Palmer 1990
Across species	Nuclear mutator genes can induce frequent mutations in cpDNA (and probably in mtDNA as well) when homozygous, in some plant species	Tilney-Bassett 1975
	Highly (50-100) accelerated mtDNA substitution rates (on par with rates in animal mtDNA) discovered for a few plant species (e.g. *Plantago, Pelargonium, Hevea*)	Luo *et al.* 1995; Palmer *et al.* 2000
	Modest substitution rate heterogeneity among angiosperms (appears to be paralleled in all three plant genomes)	Eyre-Walker & Gaut 1997; Laroche *et al.* 1997
	Studies on mtDNA sequences have suggested that generation time may account for substitution rate heterogeneity, with long-lived monocots (such as palm trees) or dicots (forest trees) having reduced rates	Bousquet *et al.* 1992; Gaut *et al.* 1992
	The *coxI* gene in long-lived gymnosperms evolves five times faster than that of annual angiosperms, due to the accumulation of T-C substitutions at edited sites. Consequently, the number of nucleotide substitutions is similar for synonymous and non-synonymous sites	Szmidt *et al.* 2001
	In conifers, there is a higher evolutionary rate in both the mitochondrial and chloroplast DNA when the organelle is inherited paternally than when inherited maternally. These results suggest that, compared with eggs, sperm tend to carry a greater number of mutations in mitochondrial and chloroplast DNA	Whittle & Johnston 2002

substitution rates. However, species with very high levels of cpDNA diversity have now been identified. An example is provided by a study of cpDNA variation in *Eucalyptus globulus* in southern Australia. A total of 105 haplotypes were detected among 270 samples (40 variable sequence characters in a 987-bp long sequence, Freeman *et al.* 2001). The level of diversity is therefore likely to be very heterogeneous across species. Unfortunately, few comparative studies exist. The diversity of techniques used and the objectives pursued explain why genome-wide estimates of nucleotide variation (the only measures truly comparable across species) are rather limited. Available estimates deal nearly exclusively with cpDNA and are often of low precision. Schaal *et al.* (1991) have reviewed 46 cpDNA studies, and summarized values of percent nucleotide change: mean values were 0.07% at the intraspecific level, 0.80% at the interspecific level and 3.40% at the intergeneric level. This means that there is on average no more than one cpDNA nucleotide site out of 1000 that differs between two conspecific individuals taken at random.

Although this compilation needs to be updated, the shift from probe-based approaches to PCR-based studies (which typically survey only a few percent of a carefully selected part of the genome) explains why progress in that direction has been quite limited, despite the increasing number of population studies that make use of sequencing.

Potential explanations for low levels of diversity in plant oDNA

A priori, the low level of diversity detected in plant oDNA was not expected. The reduced effective population sizes of organelle genes compared to nuclear genes could explain the low level of diversity observed in plant oDNA. But this argument also applies to the highly variable animal mtDNA. In fact, some of the arguments put forward to explain the *high* levels of mtDNA diversity in animals also apply to plant oDNA: for instance, free-radicals generated by electron transport chains (common to mitochondria and chloroplasts) are known to cause damage to macromolecules such as lipids, proteins, RNA and DNA and should therefore increase oDNA mutation rates. Actually, it has been suggested that the sequestration of electron transport chains into organelles is necessary to protect the nuclear genome from these free radicals (Allen & Raven 1996; Race *et al.* 1999). Furthermore, plant oDNA have a higher proportion of non-coding sequences compared to animal mtDNA. Their reduced mutation rates are therefore quite paradoxical, and may be attributed to more direct causes such as the efficiency of their DNA repair enzymes.

The reason stated above for reduced effective population sizes for plant organelle *versus* nuclear genomes may be explained as follows. In a diploid hermaphrodite species, the effective population size for organelle genes will be half that of nuclear genes, and one fourth for species with separate sexes, assuming that all individuals are homoplasmic (see 2.3) and that male and female effective population sizes are identical (i.e. perfect sexual symmetry, see Ross 1990). A higher reproductive success of males compared to females would reduce effective population sizes for both organelle and nuclear genes but the decrease would be more important for organelle genes. Indeed, it seems that a larger fraction of plants might contribute to the next generation as males at the time of flowering, whereas only a subset contribute as females at the time of seed production, at least in hermaphrodite (as opposed to monoecious) species (Oddou-Muratorio *et al.* 2001), but this needs further evaluation. In animals, the 4-fold difference may disappear in species where males have harems. In dioecious plants as well, the greater the excess of females, the lower the effective size for autosomal genes but the higher the effective size for maternally inherited organelle genes. In gynodioecious species, highly biased sex ratios in favor of females will also result in decreased nuclear effective population sizes but increased organelle effective population sizes, particularly in the case of cytoplasmic gynodioecy (Laporte *et al.* 2000). In some gynodioecious species, fairly high levels of organelle diversity have indeed been detected (e.g. Belhassen *et al.* 1993; Saumitou-Laprade *et al.* 1993; Olson & McCauley 2002). More conclusively, gynodioecious populations of *Beta maritima* were more variable and less differentiated at cpDNA markers than hermaphrodite ones (Forcioli *et al.* 1998). If the species considered are not diploid but polyploid, the effective population size will be further increased for nuclear genes relative to organelle genes. Similarly, with inbreeding and strict maternal inheritance of organelles, only the nuclear effective sizes will be reduced (Petit *et al.* 1993b).

Reduced gene flow for organelle genes compared to nuclear genes (see 3.1) should imply a better maintenance of total (species) diversity, as effective population size for a subdivided population is increased by a factor $1/(1-F_{ST})$ (Barton & Whitlock 1997). However, this holds only for the island model when there is demographic equilibrium. In highly subdivided species, local extinction (by deleting unique lineages, not found elsewhere) and range expansions (by spreading a single lineage across large areas) will in fact have the opposite effect. Preliminary data compiled by us suggest that plant species that have strong geographic structure at oDNA markers also have a reduced total diversity. A combination of high diversity and strong geographic structure – the ideal situation for phylogeographic reconstructions – is therefore not very likely in plants. Gymnosperms, characterized by a paternal mode of cpDNA inheritance, have been perceived as displaying relatively important levels of diversity. Although within population diversity is clearly higher, because of high pollen flow, it is not clear if *total* species diversity is indeed higher in these species. If so, this would support the suggestion above that high gene flow helps maintain diversity when demographic stochasticity is high.

The nature of variation in plant oDNA

Understanding of the nature of variation found in cpDNA and in mtDNA has progressed rapidly as a result of the increasing numbers of researchers characterizing the mutations identified in population and phylogeographic surveys. The types of mutations found in plant oDNA are outlined in Table 3.

Consequences of the nature and levels of variation of plant oDNA for phylogenetic reconstruction

Potentially, all of the mutations described for oDNA could generate homoplasy (i.e. represent recurrent or parallel events), complicating interpretations in phylogeographic studies. This potential would depend in part on the method used to screen mutations (sizing or sequencing) and on the targeted taxonomic level. Chloroplast microsatellites are generally considered inappropriate for phylogeographic purposes due to size homoplasy (Doyle *et al.* 1998; Lie-

Table 3. *Mutations found in plant oDNA.*

Mutation	Frequency	Source of mutation and consequences	Reference
Substitutions	Most mutations in coding sequences; 50-75% in non-coding ones	Transition/transversion ratio more balanced than in animal mtDNA, hence decreased risk of recurrent mutation	Wakeley 1996; Deguilloux, unpublished
T-C substitutions at edited sites	Most of the non-synonymous substitutions in edited sequences	Phylogenetic reconstructions remain possible (although divergence is affected) unless processed paralogs replace the original (edited) sequence	Bowe & dePamphilis 1996; Szmidt *et al.* 2001
Small indels (1-10 bp)	25-50% of mutations in non-coding regions	Slipped-strand mispairing during DNA replication or repair (inferred from association with short direct repeats)	Gaut & Clegg1993; Gielly & Taberlet 1994
Recombination between short dispersed sequences	Rare in cpDNA; more frequent in mtDNA	Associated with end points of inversions	Palmer 1992
Minute inversions caused by hairpin structures	Identified in both mt & cpDNA	Generated by short (15-30 bp) inverted sequences, as a consequence of intramolecular recombination	Golenberg *et al.* 1993; Kanno *et al.* 1993; Clegg *et al.* 1994; Dumolin-Lapègue *et al.* 1998
Variable number of tandem repeats at minisatellites	Identified in both mt & cpDNA	Replication slippage at repeats >5 bp	Sperisen *et al.* 2001; King & Ferris 2002
Variable number of tandem repeats at microsatellites	20-50 / cpDNA genome (depending on threshold considered); rarer (?) in mtDNA	Replication slippage at mononucleotide (mostly A or T) repeats	Powell *et al.* 1995; Provan *et al.* 2001; Soranzo *et al.* 1999; Sperisen *et al.* 2001; Vendramin *et al.* 2000

pelt *et al.* 2001). Similarly, the fact that small indels tend to occur at direct repeats, often clustered in some non-coding regions, and that the proportion of indels compared to substitutions decreases as the divergence between taxa increases suggests that they reoccur relatively rapidly at a restricted number of potential sites (Zurawski & Clegg 1987). Specific regions of cpDNA that have unusually high mutation rates are referred to as hotspots or hypervariable regions (e.g. the IR/LSC junction or a non-coding region near *rbc*L; Morton & Clegg 1993; Goulding 1996; Vaillancourt & Jackson 2000; see also Hipkins *et al.* 1995). For plant mtDNA, the frequent and complicated rearrangements that occur due to inverted and direct repeats can easily compromise phylogenetic interpretations when mtDNA is studied with probes (Palmer 1992; Tsumura & Suyama 1998; Wu *et al.* 1998). The maintenance of sublimons (substoichiometric molecules) at low concentrations and their rapid increase when exposed to new nuclear backgrounds or growth conditions could provide a means for rapid cryptic genome evolution (Small *et al.* 1987, 1989; Kanazawa *et al.* 1994), which may complicate phylogenetic interpretations, since the changes may be reversible. Such sublimons have recently been identified in animal (including human) mtDNA as well, where they are however less frequent than in plants (Dowton & Campbell 2001). The presence of such repeated sequences that do not sort out during mitosis (see 2.3) is akin to some form of heterozygosity.

However, in phylogeographic surveys of plants, restricting the analysis to single nucleotide changes would often result in very low resolution. Using the 'total evidence' by combining the various markers (by using different weights for phylogenetic reconstructions) seems more promising. For instance, cpSSR may be perfectly used to differentiate 'tip' haplotypes (i.e. autapomorphies). Furthermore, careful study of the various mutations identified may indicate those that may be used more readily for phylogenetic reconstructions. In particular, cpDNA indels of more than two bases that do not belong to tandem repetitions may be less prone to recurrent or parallel evolution (Gielly & Taberlet 1994). Even tandem duplications may have utility to confirm phylogenetic relationships at low taxonomic levels (Wolfson *et al.* 1991).

Sequences found in other genomes

Researchers surveying cpDNA and mtDNA variation should also be aware of the potential complications induced by the existence of sequences of cpDNA in mtDNA, and of sequences of both mtDNA and cpDNA in the nuclear genome of many plants. The following examples illustrate this point. Mitochondrial DNA of *Brassica* contains some 12 to 14 kb of cpDNA sequences distributed in over 11 locations (Hanson & Folkerts 1992). Similarly, there are 17 insertions of cpDNA sequences totaling 11 kb in the nuclear genome of *Arabidopsis thaliana* (The Arabidopsis Genome Initiative 2000). Concerning mtDNA, there are 13 small insertions totaling 7 kb as well as a large insertion of 270 kb close to the centromere of chromosome 2 (Lin *et al.* 1999). As a consequence of these transfers, nuclear mitochondrial pseudogenes (called *numts*) appear to have contaminated many PCR-based mitochondrial studies, resulting in robust, believable but incorrect phylogenies as well as false reports of heteroplasmy (Bensasson *et al.* 2001). *Numts* seem to be especially frequent in plants (Blanchard & Schmidt 1995, 1996) and while they may be interesting as molecular fossils or for studying nuclear mutation rates, measures to check and avoid *numts* have been described (Bensasson *et al.* 2001) and should be followed for phylogeographic studies based on oDNA.

2.3 Intraplant oDNA diversity and its consequences

Because organelles contain several genome copies per cell, heterogeneity of oDNA within cells (called heteroplasmy) is possible and may potentially complicate population and phylogeographic studies. The *stringent* nuclear genome has little margin for error: it is generally constrained to transmit equal number of gene copies into each daughter cell following meiosis or mitosis. By contrast, organelles are considered to have *relaxed* genomes (Birky 1983) since cells do not partition their organelles in a precise way at mitosis or meiosis. Consequently, if a cell contains more than one type of oDNA, and in the absence of selection, stochastic changes in haplotype frequencies will occur each time it divides, ultimately leading to fixation of either type in cell lineages. The rapidity of this sorting out process (also called vegetative segregation) will depend on the effective number of organelle genomes per cell. This number appears much smaller than the measured number of organelle genomes per cell (Bendich 1987; Bendich & Smith 1990; Gillham 1994). Although generally low, cases of intraplant diversity have been reported and need to be taken into account for the interpretation of population and phylogeographic studies.

Number of organelle genomes

The number of organelle genomes differs widely according to the organelle (plastid or mitochondria), the tissue or the species considered (Table 4). The amount of organelle DNA is largely controlled by the nuclear genome and seems to be related to the level of organelle gene expression (Bendich 1987; Griffin *et al.* 2001). In *Arabidopsis*, study of chloroplast division mutants has shown that the total amount of chloroplast material within a cell, rather than the number of chloroplasts, is the most important parameter for plant function (Pyke 1999). Kuroiwa (1991) has found that proplastid precursors from meristems, which are of primary importance for sorting out processes since meristems govern plant growth, contain many fewer genome copies than mature chloroplasts (Table 4).

Table 4*. Estimates of genome numbers in plastids and mitochondria.*

Organelle	Typical size of the organelle	Estimate of genome number	Estimate source	Reference
Plastid	4-6 µm Ø (lens-shaped)	1-2 genomes / proplastid precursor	*Nicotiana* meristematic cells	Kuroiwa 1991
		60 genomes (in 12 – 25 nucleoids) / chloroplast	Cells of mature leaves	Gillham 1994
		300 genomes / cell	Pea roots	Lampa & Bendich 1979
		560 genomes / cell	*Arabidopsis* leaves	Draper & Hays 2000
		1900-50 000 genomes / cell (8-23% total DNA)	Light-grown leaves	Bendich 1987
Mitochondria	0.5 x 1-2 µm	<0-1 genome equivalent / mt	Cucurbit leaves	Bendich & Gauriloff 1984
		26 genomes / cell	*Arabidopsis* leaves	Draper & Hays 2000
		100 genomes / cell	Pea roots	Lampa & Bendich 1984

Intraplant organelle diversity

Among the cases of intraplant oDNA diversity reported so far, many involve plants known to have bi-parental or paternal inheritance (Table 5), such as *Actinidia* (Chat *et al.* 2002), *Medicago* (Lee *et al.* 1988; Johnson & Palmer 1989; Fitter *et al.* 1996), or *Pinus* (Govindaraju *et al.* 1988; White 1990). Relatively stable plastid chimeras (variegated plants classified as periclinal chimeras) have long been known by horticulturists and by plant geneticists. A mutation produces a periclinal chimera if the affected cell is positioned near the apical dome of the meristem so that the cells produced by subsequent divisions form an entire layer of the mutated type (if incomplete the chimera is called mericlinal). The resulting meristem contains one layer which is genetically different from the remainder of the meristem (Kirk & Tilney-Bassett 1978; Lineberger n.d.). For instance, in *Pelargonium*, the mutant white subepidermal layer (LII), caused by plastid mutation, is sandwiched between two green layers (Baur 1909). Such chimeras breed true for LII since it is this layer that produces germ cells (as well as the white marginal tissues of the leaves). More frequently, sorting out produces unstable sectorial chimeras that yield different types of offspring depending on the position of the female flowers on the plant. An example of the latter is provided by a case of heteroplasmic sex determination in *Silene vulgaris*: gynomonoecious plants (consisting of sectors with female flowers and sectors with hermaphrodite flowers) found alongside female and hermaphrodite plants, were considered to be mtDNA chimeras since the offspring produced depend on which sector of the plant is used in the crosses (Andersson 1999).

Vegetative segregation

If an individual plant cell is heteroplasmic for organelle genomes, sorting out of organelles during mitosis will occur more or less rapidly depending on the effective number (n) of organelle genomes, assuming that selection is absent. Simple neutral models indicate that it takes about $10n$ consecutive cell divisions to reach complete homoplasmy (Michaelis 1949). The effective number of segregating plastid units in *Epilobium* (as deduced from the rapidity of sorting out) has been estimated to range from 10 to 20, and is in agreement with the number of plastids in meristematic cells (12 on average) (Michaelis 1967). Since the number of consecutive cell divisions during the life of an annual plant such as *Epilobium* is of the order of 100 (Michaelis 1967), most individuals will become homoplasmic within one generation for plastid mutations. The fact that plastids rarely fuse (Sears 1980) must limit the effective number of plastid genomes per cell. The observation that rare cases of plastid fusion seem to account for unusually slow sorting out (Vaughn 1981; Lax *et al.* 1987) supports this view. Transient tubular connections between chloroplasts within cells, which are reminiscent of bacterial pilli, have been recently rediscovered (Köhler *et al.* 1997). It has been suggested that they could play a role in establishing homoplasmicity (Shiina *et al.* 2000), but this remains to be demonstrated. Finally, maintenance of intraplant diversity may also be due to selection, as in the case of weeds heteroplasmic for mutations that confer resistance to herbicides such as triazine (Frey 1998). For plant mitochondria, Erickson & Kemble (1990) have found rapid sorting out in *Brassica napus* (completed within one generation) but intracellular selective forces seemed to be at work, so this result may not be typical. Indeed, in cybrids (plants regenerated from cells resulting from the fusion of an enucleated cell with a whole cell, thus creating a cytoplasmic hybrid), sorting out of mitochondria is generally slower than for plastids, lasting up to 4-5 generations in *Petunia* (Izhar *et al.* 1983), and is attributed to the higher number of mitochondria per cell in suspension cultures (Rose *et al.* 1990). The same result was found following paternal leakage in artificial crosses (Dulieu *et al.* 1990). Besides differences in organelle numbers, the frequent fusion of mitochondria, which may be considered as forming a syncytium (as seen in motion pictures of the cell organelles; e.g. Honda *et al.* 1964) could contribute to this difference between sorting out of plastid and mitochondrial genes, thereby reversing the expectations based on genome copy estimates.

Consequences for population surveys

Within-individual diversity is a prerequisite for recombination (see 2.5) and hence reticulate evolution (making it more difficult to reconstruct intraspecific phylogenies). Interpretations concerning genetic drift (see 2.2) and gene flow (see 3.1) for organelle genes may also be compromised in case of heteroplasmy. The possibility of having intraindividual variation should therefore be kept in mind, especially in plants where inheritance of the organelles is not strictly uni-parental. Even if no case of paternal leakage has been identified previously in the species under investigation, it would be important to look for within plant heterogeneity, since rare occurrence of intraindividual variation could precisely point to cases of bi-parental inheritance that had gone undetected with other techniques (see 2.4). Not only cpDNA, but also mtDNA should be examined in this respect, since sorting out could be slower for this organelle. The methods used to study variation should, therefore, be effective in detecting within-individual variation. PCR-based techniques, as used in many population surveys of oDNA variation, will often reveal only the most frequent type, compared to RFLP based on probes. This latter technique can indeed be quite sensitive (Larkin & Scowcroft 1981). Insertion/deletions will be especially useful compared to restriction site variation because incomplete digests will not confound interpretations, both for Southern blot-based analyses and for PCR-RFLP-based approaches. Sequencing approaches (direct or after cloning) may detect heteroplasmy if appropriate measures are taken, such as examining many clones (Chat *et al.* 2002), or looking for ambiguous nucleotide positions in electrophoregrams.

2.4 Organelle inheritance

The mode of organelle inheritance has a major effect on the distribution of oDNA diversity and is probably the single most important factor determining the level of geographic structure in plants, as shown by surveys in Pinaceae where cpDNA and mtDNA are transmitted through gametes of opposite sexes (Table 5). Furthermore, even low levels of paternal leakage may have disproportionately large effects (see 3.1), and its occurrence implies that the geographic patterns unraveled are no longer the outcome of colonization by seeds only. This should be borne in mind, because reports of bi-parental or even of paternal inheritance for both cpDNA and mtDNA are relatively frequent in plants. Consequently, determining the mode of transmission of organelle genomes is of utmost importance before investing in any thorough population survey (e.g. Cruzan *et al.* 1993), although this phase is often omitted. Here, we provide a short overview of methods that may be used for that purpose and discuss their respective advantages. We also give some information on the mechanisms involved and consider the critical assumption of a uniform organelle transmission rate across a species' range.

Plastids

One of the most complete reviews of cpDNA inheritance in angiosperms is that of Harris & Ingram (1991). Bi-parental transmission was found in 27% of the 88 families, 21% of the 233 genera and 27% of the 398 species investigated. There was evidence for variation in mode of inheritance at all taxonomic levels, including among plants of the same species, indicating that bi-parental plastid transmission is not a rare phenomenon in angiosperms (Table 5) and that uni-parental transmission should no longer be assumed. In some cases, studies involving chlorophyll deficiency or herbicide resistance have examined a large number of progenies, allowing detection of rare cases of paternal transmission, sometimes as low as 0.05%. In other cases, however, maternal inheritance appears to be close to absolute. In conifers, there are now numerous records of paternal inheritance of cpDNA, whereas a few cases of paternal or predominantly paternal inheritance of cpDNA have now been reported in angiosperms (see Table 5).

Mitochondria

Studies of mitochondrial inheritance in plants have been much more limited, and the situation is improving only slowly (Smith 1989; Reboud & Zeyl 1994;

Table 5. *Some examples of oDNA transmission patterns in plants.*[1]

Family	Species	cpDNA[2]	mtDNA[2]	References
Angiosperms – Dicotyledons				
Actinidiaceae	*Actinidia deliciosa*	P(m)	M	Chat *et al.* 1999, 2002
Asteraceae	*Helianthus annuus*	M	M	Rieseberg *et al.* 1994
Brassicaceae	*Arabidopsis thaliana*	M	M	Röbbelen 1966; Martìnez-Zapater *et al.* 1992
	Brassica napus	M	Mp	Corriveau & Coleman 1988; Erickson & Kemble 1990
Convolvulaceae	*Pharbitis nil/P. limbata*	Pm	?	Hu *et al.* 1996
Cucurbitaceae	*Citrullus lanatuss*	M	M	Havey *et al.* 1998
	Cucurbita pepo	M	M	Havey *et al.* 1998
	Cucumis melo	M	P	Havey *et al.* 1998
	Cucumis sativus	M	Pm	Corriveau & Coleman 1988; Havey 1997
Fabaceae	*Lens culinaris*	Mp	Mp	Rajora & Mahon 1994, 1995
	Medicago sativa	PM	M	Lee *et al.* 1988; Masoud *et al.* 1990; Forsthoefel *et al.* 1992
Fagaceae	*Quercus robur*	M	M	Dumolin *et al.* 1995; Dumolin-Lapègue *et al.* 1998
Geraniaceae	*Pelargonium zonale*	MP	MP	Baur 1909; Nagata *et al.* 1999
Salicaceae	*Populus* spp.	M	M	Mejnartowicz 1991; Rajora & Dancik 1992; Radetzky 1990; Rajora *et al.* 1992
Solanaceae	*Petunia hybrida*	M(p)	M(p)	Dulieu *et al.* 1990
	Nicotiana tabacum	M(p)	M	Medgyesy *et al.* 1986
Turneraceae	*Turnera ulmifolia*	PM	?	Shore & Triassi 1998
Apiaceae	*Daucus carota*	M	M	Steinborn *et al.* 1995; Thompson 1961
Zygophyllaceae	*Larrea* spp.	P	?	Yang *et al.* 2000
Angiosperms – Monocotyledons				
Musaceae	*Musa acuminata*	M	P	Fauré *et al.* 1994
Poaceae	*Secale cereale*	MP	MP	Fröst *et al.* 1970; Soliman *et al.* 1987
Gymnosperms				
Araucariaceae	*Agathis australis*	P	P	Owens *et al.* 1995
Cephalotaxaceae	*Cephalotaxus drupaceae*	M	M	Mogensen 1996
Cupressaceae	*Calocedrus decurrens*	P	P	Neale *et al.* 1991
	Chamaecyparis spp.	P	P	Kondo *et al.* 1998
	Cunninghamia sp.	M	?	Lu *et al.* 2001
Ephedraceae	*Ephedra*	M	M	Mogensen 1996
Pinaceae	*Picea abies*	P	M	Grivet *et al.* 1999; Sperisen *et al.* 2001; Sperisen & Vendramin, unpublished data
	Pinus taeda	P	M	Neale & Sederoff 1989
	Pinus contorta	P	Mp	Wagner *et al.* 1987, 1991a
Taxaceae	*Taxus baccata*	P	M	Mogensen 1996
Taxodiaceae	*Sequoia sempervirens*	P	P	Neale *et al.* 1989

[1] This is a non-exhaustive list focusing on species where both cpDNA and mtDNA inheritance have been reported and/or where unusual modes of inheritance were found, and is meant to illustrate the diversity of transmission modes observed in seed plants.

[2] M: Maternal, P: Paternal, MP, PM: bi-parental but predominantly maternal or paternal, Mp, Pm: uni-parental with some leakage, M(p), P(m): predominantly uni-parental with exceptional cases of leakage.

Röhr *et al.* 1998; and see Table 5). Until recently, maternal inheritance of mtDNA was virtually axiomatic in angiosperms (e.g. Erickson & Kemble 1990; Mogensen 1996). Well-studied cases of bi-parental cpDNA transmission associated with strict maternal inheritance of mtDNA (e.g. in *Medicago* or in *Pinus*) may have reinforced this impression. Most data have been based on RFLPs, and given the sample sizes typically used, these would not detect paternal contribution lower than 1-10%. Although cytoplasmic male sterility (CMS), generally attributed to mtDNA (see 3.4), has been studied in over 150 species (Hanson 1991), and generally found to be maternally inherited, its obvious effects on fertility (rendering reciprocal crosses impossible) and the possibility of reversion have limited its usefulness for mtDNA inheritance studies. The report of strict paternal inheritance of mtDNA in some families of conifers, as well as more recent findings of angiosperms showing paternal or bi-parental transmission of mitochondria, but maternal inheritance of plastids (in *Musa*, *Brassica*, *Cucumis*; see Table 5) indicate that the modes of transmission of the two organelles are not necessarily coupled, and that the impression of greater maternal bias for mitochondria may have been premature.

Cytological and genetic determination

The extensive surveys based on genetic and cytological analyses during the first three quarters of the 20th century were completed by studies relying on molecular investigations and by new cytological approaches, especially based on epifluorescence microscopy (Corriveau & Coleman 1988; Nagata *et al.* 1999). The increase or decrease in DNA content of mtDNA or cpDNA in the reproductive cell of pollen grains seems to be well correlated with patterns of organelle transmission, as inferred using genetic markers, but subsequent elimination of male organelles or oDNA in the embryo remain possible and paternal and bi-parental inheritance cannot be distinguished from each other with such methods. Molecular methods are in principle more general and more precise, but they also have their limitations. Given the low levels of diversity of oDNA, the identification of molecular differences between the parents, especially in intraspecific crosses, may remain difficult. Highly polymorphic markers such as cpSSRs (see 2.2) may help overcome this problem. Contamination by foreign pollen in controlled crosses can also be monitored more systematically using nuclear markers and this has become standard practice (e.g. Wagner *et al.* 1989; Chat *et al.* 1999). For all these studies, it is essential to include as many crosses as possible since genetic variation may exist for the control of oDNA inheritance (see 2.2). Use of serial backcrosses (Lansman *et al.* 1983) and of sensitive PCR techniques (e.g. allele-specific amplification) are also recommended.

Determination of organelle transmission in nature

Another advantage of molecular markers is that they allow the determination of transmission patterns of organelles in nature, without the need to produce controlled crosses. Indeed, some wild plant species may be difficult to cross. In such cases, convincing evidence for the predominant mode of inheritance may nevertheless be obtained in the field, provided that within population oDNA variation exists and information on mating system is available, as shown in Oddou-Muratorio *et al.* (2001). When half-sib progenies are collected, and oDNA variation is studied in both the parents and offspring, only a subset of the offspring is expected to be informative, compared to controlled crosses, as selfing may occur or the father may have the same oDNA haplotype as the mother. The number of informative crosses, for which the two parents are expected to harbor different cpDNA haplotypes, may be estimated by assuming that plants either self or mate at random in the population:

$$N_{\text{inf}} = \sum_i N\,(1-p_i)\,(1-s),$$

where p_i is the frequency of haplotype i in the population, s is the selfing rate and N is the number of seedlings analysed. In controlled crosses, when all seedlings have an oDNA haplotype identical to that of one of the two parents, the maximum rate of transmission (P) from the other parent is given by the binomial model: $P = 1-(1-\beta)^{1/N}$ where N is the total number of seedlings assayed and β is the statistical power (Milligan 1992). For half-sib progeny, N_{inf} is simply substituted for N in the equation. Full paternity analysis, if available, will eliminate the need for assumptions of random mating and should yield the same precision as regular artificial controlled crosses. Such methods measure *real* levels of bi-parental transmission in nature.

Mechanisms of organelle transmission

In plants, the mechanisms of inheritance of organelle (especially plastid) genes have received much attention (Whatley 1982; Connett 1987; Hagemann & Schröder 1989; Reboud & Zeyl 1994; Mogensen 1996; Birky 2001). They are extremely varied, including simple differences in quantitative inputs of organelles between egg and sperm cells (Russell 1987), pure exclusion of all organelles originating from the sperm cell during syngamy (Mogensen 1988), selective increase or decrease of oDNA in the generative cell of pollen grains (Nagata *et al.* 1999) or active elimination of part or all of the organelles in the egg cell of conifers (Chesnoy 1987). In species showing regular bi-parental transmission of plastids (such as *Pelargonium*, *Oenothera* or *Medicago*), rates of paternal transmission are variable and are determined by environmental (Yu & Russell 1994, but see Tilney-Bassett 1970) and genetic factors, including the nuclear genome, the oDNA genomes and their interactions (Table 6).

Table 6*. Evidence for different factors governing rates of paternal transmission of plastids.*

Factor	Example	References
Environmental		
	In *Medicago*, the number/distribution of plastids from each parent during early embryo development influences the degree of plastid transmission following hybridization	Zhu *et al.* 1993
Genetic		
Nuclear genome	Capacity to transfer plastids via pollen in *Petunia* mutant is under the control of two nuclear genes acting at the male gametophytic level	Derepas & Dulieu 1992
	In *Pelargonium* two complementary genes acting at the level of the female sporophyte affect plastid transmission	Amoatey & Tilney-Bassett 1994
	In *Oenothera* the genotypes of both parents influence the degree to which plastids are transmitted to offspring: flowers with long styles/pollen with low growth capacity result in reduced frequency of paternal transmission as the time interval between pollination and fertilization is lengthened, resulting in more plastid degeneration in the sperm cell	Chiu & Sears 1993
Organelle genome	Plastid genomes at a numerical disadvantage at fertilization may come to predominate in mature plants due to differential multiplication rates e.g. *Medicago sativa, Oenothera, Pelargonium*	Fitter & Rose 1993; Schötz 1975; Tilney-Bassett 1975
Nucleo-cytoplasmic interactions	In *Oenothera* relative efficiencies of plastome transmission coincide with the relative compatibilities between each plastome and the nuclear background of the progeny	Chiu & Sears 1993

The fact that nucleo-cytoplasmic interactions can play a role suggests that patterns of inheritance in wide crosses (e.g. involving different species) may be atypical, so that transmission rates measured using such crosses may differ from those involving more closely related individuals. The mechanisms involved in organelle transmission can have similar effects on both types of organelles, or may act independently on cpDNA and mtDNA. For instance, Nagata *et al.* (1999) showed that the increase or decrease of oDNA in young generative cells determines the mode of inheritance in the species studied and is *independent* for cpDNA and mtDNA, whereas some of the mechanisms described by Hagemann & Schröder (1989) should result in similar transmission patterns for both types of organelles. One may therefore expect a positive but weak correlation between the mode of transmission of cpDNA and mtDNA in plants, although this has not been tested so far. Also unknown is the mode of transmission of oDNA into the triploid endosperm of angiosperms, which could differ from that of the embryo. Indeed, the two sperm cells are dimorphic in some plant species differing notably by their organelle content (Russell 1991).

Consequences of bi-parental inheritance

By repeatedly mixing oDNA lineages within cells, bi-parental inheritance could have several consequences of importance for population and phylogeographic studies. Compared to strict maternal inheritance, bi-parental inheritance will increase gene flow (see 3.1), may allow recombination (see 2.5) and could facilitate the invasion of selfish DNA elements (see 3.5). It could also modify genome structure and size in an unpredictable fashion. For example, by repeatedly mixing different organelle genomes into the same cells, bi-parental inheritance could favor intracellular forces, such as *replication races* between organelle genomes, leading to slimmer genomes (Walsh 1993; Selosse *et al.* 2001). The evolutionary history of *organelle sex* is full of reversals and parallel changes (Birky 1995), suggesting that uni-parental inheritance is not consistently advantageous or detrimental for a species, and that transient selection pressures may be at work. The observed correlation between plastid inheritance and mating system, with most cases of bi-parental or paternal inheritance found in allogamous species (Reboud & Zeyl 1994), is noteworthy, since it is in allogamous species that the impact of paternal leakage is greatest (see 3.1). This observation points to some active 'Red Queen' process where organelle and nuclear genes are engaged in an evolutionary contest to modify the extent of cytoplasmic inheritance. Characterizing differences in rates of paternal transmission across a species' range would allow a better understanding of the mechanisms shaping this intriguing component of the genetic system and would help interpret phylogeographic surveys.

2.5 Recombination

Here we discuss the evidence that has been produced to substantiate the existence of cases of oDNA interparental recombination in nature and discuss its consequences for phylogeographic studies.

Evidence

A prerequisite for interparental recombination is the existence of heteroplasmic cells, following bi-parental inheritance, which has been reported several times (see 2.2 and 2.4). However, only indirect evidence supports interparental oDNA recombination under natural conditions (e.g. Dally & Second 1990; Govindaraju *et al.* 1989). On the contrary, the study of somatic hybrids produced by fusion of protoplasts has established that plant mtDNA but not cpDNA will readily recombine when present in the same cell (Belliard *et al.* 1978, 1979, reviewed in Rose *et al.* 1990). For instance, Donaldson *et al.* (1994) found 95% of rearranged mtDNA in somatic hybrids of *Nico-*

tiana, but independent segregation for the parental cpDNA types, whereas Cardi *et al.* (1999) report up to 75% of non-parental mtDNA restriction patterns in potato somatic hybrids and again no cpDNA recombinant. For cpDNA, recombination has been detected in *Nicotiana* cybrids, but only following stringent selection procedures (Medgyesy *et al.* 1985; Thanh & Medgyesi 1989), resulting in a fine mosaic of parental plastid molecules (Fejes *et al.* 1990). The few direct attempts to identify recombination by investigating cpDNA in the progeny of normal crosses (e.g. Chiu & Sears 1985) have failed, probably as a consequence of the rarity of plastid fusion (Sears 1980). Until more data becomes available, one may tentatively suggest that, in nature, interparental recombination involving mtDNA would be limited by the rate of bi-parental transmission, whereas for cpDNA bi-parental inheritance would be a necessary but not a sufficient cause for interparental recombination.

Consequences of recombination

The occurrence of interparental recombination, even at frequencies lower than mutation rates, could dramatically affect intra- or interspecific phylogenies and hence phylogeographic hypotheses, and will increase levels of oDNA diversity. Indeed, in the absence of recombination, M mutations will generate M+1 haplotypes, assuming no homoplasy (each new mutation adds a new variant), but if recombination occurs, there will be 2^M possible haplotypes and a fully reticulating phylogenetic tree. In the complete absence of recombination, there is in fact a risk of mutational meltdown (Lynch *et al.* 1993). This is due to the irreversible accumulation of slightly deleterious mutations, a process called Muller's ratchet (Muller 1964), acting synergistically with low population sizes. This problem needs to be considered with particular attention in plants, because they have greater coding capacity than animals in their cytoplasm (mtDNA and cpDNA). Furthermore, lack of recombination implies complete linkage. If there is an advantageous mutation (the driver) which goes to fixation, all other mutations not present on this genome will become extinct, with only those variants associated with the driver (the hitchhikers) remaining (see 3.4).

2.6 Dynamics of dicytoplasmic systems

The nuclear and organelle genomes, being transmitted differently, are generally expected to be in linkage equilibrium, except in the case of the recent admixture of differentiated populations or species (Asmussen & Arnold 1991) or when there is strong epistasis (nucleo-cytoplasmic interaction, see 3.4). On the contrary, in a *dicytoplasmic* system where both organelle genomes are normally transmitted by the same parent, full disequilibrium is expected.

Paternal leakage of one of the two organelles (or maternal leakage if cpDNA and mtDNA are both paternally transmitted) should have a major effect on the combined history of the two organelle lineages. Indeed, if such events occur, even very rarely, the two genomes will no longer tell the story of a single 'combined' lineage, and could be considered as two distinct cytoplasmic 'loci'.

Evidence from artificial crosses and *in vitro* manipulations

Breeders have attempted to create new organelle associations by taking advantage of rare paternal transmission, in order to combine useful features of cpDNA (such as herbicide resistance) and of mtDNA (CMS) (Erickson & Kemble 1990). The study of somatic hybrids differing at both organelles has shown that cpDNA and mtDNA generally segregate independently (Rose *et al.* 1990). Note however that direct interactions between the two organelles may exist (see 3.4), as well as indirect ones through the association with the nuclear genome (see 3.5). Because of the high frequency of mtDNA recombination, the only way to obtain new association of (intact) parental cpDNA and mtDNA is through paternal leakage of chloroplasts (and not of mitochondria) in crosses (Medgyesy *et al.* 1986). On the other hand, when maternal inheritance is the rule, both genomes can be used interchangeably as markers. For instance, in studies on male sterility in cultivated species, cpDNA markers have sometimes been used to distinguish between different male sterile cytoplasms, despite the fact that CMS is generally attributed to mtDNA (see 3.4).

Population studies

In natural populations, if cp and mtDNA are not strictly co-transmitted, then the number of genetically distinct 'cytoplasms' will quickly multiply. If both oDNA are strictly co-transmitted, they will represent a single (maternal) lineage and most mutations in one genome will match a corresponding lineage in the other. Dumolin-Lapègue *et al.* (1998) tested the association between the two organelles in oaks (where both genomes are predominantly maternally transmitted), and indeed demonstrated that in these species both organelles seem to be fairly strictly co-transmitted across generations (i.e. are in full linkage disequilibrium, except for recurrent or parallel mutation events). Subsequent studies reached the same conclusion in *Beta maritima* (Desplanque *et al.* 2000), *Silene vulgaris* (Olson & McCauley 2000), evergreen oak species (Belahbib *et al.* 2001) and *Cycas taitungensis* (Huang *et al.* 2001). Future studies with other species should provide evidence for partial or complete uncoupling of the two lineages, given that bi-parental inheritance exists in some taxa.

3. Determinants of the geographic structure of organelle genes

Probably few higher organisms follow Beijerinck's Law, which states that *everything is everywhere but the environment selects* (cited in Sauer 1988). Indeed, although the physical and biotic environment may control plant invasions, seed dispersal has intrinsic limitations. Plants have their *Drosophila* (broad dispersal) as well as their salamanders (highly restricted dispersal). However, they are often strongly structured at maternally inherited organelle genes, implying that seed dispersal is often limited. Currently, a great deal is known about the organization of diversity at nuclear *versus* organelle genes in plants, a side-effect of the low level of oDNA diversity that incited focus on a few markers but with large sample sizes. In addition to a review of this, we also provide some elements of theory that helps interpret patterns of diversity in terms of past and ongoing gene flow. Such theory may support the use of organelle genomes in phylogeography, inspire new investigations, and guide experiments by pointing to the most important parameters to be measured (Birky 1991). The mechanisms underlying seed flow are then discussed, followed by consideration of other processes that can shape geographic variation of organelle lineages within species, such as history (Quaternary history, plant culture and domestication), selection, interspecific gene flow or even horizontal transfers.

3.1 Intraspecific gene flow

In phylogeographic studies, better historical inferences will be possible whenever there is a strong geographic structure coupled with high levels of diversity. In recent years (see 1), together with growing appreciation that intraspecific diversity can indeed be identified, especially in cpDNA (see 2.2), theoretical approaches (e.g. Petit *et al.* 1993b; Ennos 1994) have promoted the appreciation of the potential for organelle markers to detect geographic structure, compared to nuclear markers. They also showed that these markers, in combination with nuclear markers, may be very useful for gene flow studies in plants (McCauley 1995).

Dispersal *versus* gene flow

Seeds or plant segments (vegetative organs more or less specialized in dispersal) generally play a greater role in local recruitment or *colonization* of new, empty habitats and a more limited role in *gene flow sensu stricto* (the movement of genes among established populations), unless population sizes are low and there are frequent colonization and extinctions of local populations (as in metapopulation models: e.g. Husband & Barrett 1996; Oddou-

Muratorio *et al.* 2001). As a consequence, the maternal genetic structure established during colonization will generally persist for a long time (e.g. Petit *et al.* 1997).

Organization of diversity in organelle and in nuclear genes (theory)

In plants, subdivision of diversity is expected to be much stronger for maternally inherited organelle genes than for nuclear genes. Reasons include the lower effective population sizes for organelle *versus* nuclear genes (see 2.2) and the contrasting levels of gene flow experienced by each genome due to differential inheritance. Indeed, nuclear genes are dispersed in pollen (half of the time, when they are transmitted through the male gamete) and in seeds (in all cases, whether they are transmitted from the male or from the female gamete), whereas maternally inherited organelle genes are dispersed only in seeds. The situation is therefore intrinsically asymmetric, and even if pollen flow is very low or absent (autogamous or apomictic species), organelle genes will never experience more gene flow than nuclear genes. Actually, assuming *equal* seed and pollen flow, maternally inherited genes are expected to partition up to three times more diversity among populations, compared to nuclear genes (Petit *et al.* 1993b and unpublished data). Finally, gene flow by pollen is often more important than by seeds (e.g. Levin & Kerster 1974; Ellstrand 1992). The relative importance of seed and pollen flow based on information from organelle and nuclear genes was first studied theoretically by Petit (1992) using the simple island model of population structure and assuming equilibrium between gene flow and drift. Effective migration rates for plants are more complex than those for animals, due to the asymmetry between male and female transmission. If α is the rate of transmission through female gametes, the effective migration rate is $m_e = \alpha m_s + (1-\alpha)(m_p + m_s - m_p m_s)$ for plants (where s stands for seeds, p for pollen), compared to $m_e = \alpha m_{♀} + (1-\alpha) m_{♂}$ for animals. For plants, this expresses the fact that genes transmitted through male gametes can migrate with seeds, with pollen or with both, whereas for animals genes may migrate with males or females exclusively each generation. If genetic differentiation estimates are available for both organelle (G_{STc}) and nuclear markers (G_{STn}), the migration ratios $m_{♀} / m_{♂}$ and m_p / m_s are given by:

$$r = \frac{8\alpha(1-\alpha)(1/G_{STc}-1) - \alpha(1/G_{STn}-1)}{(1-\alpha)(1/G_{STn}-1) - 8\alpha(1-\alpha)(1/G_{STn}-1)}$$

for animals with separate sexes, assuming a sex ratio of one, and

$$r = \frac{2(1/G_{STc}-1) - (1/G_{STn}-1)}{(1-\alpha)(1/G_{STn}-1) - (1/G_{STc}-1)}$$

for hermaphrodite plants, assuming equal number of effective males and females (i.e. no sexual asymmetry) and no selfing. The relationship between nuclear and organelle G_{ST} values can then be deduced for the complete range of pollen to seed ratios (Figure 3).

In animals, G_{ST} for organelle markers will always be higher than for nuclear markers, regardless of the mode of inheritance. In plants, this is also true for maternally inherited genes. For paternally inherited genes, genetic differentiation should be equal to or slightly higher than for nuclear genes, since both are transmitted through seeds and pollen (but not with the same frequency).

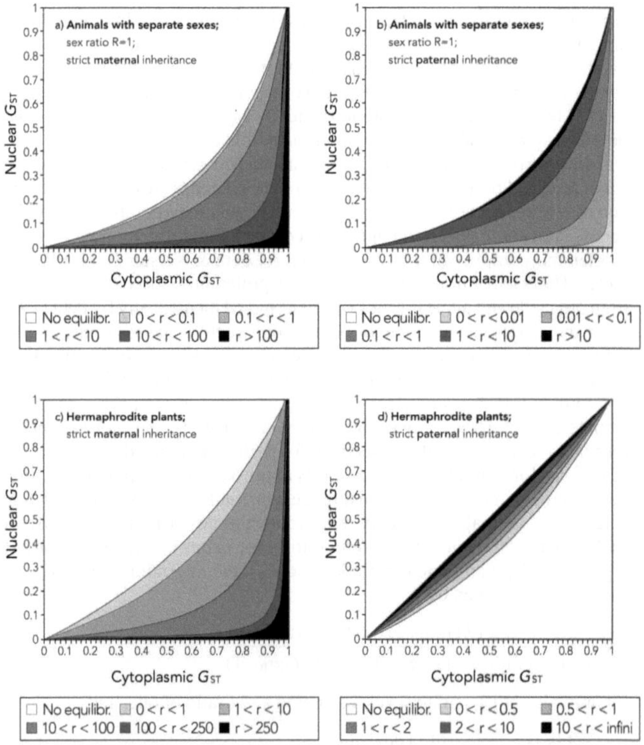

Figure 3*. Parameter space for nuclear* G_{ST} *versus organelle* G_{ST} *at equilibrium in an island model of population structure, for different values of the migration ratio* r. *After Petit (1992).*

Organization of diversity for organelle and nuclear genes (experimental data)

Despite a number of unrealistic assumptions, these theoretical predictions have been roughly confirmed by experimental results from population surveys. Table 7 provides mean values of G_{ST} from an unpublished literature review (Petit *et al.* 2005), comparing the results of angiosperms with those for conifers. All conifer species studied are characterized by contrasting modes of inheritance (cpDNA being predominantly paternally inherited and mtDNA predominantly maternally inherited). The effect of the mode of transmission is particularly clear in conifers (mean G_{ST} is 0.76 for the maternally inherited mtDNA and only 0.16 for the paternally inherited cpDNA), since the comparison is between two organelle genomes that differ only by their mode of transmission. Except for the paternally inherited cpDNA of conifers, there have been no other reports where $G_{STn} > G_{STc}$ (unpublished compilation; see also Ennos 1994; El Mousadik & Petit 1996; Ouborg *et al.* 1999). The dispersion of values is very large and G_{STc} in particular can range from < 0.1 to close to fixation.

Table 7*. Subdivision of diversity at organelle and nuclear genes in angiosperms and conifers. N equals the number of studies.*

Group	Genome	N[1]	G_{ST} *mean*	G_{ST} *min*	G_{ST} *max*
Angiosperms	Chloroplast	69	0.71	0.07	1.00
	Mitochondria	11	0.61	0.08	0.97
	Nucleus	42	0.19	0.02	0.67
Conifers	Chloroplast	24	0.16	0.00	1.00
	Mitochondria	10	0.76	0.23	0.97
	Nucleus	21	0.11	0.02	0.38

[1] Number of studies.

Possible complications: leakage during transmission and assumption of equilibrium

For most of the species surveyed so far, the mode of inheritance has not been determined precisely. When pollen flow is much higher than seed flow (as in many wind-pollinated species), even a low level of paternal leakage could substantially affect the distribution of diversity. To illustrate this, consider a species with a high pollen: seed migration ratio ($m_p = 0.225$, $m_s = 0.0004$, $N = 100$; Petit 1992). Figure 4 provides G_{STc} values for a range of paternal leakage rates using these parameters. The results indicate that even rates below 1% can significantly affect the distribution of diversity. Note that the converse is not true for paternally inherited genes: a low level of maternal leakage will not significantly affect the distribution of oDNA diversity, even if seed flow is much higher than pollen flow. This is because paternally inherited genes are always moved in seeds as well as in pollen. So far, no study has attempted to contrast subdivision measured with each organelle when one of them presents some paternal leakage and the other is strictly maternally inherited, even if phylogeographic or population surveys have included species where cpDNA inheritance is not strictly uni-parental (e.g. Cruzan *et al.* 1993; Dong & Wagner 1994; Mason *et al.* 1994). Another caveat is the assumption of equilibrium, and the fact that these estimates may be scale-dependent (McCauley 1995, 1997). In fact, recent studies have questioned the usefulness of obtaining such estimates (Whitlock & McCauley 1999). However, new approaches have helped resolve some of the criticisms raised by previous models. In particular, indirect estimates can be compared to direct estimates (Latta *et al.* 1998) and other models than the very simplified island model can now be considered, including the stepping-stone and isolation by distance models (Hu & Ennos 1997; Rousset 1997; Oddou-Muratorio *et al.* 2001). Consequently, scale-specific estimates of gene flow may be obtained. An adaptation to oDNA markers of recent spatially realistic metapopulation models (Hanski 1999), that account for population dynamics within landscape contexts, will probably further improve gene flow estimates (Hanski 1999; Sork *et al.* 1999).

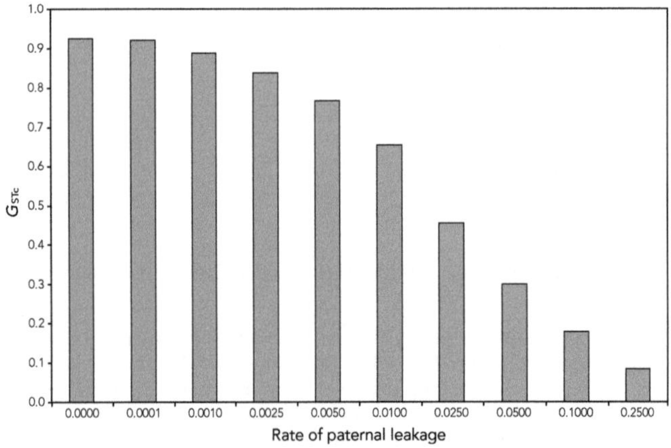

***Figure 4**. Effect of increasing levels of paternal leakage on oDNA population structure: case of asymmetric pollen and seed flow (see text).*

Alternative methods relying on joint nucleo-cytoplasmic data have also been proposed (Orive & Asmussen 2000). Comparing estimates of clinal width based on markers having different modes of inheritance may also provide insights into relative pollen and seed flow. For instance, in *Abies alba*, genetic surveys based on mtDNA and cpDNA (characterized respectively by maternal and paternal modes of inheritance) indicate that stepwise gene flow through pollen rather than through seeds has resulted in genetic exchanges among formerly separated Ice-age refugia (Liepelt *et al.* 2002).

Consequences for sampling

For maternally inherited organelle genomes, the generally high level of fixation of oDNA diversity within populations is a great opportunity for phylogeographic surveys. It means that instead of focusing efforts on intrapopulation sampling, in order to determine precisely haplotype frequencies, researchers should instead sample more populations with lower sample sizes in each (Pons & Petit 1995). Note that the partitioning of diversity may still be determined with sample sizes as low as 2-3 individuals per population, provided that there are many populations (e.g. Tremblay & Schoen 1999). With many points sampled across the range, the spatial distribution of haplotypes will be better reconstructed, allowing improved reconstruction of past expansions (see discussion in Petit & Grivet 2002). Another decision that has to be taken in phylogeographic surveys is the relative amount of effort allocated to sample many informative characters (such as nucleotide sites) *versus* analyzing many individuals. In plant organelles, sampling of many characters may become very costly, if informative sites are distributed sparsely, so that comparatively longer sequences need to be investigated. A solution is to identify the polymorphism first and to find a way to screen these rapidly. Attention must be given to the way in which variation is identified (i.e. use of a representative sample), otherwise biased results may be obtained.

Problems of inference in phylogeography

Although surveys based on organelle genes often detect higher levels of geographic structure than those based on nuclear genes, it is often considered that estimates of gene flow based on oDNA markers will provide low resolution because they are inherited as a single locus. Actually, as colonization occurs by seeds only, there is essentially a single history to describe. The videotape of colonization cannot be rerun again and again to obtain replicates of that (unique) history. With nuclear genes, however, all loci segregate largely independently, and the stochasticity of the genetic processes may be better evaluated, but the history being described is now that of colonization and gene flow in combination. Indeed, this history will more likely reflect the vagaries of pollen flow than the actual colonization history of the species. It is therefore important to acknowledge the truly phenomenological nature of any colonization pattern. At local or regional scales, replicates may be provided not across loci (since only one locus is available, even when both oDNA genomes are investigated, see 2.6) but across regions. At the range-wide scale, the analysis of several codistributed species, as well as simulation studies, may help shift attention to actual processes.

3.2 Ecology of dispersal

The unit of dispersal is often called a 'diaspore', short-hand for a spore, seed, fruit or portions thereof. Diaspores are the only vehicle for dispersal *sensu stricto* in many plant species that have no mechanism of vegetative dispersal. Dispersal determines the spatial arrangement and physical and biotic environment of seedlings and is therefore a crucial stage in the life cycle of plants. If organelle genes are maternally inherited, dispersal will condition their geographic spread and hence ultimately their phylogeographic structure. The diversity of the mechanisms of diaspore dispersal is considerable (van der Pijl 1972; and see below), and whenever a phylogeographic survey of a plant species is planned, it is advisable to carefully evaluate evidence concerning the mechanism(s) involved, as well as the existence of vegetative dispersal.

Modes of diaspore dispersal

Dispersal mechanisms have been classified according to the nature of the vector (van der Pijl 1972): autochory (inherent plant mechanisms such as explosive fruits), anemochory (wind), barochory (gravity; this category may be a fallback option), hydrochory (water), various forms of zoochory (animals), such as epi- or endozoochory. These categories have been used for classifications of floras (e.g. Mori & Brown 1994) but for several reasons are likely to be only weakly correlated with dispersal ability: i) Within any of these broad categories, large differences in dispersal abilities exist due to variations in seed morphology and physiology (see following section). ii) Many seeds may be dispersed by multiple vectors, or secondary dispersal may occur (Chambers & MacMahon 1994; Green & Johnson 1995). iii) Factors related to plant life history, such as fecundity and generation turnover additionally affect the dispersal potential and dynamics of species (Higgins & Richardson 1999; Hampe & Bairlein 2000). iv) Dispersal through time (e.g. existence of a seed bank or cumulated recruitment over several years for perennial species) may play an important role in determining spatial genetic structures (Levy & Neal 1999; Caron *et al.* 2000). v) The processes associated with dispersal are numerous and include fruit/seed removal, seed delivery, seed predation, seed bank dynamics, germination and seedling establishment, all necessary to link the reproductive cycle of adults with the establishment of offspring, thereby closing the 'seed dispersal loop' (Fenner 2000; Wang & Smith 2002). Among dispersal syndrome groups, seedling functional types differ, which may play a role in establishment (Ibarra-Manríquez *et al.* 2001). Factors acting at any of these stages of plant recruitment may constrain and overwhelm patterns generated by seed movements' *stricto sensu*. For instance, many plant species (e.g. orchids, as well as many members of the Ericales and Gentianales) produce an extremely high number of very small 'dust seeds', which disperse very well but depend on the presence of specific mycorrhizas or plant hosts for establishment (e.g. McKendrick *et al.* 2000). vi) Finally, the importance, frequency and geographic scale of perturbations of plant populations (Clark *et al.* 1998a,b; Kollmann 2000) and the degree of range fragmentation (Petit *et al.* 2002a) will also condition the stability of the spatial genetic structure established during colonization. Consequently, even good colonists may retain strong geographic structure at the broadest scale.

Other factors affecting seed dispersal

There are other important factors influencing dispersal capability within these broad categories. Within the animal-dispersed category, seeds may be ingested and defecated or regurgitated (fleshy fruits), transported and cached for later use (nuts), or attached with the help of barbs, hooks, or spines, potentially resulting in very different dispersal efficiency. Furthermore, the vagility of animal vectors may differ considerably. For wind-dispersed seeds, tiny cottonous, plumose or dust seeds may be carried at much longer distances than samaras, for instance. This is reflected in the level of subdivision of cpDNA diversity: 0.07 to 0.20 for *Populus tremula*, characterized by light cottonous seeds, or 0.21 for *Epipactis helleborine*, an orchid characterized by dust seeds, compared to 0.97 for *Carpinus betulus*, characterized by heavier winged seeds (Salvini *et al.* 2001; Squirrell *et al.* 2001; Grivet & Petit 2002b). Within a subcategory, large differences can exist: winged seeds may be very small, as in birch (*Betula*), a pioneer species, or very large in some late-successional tree species, potentially resulting in very contrasting dispersal patterns. Even between closely related species with similar seeds, small differences can matter: for instance, acorns of *Quercus petraea* are less visible in the canopy (absence of peduncle), are slightly smaller, less elongated, richer in tannins, less resistant to immersion and have delayed germination compared to acorns of *Q. robur*, a closely related, interfertile species. Consequently, acorns of *Q. petraea* are not as readily selected by jays (Bossema 1979) or transported in rivers. Furthermore, *Q. petraea* seedlings cannot withstand concurrence in more open sites. All these factors result in reduced long-distance seed dispersal for this latter species. This is reflected in the partitioning of cpDNA diversity: in Europe, N_{ST} (a measure of subdivision of diversity that takes into account similarities between haplotypes) for *Q. petraea* is 0.891 (with a standard error of 0.010), compared to only 0.809 (0.010) for *Q. robur* (Petit *et al.*

2002b). This trend was verified separately in the Alpine region, in France, central Europe, the United Kingdom, North Balkans and northern Europe (Petit *et al.* 2002b).

Stratified seed dispersal

Reid's paradox (Reid 1899; Clark *et al.* 1998b) posits that directly measured seed dispersal events away from the mother plant cannot account for the rapidity of the spread of trees and herbs during the Holocene. The only solution to this paradox is the existence of long-distance jumps across the landscape. Darwin (1859) was particularly interested in accidental means of long-distance transport, for instance with oceans, or in the mud attached to the feet of vertebrates, or in the digestive tracts of birds or mammals. Ant-dispersed temperate woodland herbs that have colonized Europe during the Holocene provide a striking example; dispersal distances by ants are generally <2 m, with a record at 35 m. Yet dispersal values two or three orders of magnitude higher are needed to account for their postglacial colonization rates: clearly, they had to disperse by other means (Cain *et al.* 1998). Higgins & Richardson (1999) and Cain *et al.* (2000) reviewed the importance of long-distance seed dispersal in plant populations and insisted that despite the difficulty of studying them they are so essential that attempts to account for the tail of the dispersal curve are obligatory for understanding the key processes in plant dynamics and phylogeography.

Seed heteromorphy illustrates the duality of selection pressures on dispersal in plants (Imbert 2002). A fascinating example is given in Sauer (1988), where a thorough discussion of many examples of plant dispersal and migrations may be found: in the sea rocket (*Cakile* spp.), a quick growing annual that colonizes ocean beaches, "*when the seeds are ripe, the indehiscent pod breaks in two. The bottom segment with half the seeds remains attached to the dying mother plant, which commonly gets buried in the sand; the top segment with the rest of the seeds is commonly washed away by storm waves to drift in ocean currents.*" Similarly, in species having cleistogamous flowers (flowers that do not open, resulting in obligatory self-pollination), two modes of dispersal generally coexist, with the outcrossed progeny of chasmogamous flowers being dispersed and the selfed progeny of cleistogamous flowers, genetically more similar to their parent, remaining close to it (Schoen & Lloyd 1984). Extreme examples of this strategy can be found in the 50 known amphicarpic plant species (amphicarpy is a phenomenon in which a plant produces aerial as well as subterranean fruits from cleistogamous flowers situated on subterranean shoots) (Lev-Yadun 2000).

In general, directed seed dispersal by animal vectors results in a seed shadow with many seeds concentrated close to and some seed accumulation far away from the initial seed source (for instance at another distant food source, under perches or in latrines). An example illustrating the multiplicity of vectors and their consequences for plant establishment is provided by Wenny & Levey (1998) for a neotropical tree species, *Ocotea endresiana*. Five bird species are involved in the seed dispersal of this tree. The resulting distribution is bi-modal: seeds are dispersed either near the fruiting tree, by four of the five bird species, or under song perches in favorable gaps, by the male of the fifth bird species. Recruitment primarily occurs in this latter habitat, illustrating the importance of directed dispersal to existing canopy gaps, whereas the other species provide dispersal conducive to random colonization of future gaps. Probably the largest existing data set illustrating this disperser-community effect on seed dispersal patterns is provided by Jordano & Schupp (2000) for the tree *Prunus mahaleb* in southeastern Spain.

Oaks also illustrate the importance of mixed dispersal strategies, involving gravity and rodents at short distances (10 - 100m), and birds (corvids) and rivers at medium (0.5-5 km) and long (>10 km) distances. The longest and hence rarest dispersal events, which cannot be directly studied, actually determine the spatial pattern established, the rapidity of the spread and the maintenance of diversity during colonization (Le Corre *et al.* 1997a, Petit *et al.* 1997, 2001). Even more surprising is the suggestion that the main agents of long-distance dispersal for 'wind' dispersed seeds are in fact birds (Wilkinson 1997). Other 'freak' events are described by Higgins & Richardson (1999), illustrating the difficulty to predict the nature and frequency of long-distance seed dispersal events from direct observation of the source populations; considering the sink populations, which were actually established through long-distance seed dispersal events, may actually be more informative, along with knowledge of actual migration rates and genetic patterns.

Potential role of pollen movements during colonization

Although, as a rule, *"only the sporophytic phase of plants can prospect new environments"* (Harper 1977), the male gametophyte may play some role during colonization. First, especially in allogamous species, colonists that have arrived by seeds may leave progeny only insofar that they can capture pollen (thereby rescuing the population established by long-distance seed dispersal; Richards 2000; McCauley *et al.* 2001). Consequently, the final configuration of successful populations across the landscape may eventually be shaped by pollen movements as much as by seed (or plant part) movements. Second, species may actually colonize new sites by pollen, taking advantage of female flowers of other related species to establish their own genes through hybridization and introgression (see 3.5).

Clonal spread and vegetative propagation

In addition to seed dispersal, vegetative dispersal may be important in shaping the geographic structure of oDNA in some plant species. One mechanism of vegetative dispersal is maternal apomixis, i.e. vegetative reproduction through seeds. The above discussion on seed dispersal mechanisms also applies to maternal apomixis. It is a comparatively common phenomenon, being found in some 40 angiosperm families (Vielle Calzada *et al.* 1996; Richards 1997). Interestingly, a case of paternal apomixis has been recently discovered in *Cupressus* (Pichot *et al.* 2001), where both organelles are paternally inherited (Chesnoy 1987). Other mechanisms of vegetative propagation include local multiplication of ramets that remain in contact with the first individual producing them (or eventually become detached). Such dispersal is restricted to local filling by clones, and will be important for colonization in combination with other long-distance processes (by seeds or by other means). Plant parts can also move and establish new populations. These parts may be small or unspecialized or on the contrary be very specialized and able to move long distances (van der Pijl 1972). For mixed strategies, involving seed dispersal and vegetative dispersal, the use of appropriate markers (nuclear multi-locus genotypes in particular) should allow identification of the main components of dispersal and gene flow (seed, pollen and vegetative), but this remains a largely untouched topic. In strictly clonal species, the study of colonization will be facilitated if there is more than one clone involved. In fact, there have been attempts to detect migration routes and barriers by examining the distribution of species (rather than molecular markers) in apomictic species complexes (Tyler 2000). On the other hand, cpDNA markers have been used to better describe apomictic dandelion species and their genetic relationships, pointing to significant gene flow between sections in which sexual reproduction is not currently known (Wittzell 1999). Similarly, a combination of cpDNA and nuclear markers has allowed to clarify the relative importance of sexual and asexual reproduction within *Allium vineale* (Ceplitis 2001) and to demonstrate the recurrent formation of polyploid and/or apomicts in *Arabis hoelboellii* (Sharbel & Mitchell-Olds 2001).

3.3 History

"...there is no more positive guide to the past occupation of any area by a particular species [than the discovery of fossils]. Nevertheless, we may garner a great deal of information from autecological and genecological studies of well-chosen species..." (Baker 1959).

When potential for interdisciplinary approaches exist, for instance by combining a phylogeographic survey based on organelle markers with direct historical information (fossils and other historical accounts), considerable progress may be made in the understanding of a flora. We consider here the importance of the Quaternary history for interpreting phylogeographic data, discuss the prospects offered by paleogenetics, and consider the special case of cultivated plant species.

Quaternary history

Current gene flow is not the sole factor determining genetic structure of plants, a fact that most phylogeographic surveys readily demonstrate. This is particularly obvious in temperate forest tree species whose history has been independently recorded in peat bogs and lakes (MacDonald 1993). At the range-wide scale, it is clear that the maternal genetic structure of these species is largely determined by the past distribution of populations during the last Ice Age (in refugia: locations where species survived during the harsher climatic episodes) and by subsequent colonization routes (e.g. Dumolin-Lapègue *et al.* 1997b; Petit *et al.* 2002b,c). High-Alpine plants offer another type of opportunity: here, the objective is to distinguish between the '*tabula rasa*' or the 'nunatak' hypotheses (Stehlik *et al.* 2001). For arctic plants, the huge circumpolar current ranges points to efficient dispersal abilities in these cold landscapes (Tremblay & Schoen 1999). Areas that remained unglaciated within the Arctic, such as eastern Beringia, may have played the role of refugia, like in temperate regions (Abbott *et al.* 2000). For tropical plants, climate changes were not as drastic as at higher latitudes (Dynesius & Jansson 2000) but changes in precipitation (especially periods of relative drought) may have elicited range contractions and subsequent expansions (e.g. Muloko-Ntoutoume *et al.* 2000). Clearly, each plant phylogeographic study will have to be considered within its own geographical and historical context.

The abundance of fossil material for some (but unfortunately not all) plant taxa offers the possibility to quantitatively combine historical information, such as age of arrival of a species in a region, with genetic data (e.g. Comps *et al.* 2001; Petit *et al.* 2001). Although plants are priviledged in this respect, oDNA sequences have rarely been used for studies of molecular clock calibrations to infer coalescent times within species, probably because of the (often) limited level of DNA variation and the lack of established calibrations. An exception is the study by Comes & Abbott (2001) of Mediterranean *Senecio* species, where intraspecific cpDNA lineages were shown to have survived for ca. 0.4-1.0 million years. As the importance of the historical frame for phylogeographic survey in plants has been reviewed recently (Cruzan & Templeton 2000; Petit *et al.* 2001) and also holds for other organisms, it will not be developed further here.

Paleogenetics and phylogeography

Population studies in plants have been based so far on contemporary material; i.e. DNA isolated from fresh material (generally leaves or buds). Although a great deal may be learnt about the spatio-temporal diversity dynamics of a species using samples from existing populations, a lot of genetic information is buried with fossils. The direct analysis of DNA present in plant fossils therefore offers considerable prospects. Indeed, macro- and micro-remains of plants are very numerous, owing to the greater biomass that plants represent in terrestrial ecosystems. Pollen grains and wood are especially abundant, whereas seeds are also frequently unearthed during archeological surveys. Ancient DNA (aDNA) studies on such material remain scarce, however. DNA isolation in hard and dry tissues of plants is complicated by the presence of polysaccharides or phenolics, adding to the already considerable difficulties in accessing highly degraded aDNA. Some early papers, such as that by Golenberg *et al.* (1990), which reported the amplification of Miocene sequences of plant tissues, were subsequently strongly criticized (e.g. Sidow *et al.* 1991), raising doubts on much of the early research in the field. However, several ongoing projects have now clearly shown that sequences present in high copy number (such as cpDNA or mtDNA) may still be accessible in (younger) plant remains. This is the case for dry oak wood, up to at least a few centuries (Dumolin-Lapègue *et al.* 1999a; Deguilloux *et al.* 2002). In addition, evidence for the persistence of cpDNA in cones from the last Ice Age (Kobayashi *et al.* 2000) and in pollen grains of *Abies* from the Pleistocene (150 000 BP) (Suyama *et al.* 1996) have been reported (note that both cpDNA and mtDNA are present in mature pollen grains of conifers, despite discordant inheritance; Wang *et al.* 1996). These two latter paleogenetic studies simply confirmed species identification using aDNA, but could in principle be extended to survey intraspecific organelle lineages. Work is in progress with European oaks in order to identify cpDNA haplotypes in fossils and to compare them with the existing phylogeographic structure. Archaeological remains of cultivated plants (mostly seeds so far) have also been successfully investigated for the presence of aDNA a few thousand years old (Goloubinoff *et al.* 1993; Rollo *et al.* 1994).

Origin of cultivated species and human impact on geographic structure

Many important domesticated plant species (which include some 230 species and 180 genera out of >300 000 species and >13 500 genera of angiosperms) have now been the subject of at least preliminary investigations using oDNA markers, to understand the process of their domestication (when and where) and to identify their wild progenitors and the origin of their cytoplasm in allopolyploid species. Simultaneously, phylogeographic studies of their wild relatives are carried out to characterize these important genetic resources. Most examples studied so far have reported considerable bottlenecks for organelle genes in domesticated plants compared to their wild progenitors (Doebley, cited in Gepts n.d.), illustrating the usefulness of these markers. Phylogeographic studies of domesticated plants are often complicated by the important seed dispersal ability of these species, at least prior to domestication. Indeed, selection by man sometimes results in the loss of dispersal ability by natural means and – in extreme cases – in the incapacity to survive in the wild, but this is then more than compensated by direct dissemination by humans. Another difficulty for phylogeographic surveys with wild relatives of domesticated species is the existence of ongoing gene flow (including seed flow, i.e. escape from cultivation) between the wild and domesticated populations (Doebley 1990; Zohary & Hopf 2000; Muller *et al.* 2001). Archaeological, historical and genetic evidence is needed to decipher such complex histories (Gepts n.d.). Phylogeographic studies of cultivated (but not yet domesticated) species (including for instance ornamental species, or forest trees) may also be complicated by extensive seed dispersal and plantations involving exotic material.

The spread of agriculture may have also affected the phylogeographic structure of some species, both in herbs (e.g. *Lolium*, Balfourier *et al.* 2000) or in trees such as *Castanea* (Fineschi *et al.* 2000). Cryptic invasions by exotic populations of the same species are also possible, as in the common reed in North America, potentially disrupting the preexisting phylogeographic structure (Saltonstall 2002). The comparison of oDNA geographic structure in introduced and native parts of the range is likely to be informative; evidence is accumulating that show that introduced populations have not always experienced strong founding events (Squirrell *et al.* 2001). In general, human influences, such as heavy afforestation with exotic material, will tend to blur a preexisting genetic structure because they are akin to gene flow, as shown in European oaks (König *et al.* 2002).

3.4 Selection

One potential caveat in the interpretation of oDNA population surveys is the assumption of selective neutrality. Mitton (1994), McCauley (1995) and Rand (2001) have discussed the possibility that selection can play a role in the level and distribution of oDNA diversity. However, phylogeographic studies based on oDNA are often disconnected from more classical studies of organelle genetics where the role of selection is discussed. In fact, the possibility that selection acting directly or indirectly on cytoplasmic genomes has shaped geographic patterns of oDNA variation has rarely been considered seriously, compared to other parameters such as gene flow or past demography. Except for the detailed studies on *Oenothera* and *Epilobium* cited in Part 1, which predate the rise of molecular markers, the bulk of the data on selection on oDNA comes from cultivated plants, where the analysis of reciprocal crosses and of maternal effects has a long tradition. To help bridge this gap, we consider here the major traits encoded by organelle genes, including cytoplasmic male sterility, and then discuss the importance of deleterious mutations affecting oDNA as well as the indirect selective effects due to the association between organelle genes and nuclear genes.

Major traits encoded by organelle genes

The analysis of alloplasmic lines, i.e. pure lines that differ only by their cytoplasmic genomes, which are readily produced in self-compatible plants, can reveal much about the selection pressures acting on organelle genomes. A good example is that of wheat and related species (Wang *et al.* 1997). In this genus, alloplasmic lines differ by many characters, either vegetative or reproductive. Detailed molecular screenings of cpDNA and mtDNA (both strictly maternally inherited in this group) have allowed the detection of much variation. Polymorphisms located in the cpDNA were preferentially related to growth characteristics, whereas those located in mtDNA were correlated mostly with reproductive traits (including but not only pollen production). This seems to fit with the function of these two organelles, which are expected to affect fitness rather than morphology, being specialized in the control of photosynthesis and respiration, as opposed to morphogenesis (Birky 1988).

Organelle genes have been shown to be involved in plant sensitivity or resistance to several fungi (Grun 1976). Examples include mildew (*Erysiphe* sp.) resistance in *Epilobium* (Michaelis 1935), as well as southern corn leaf blight (*Helminthosporium maydis*) and yellow leaf blight (*Phyllosticta zeae*) resistance in corn. Sensitivity of mtDNA to fungal toxins seems to be involved in these cases (Newton 1988). Similarly, sensitivities to other fungal toxins (e.g. tentoxin) and to herbicides (notably atrazine) are encoded by the chloroplast genome (e.g. Glimelius *et al.* 1981; Erickson & Kemble 1990).

CMS

Plant mitochondria have a special role in pollen and in chloroplast development. The latter is testified by the existence of heteroplasmic maize with green and non-green striped phenotypes inherited cytoplasmically and caused by mtDNA lesions (Newton 1988), and by the observation that specific alterations of mtDNA show deleterious effects on chloroplast structure and function (Roussell *et al.* 1991). Mitochondrial genes have a particularly important role in pollen development, as indicated by numerous reports of cytoplasmic male sterility (CMS). The first demonstration that CMS was of mitochondrial origin is by Belliard *et al.* (1979), by means of protoplast fusion and the production of somatic hybrids. Generally CMS plants are vigorous and normal but lack pollen (Hanson & Conde 1985). In fact, pollen production by anthers is probably one of the most energy-demanding stages of plant's life, which may partly explain this pattern (Morand *et al.* 2001). Furthermore, in pollen mother cells, the energy (in the form of ATP) is produced mostly by mitochondria; as a consequence, any mitochondrial impairment should alter the development of pollen grains. From an evolutionary perspective, CMS is attributed to the differential transmission of organelle and nuclear genes and is probably the best known example of genomic conflict in plants. It has also considerable economic importance for the production of hybrid seeds. In most cases examined in detail, CMS results from expression of novel mitochondrial polypeptides (Leon *et al.* 1998). However, on rare occasions, cytoplasmic male sterility has also been attributed to plastid genes, for instance in *Oenothera* with pollen characterized by abnormal starch grain shapes (Stubbe 1960). Until recently, no similar effect on male fertility had been reported for animal mtDNA, but recently effects on sperm mobility and male fertility were reported in humans as well as in other vertebrates (Ruiz-Pesini *et al.* 2000; Gemmel & Allendorf 2001). Plant mitochondrial DNA diversity has been studied in natural populations of some gynodioecious species, in conjunction with the dynamics of CMS and restoration by nuclear genes (e.g. Belhassen *et al.* 1993; Saumitou-Laprade *et al.* 1993; Manicacci *et al.* 1996; Tarayre *et al.* 1997; de Haan *et al.* 1997; McCauley 1998; Olson & McCauley 2002). In such models, it is expected that selection acting on gender could influence the distribution of oDNA diversity, as confirmed by theoretical approaches (reviewed in Werren & Beukeboom 1998). Cytoplasmic male sterility may affect organelle and nuclear gene flow not only at the intraspecific but also at the interspecific level (see 3.5).

Accumulation of deleterious mutations in organelle genomes

The accumulation of deleterious mutations is a general feature of organelle genomes (e.g. Lynch & Blanchard 1998). In phylogeographic surveys based on animal mtDNA, there is generally an excess of singleton haplotypes, compared to the neutral model, pointing to the existence of slightly deleterious mutations that never 'catch up' in the population (Rand 2001). Consequently, divergence between species is not simply the extrapolation of divergence within species. We have obtained preliminary data (Grivet 2002) that suggests that this pattern also holds for cpDNA, by pooling the results from many species, since single species estimates are difficult to obtain given the low level of intraspecific diversity. A specificity of plant organelles, their extremely low substitution rates, may actually represent their solution for coping with Muller's ratchet, by slowing it down as much as possible. Episodes of biparental inheritance and rare recombination during seed plant evolution may have been sufficient to rescue organelle genomes from mutation meltdown. Cryptic heterozygosity in plant mtDNA may also play a role (see 2.1). Compensatory mutations in the nuclear genome are also very important, as in the restoration of CMS. A solution used by plants but not by animals is the edition of transcripts. RNA editing (posttranscriptional alterations of single nucleotides within a mRNA) occurs in most land plants, and involves specific changes (mostly from C to U) in RNAs transcribed from the organelles. Edited triplets specify a different amino acid from their unedited (genome-encoded) counterpart, thereby restoring the function of the protein. This editing [first discovered in plant mtDNA (Covello & Gray 1989; Gualberto *et al.* 1989) and subsequently in cpDNA, where it is less frequent (Gray 1996)] is therefore vital for plants. It may be viewed as a way for the nuclear genome to cope with deleterious mutations in mt or cpDNA, by curing them indirectly. This is apparently a very costly and inefficient system, as each editing event is site-specific (there are more than 400 edited nucleotides in plant mtDNA; Hermann & Bock 1999) and requires a special machinery – an *editosome* – for processing, but it fits with the fact that most regulation is posttranscriptional in both mitochondria and chloroplasts (Leon *et al.* 1998).

Indirect selection

The absence or rarity of interindividual recombination in oDNA means that the whole genome (or indeed both organelle genomes) will be selected as a unit (2.5). It is therefore difficult to identify the precise oDNA polymorphism responsible for a given adaptive trait. This problem is magnified in predominantly selfing species, where association between nuclear loci and cpDNA genotypes is expected (Saghai Maroof *et al.* 1992). In outbreeding species, founding events, to which maternally inherited genes are so sensitive (Wade *et al.* 1994), may result in another form of disequilibrium between the nuclear and the organelle genomes (Asmussen & Arnold 1991). Multiple Mantel tests can be used to study the relationship between population divergence at organelle and nuclear markers, a method that may be more operational than the analysis of disequilibria at the individual level (Le Corre *et al.* 1997b). If differentiated populations exist in refugia, and habitat preferences evolved during isolation, colonization routes and contact zones that have been identified may not reflect solely the consequence of chance dispersal events into newly available habitats. Instead, selection during colonization processes may have shaped the present distribution of organelle lineages. Petit *et al.* (2002c) suggested that in European oaks, the distribution of some cpDNA lineages has been constrained by their association with oak species having different ecological requirements at the outset of postglacial recolonization, an association that subsequently disappeared as a consequence of introgression. More generally, the interplay between adaptation and selection needs to be taken into account (Davis & Shaw 2001) and could ultimately affect phylogeographic patterns. Another form of selection that should be considered in phylogeographic surveys of temperate species is selection for dispersal ability during range expansion (e.g. Cwynar & MacDonald 1987; Hampe & Bairlein 2000). This could affect the distribution and structure of diversity across the range compared to neutral models of expansion. Mode of oDNA transmission is also controlled genetically (see 3.3), and selection for increased or decreased levels of paternal leakage would also affect geographic structure.

3.5 Interspecific gene flow

Plants have long been known for their unusually high levels of interspecific gene flow (e.g. Stebbins 1950). The frequent occurrence of plant species having gained their organelles from other species may compromise not only interspecific but also intraspecific phylogenies. Simultaneously, plant organelles may interact in such a way with nuclear genes (especially with those of other species) so as to block or reduce interspecific gene flow. These two somewhat contradictory processes, if undetected, may easily compromise the interpretation of phylogeographic surveys.

'Cytoplasmic captures'

The first detailed review on the introgression of organelle genes across species (Rieseberg & Soltis 1991) reported many cases of plants with the nuclear genome of one species and the cytoplasmic genome of another one. Such reports became known as 'cytoplasmic captures'. The authors concluded that: *"cytoplasmic exchange appears to have occurred in the absence of significant gene flow and sometimes the native cytoplasm has been completely displaced by an alien one."* With the exception of one case where nuclear gene flow seemed to predominate over cpDNA gene flow (in pines, where this genome is paternally inherited; Wagner *et al.* 1987), all other cases were interpreted as indicating faster 'transfer' for the chloroplast genome than for the nuclear genome across species borders. Several explanations were proposed to account for these findings. Organelle genomes, present in multiple but identical copies in all cells, are effectively haploid, and therefore more exposed to drift (see 2.2). After hybridization between two species, one cytoplasmic genome would be quickly eliminated by drift, whereas the nuclear genome of one of the two species would recover through disruptive selection on coadapted gene loci. Actually, since hybrids and their offspring are often male-sterile (see below), they may transmit organelle genomes without diffusing too many nuclear genes, incorporating alien cytoplasms in a given population without altering much the nuclear gene frequencies.

Pollen swamping

A completely different scenario for interspecific gene flow was proposed for *Eucalyptus* in Tasmania (Potts & Reid 1988) and for oaks in Europe and North Africa (Petit 1992; Bacilieri *et al.* 1996; Petit *et al.* 1997; Belahbib *et al.* 2001). This scenario considers that species may be able to colonize a site already occupied by related species by pollen rather than by seeds (i.e. through pollen swamping, following hybridization). Indeed, once a given species has established itself, a large number of female flowers are produced, which are partly receptive to the pollen of the other species. Given the asymmetry of pollen and seed flow (see 3.2), this would result in hybrids, and eventually in backcrosses and other hybrid-derivatives characterized by the original (maternal) cpDNA structure but an increasing proportion of the nuclear genome of the second species. The species expanding later relies therefore not on seeds but on pollen for its dispersal into new environments. Petit *et al.* (2001) suggested that this dispersal scenario could be the rule, rather than the exception.

Pollen swamping implies that the maternal lineage is static (i.e. there is little seed flow) but that pollen movements are much more extensive, so in a sense the term 'nuclear capture' would better reflect the actual process (i.e. repeated backcrosses with incoming pollen). This highlights the danger of considering 'cytoplasmic gene flow', a concept with no reality (only seeds and pollen actually move). Similarly, 'cytoplasmic capture' is a term coined by taxonomists interested in phylogenetic reconstruction, but which is misleading. To our knowledge, the only case of 'chloroplast capture' by a resident nucleus has been described in kiwifruit, where cpDNA is paternally inherited (Chat *et al.* 2003). Indeed, gynogenetic kiwifruit plants can be produced that receive their chloroplasts as the sole contribution from the father, i.e. without concomitant transfer of paternal nuclear DNA, as if cpDNA had been cloned from one individual into another one, something that most researchers would have deemed impossible except through artificial cloning procedures (e.g. Verhoeven & Blass 1988; Eigel & Koop 1992).

In principle, convergent evolution or incomplete lineage sorting could also account for patterns of shared cytoplasmic variants. In fact, because of the particular mode of mtDNA evolution in plants (see 2.2), convergence at the molecular level was suggested to explain cytoplasmic sharing in two conifers (supposedly more refractory to hybridization): American pines (Wu *et al.* 1998) and Japanese firs (Tsumura & Suyama 1998). We suggest that past or ongoing introgression is more likely in these examples as well, as in most plants studied to date.

Quantification of cytoplasmic sharing

Regardless of the actual process involved, and given the observed frequency of interspecific cytoplasm sharing, it is now common practice to sample related congeneric species (in some groups, sampling of related genera may even be necessary; Soltis *et al.* 1991). The propensity of different species to share similar organelle variants when in sympatry can now be measured and tested statistically (Belahbib *et al.* 2001), provided that pairs of sympatric populations are sampled across the range. Dumolin-Lapègue *et al.* (1999b) and Petit *et al.* (2002b) have proposed related measures.

Nucleo-cytoplasmic barriers to gene flow

Nucleo-cytoplasmic interactions have long been suspected to impede interspecific gene flow (Grun 1976). Indeed, the outcome of interspecific crosses often depends on which plant is used as female. The numerous cases of cytoplasmic male sterility exemplify these (partial) fertility barriers. Studies in Germany with *Oenothera* and *Epilobium*, as well as in the USA with *Solanum*, have made use of systematic crosses to detect these interactions, both at the intra- and interspecific level (Grun 1976), and may be considered as the ancestors of the current geographic surveys of cytoplasmic variation, except that organelle genes were identified through their phenotypic effects. Although these deleterious interactions are not limited to interspecific crosses, they are clearly more frequent and of larger magnitude when parents do not belong to the same species. For instance, in a review of the genetic basis of CMS, Grun (1976, p. 298) showed that in cases of male sterility resulting from interspecific crosses, there are more nuclear genes involved than in cases resulting from intraspecific crosses. The genes involved were also more often dominant (60% compared to 20%). Finally, gametophytic determinism of male sterility (which is less stringent than sporophytic determinism) seems to be restricted to intraspecific crosses.

The idea that organelle and nuclear genomes coevolve following speciation and cannot be freely exchanged, even between closely related species, has been suggested several times (references in Hupfer *et al.* 2000). This might be due to cytoplasm exerting a selection pressure on foreign nuclear genome, a feature that has been empirically demonstrated in both natural and synthetic allopolyploids in *Brassica* (Song *et al.* 1988, 1995). In phylogeographic surveys, however, there are few examples where such interactions have been suspected, although most cases of interspecific gene flow of organelle genes are highly asymmetric, which is compatible with the hypothesis that nucleo-cytoplasmic interactions are in fact involved. In a hybrid zone between *Pinus banksiana* and *P. contorta*, Wagner *et al.* (1991b) showed that (paternally inherited) cpDNA markers remain closely associated with morphological characters distinguishing the two species, thereby limiting interspecific gene flow. It would therefore be of interest to combine phylogeographic surveys with large-scale studies of nucleo-cytoplasmic interactions. This may help understand whether nucleo-cytoplasmic interactions have limited the mixing of expanding organelle lineages and maintained sharp transitions between these.

In somatic hybrids, the segregation of mtDNA and cpDNA often depend on the nuclear genome, at least in distant (e.g. intergeneric) crosses (e.g. Bonnett & Glimelius 1990). Such studies indicate that incompatibility may be greater between the nuclear and mitochondrial genomes than between the nuclear and the chloroplast genome. Indeed, mitochondria are essential for meristematic activity and early cellular differentiation, but plastids do not play a vital role at this early developmental stage.

3.6 Horizontal transfers

Plant organelles have been shown to harbor sequences (mitochondrial plasmids, mobile introns, and perhaps viruses) that behave differently from the rest of the organelle genome, especially concerning their mode of transmission. In particular, some of them may be involved in horizontal gene transfers *sensu stricto*, i.e. gene flow between organisms that do not mate. This is of great potential for population and phylogeographic studies, as these sequences have therefore their own unique history, different from that of other genes in the organelle genomes. Empirical studies would be well inspired to take advantage of such diversity.

Mitochondrial plasmids, graft-transmitted elements and a mobile intron

Plasmid-like DNA or RNA, either circular or linear, with sizes varying between 1 and 11 kb, have been found in plant mitochondria from a wide range of angiosperm species (e.g. *Beta*, *Brassica*, *Chenopodium*, *Helianthus*, *Oryza*, *Phoenix*, *Silene*, *Sorghum*, *Vicia*, *Zea*). They lack homology with the main mtDNA and are thought to be autonomously replicating molecules. They are known in mtDNA from fungi but not from animals and are also absent in cpDNA. The distribution of these mtDNA plasmids within plant species is very irregular, and they may be present or absent even among individuals belonging to a single natural population (Saumitou-Laprade *et al.* 1989). This indicates that they are dispensable and that they are not inherited strictly maternally, as shown in crosses of *Beta* (Saumitou-Laprade *et al.* 1989) or *Brassica* (Palmer *et al.* 1983; Erickson *et al.* 1989). Models indicate that some (even limited) level of bi-parental transmission is necessary for their spread, although their long-term evolution is more likely strict maternal inheritance (Bengtsson & Andersson 1997). They might be sometimes transmitted horizontally, as in the fungus *Podospora anserina* (Kempken 1995). Mitochondrial plasmids may therefore present all intermediate dynamics between that typical of a virus and that typical of mtDNA. Furthermore, selection against such plasmids may affect their mitochondrial host, especially if their mode of transmission is partly coupled. A number of reports indicate that male sterility determinants could be transmitted through grafting. They are either transmitted through the egg only or through both the egg and the sperm (Grun 1976). They have been identified as viruses whose behavior in populations should be similar to that of mtDNA plasmids.

Another example of a selfish genetic element found in plant organelles is a mobile mtDNA intron situated in the *cox*1 gene. It is a *group I* intron; that is, it can direct its own 'homing' to homologous genes lacking it through the aegis of an intron-encoded endonuclease (Dujon *et al.* 1989). The intron of *cox*1 is an extraordinarily invasive mobile element, which has probably been acquired over 1000 times separately during angiosperm evolution, via a recent wave of cross-species horizontal transfers, within the last 10 million years or so. Such invasions may still be ongoing and are facilitated by the very conservative sequence evolution of mtDNA, which preserves the site necessary for its integration (Vaughn *et al.* 1995; Cho *et al.* 1998; Palmer *et al.* 2000). This intron is now present in some 17% of angiosperms tested (48 species out of 281, Cho *et al.* 1998). Inter- and intraspecific phylogeographic studies may illuminate the actual mechanisms at hand.

Bengtsson and Andersson (1997) have suggested that several reproductive features of plants (such as uni-parental inheritance of organelles or evolution of apomixis) may derive from adaptive responses to such parasites.

Conclusion: new prospects for population and phylogeographic studies

Since phylogeography is essentially an interdisciplinary endeavor, intersecting population genetics, phylogenetics and (paleo) ecology, we have attempted here to merge information originating from disparate fields (see also Avise, this volume). We hope that this eclecticism will stimulate new, non-conventional approaches, but realize the difficulty in integrating fields that seldom communicate. For instance, plant molecular geneticists have made considerable progresses in organelle genetics in an experimental, often highly artificial context, by relying on *in vitro* methods and cybrid production. This may seem at odds with the preoccupation of population geneticists trying to unravel the phylogeography of plant species in the wild. However, such studies provide basic knowledge on recombination, mutation, and selection of organelle genomes and on their compatibility with the nuclear genome, which may be important for understanding population genetics and phylogeographic surveys based on oDNA. The classic results of plant organelle inheritance and geographic variation inferred from patterns of compatibility in crosses (nicely summarized in Grun 1976) seem to have been largely forgotten by the new wave of researchers conducting phylogeographic studies. Furthermore, recent studies of molecular evolution of oDNA (macroevolution) have been largely disconnected from microevolutionary investigations, although the former is a natural extension of the latter. Theoretical and experimental progress on levels of selection and on the interplay of cooperation and competition in shaping inter- and intraspecific biotic interactions (Maynard Smith & Szathmáry 1995; Taylor *et al.* 2002) should provide a unified and integrated approach particularly well suited to the study of organelle variation in plants.

The two plant organelle genomes, despite being embedded in and serving the same organism and experiencing the same history (at least when they are both similarly transmitted from mother to offspring), have evolutionary dynamics that are incredibly different from each other and from that of organelles in other organisms. This illustrates the many solutions that intracellular mutualist symbionts may find for survival in host cells. For phylogeographic surveys, the low levels of sequence diversity of cpDNA and mtDNA oblige investigators to look more closely at the mechanisms involved in generating mutations and how these are distributed in these comparatively large and complex genomes. The skepticism of early investigators about the usefulness of cpDNA variation for population and phylogeographic surveys has now been largely dispelled. However, the idea that plant mtDNA cannot provide useful markers for phylogeographic surveys persists. Yet, with current techniques, many species have been shown to have sufficient variation in the two organelle genomes for phylogeographic surveys, at least at broad geographic scales, although a few species may lack variation across parts of their ranges. One reason why it is of interest to focus on both cpDNA and mtDNA is the variety of patterns of inheritance and the fact that transmission modes do not necessarily align for the two organelles. More detailed studies of organelle inheritance and its consequences for mono- and dicytoplasmic phylogeographic surveys are needed to take full advantage of these particularities. More generally, it appears crucial to reevaluate a number of biological assumptions concerning plant organelle genetics; otherwise they tend to become implicit and are used to interpret all subsequent observations (Dowton & Campbell 2001).

We have seen that, for oDNA, there are in principle no *replicates* of the phylogeographic patterns observed. Hence there is a potential advantage of the existence of different cytoplasmic genomes or sequences (such as selfish DNA elements) in plant cells, especially if their transmission across generations is not fully coupled. By analogy, the comparative phylogeography of plants and their specialized pathogens (Burban *et al.* 1999) or symbionts might provide independent evidence on the history of the host plants. Finally, the comparative phylogeography of codistributed seed plants (e.g. Soltis *et al.* 1997; Taberlet *et al.* 1998) may enhance the appreciation of the stochasticity of colonization processes.

Another message is that interspecific gene flow has to be understood when studying the phylogeography of a group, given the propensity of plants to hybridize and introgress. Schaal *et al.* (1998) correctly pointed out that plants are singular in their ability to repeatedly violate the assumption of non-reticulating lineages. The study of oDNA variation (producing non-reticulating gene trees both within and across species) should help free scientists from preconceived species-centered attitudes, with obvious consequences when sampling material at the outset of a phylogeographic survey.

Finally, phylogeography should not be limited to purely phenomenological description; it can provide essential background information to disentangle current from past processes, understand the consequences of crucial events such as colonization in the life span of plant species, and may allow researchers to test adaptive hypotheses. To achieve this, it is essential to consider life history traits (such as seed dispersal) and the genetic system itself (such as transmission rates of organelles through male and female gametes or recombination rates) not as fixed components but as evolving ones.

Plant phylogeography should therefore provide fertile grounds for innovative investigations in evolutionary biology. Such studies are also likely to increase with the emergence of new applications. In conservation and management of genetic resources, phylogeographic studies may help identify native populations or key regions deserving conservation or management priority (König *et al.* 2002; Saltonstall 2002). They may also clarify the origin and status of rare endemics (Feliner *et al.* 2002; Szalanski *et al.* 2001; Widmer & Baltisberger 1999). Phylogeographic survey will provide the basis for the tracing of wood (or other plant products), providing tools to limit illegal logging or to label products originating from sustainably managed regions (Deguilloux *et al.* 2002). Archaeology will also benefit from those developments. Finally, the necessary background on seed flow and inheritance patterns to emerge from phylogeographic surveys will be important in evaluating risks associated with the use of transplastomic plants (Cummins 1998; Daniell *et al.* 1998) and will help study plant migration and its genetic consequences following expected climate changes (Pitelka 1997).

Acknowledgements

We are very grateful to those colleagues that sent some comments on a previous version of the manuscript: Arndt Hampe, Stephen Harris, Xavier Reboud, Haya Anne Tsitrone and Fernand Vedel. Daniela Salvini and Silvia Fineschi helped compile some of the data presented in Figure2 and in Table7, and Kate Porter did the literature survey whose results are shown in Figure 1. She also considerably helped organize the text and polish the style. Steven Weiss also greatly helped edit the paper. Jérôme Duminil, Pierre Saumitou-Laprade, Sascha Liepelt and Birgit Ziegenhagen pointed out some useful references. A grant from the Commission of the European Communities, Agriculture and Fisheries (FAIR) specific RTD programme (CT97-3795, «CYTOFOR») was instrumental in helping to initiate this review.

References

Abbott RJ, Smith LC, Milne RI, Crawford RMM, Wolfe K, Balfour J (2000) Molecular analysis of plant migration and refugia in the arctic. ***Science* 289**, 1343-1346.

Albert B, Godelle B, Atlan A, De Paepe R, Gouyon PH (1996) Dynamics of plant mitochondrial genome: model of a three-level selection process. ***Genetics* 144**, 369-382.

Allen JF, Raven JA (1996) Free-radical induced mutation vs redox regulation: costs and benefits of genes in organelles. ***Journal of Molecular Evolution* 42**, 482-492.

Amoatey HM, Tilney-Bassett RAE (1994) A test of the complementary gene model for the control of bi-parental plastid inheritance in zonal pelargoniums. ***Heredity* 72**, 69-77.

Andersson H (1999) Female and hermaphrodite flowers on a chimeric gynomonoecious *Silene vulgaris* plant produce offspring with different genders: a case of heteroplasmic sex determination? ***Journal of Heredity* 90**, 563-565.

Asmussen A, Arnold J (1991) The effects of admixture and population subdivision on cytonuclear disequilibria. ***Theoretical Population Biology* 39**, 273-300.

Atchison BA, Whitfeld PR, Bottomley W (1976) Comparison of chloroplast DNAs by specific fragmentation with EcoR1 endonuclease. ***Molecular and General Genetics* 148**, 263-269.

Avise JC (2000) ***Phylogeography: The History and Formation of Species***. Harvard University Press, Cambridge.

Avise JC, Giblin-Davidson C, Laerm J, Patton JC, Lansman RA (1979) Mitochondrial DNA clones and matriarchal phylogeny within and among geographic populations of the pocket gopher, *Geomys pinetis*. ***Proceedings of the National Academy of Sciences, USA* 76**, 6694-6698.

Avise JC, Shapira JF, Daniel SW, Aquadro CF, Lansman RA (1983) Mitochondrial-DNA differentiation during the speciation process in *Peromyscus*. ***Molecular Biology and Evolution* 1**, 38-56.

Avise JC, Arnold J, Ball RM, Bermingham E, Lamb T, Neigel JE, Reeb CA, Saunders NC (1987) Intraspecific phylogeography: the mitochondrial DNA bridge between population genetics and systematics. ***Annual Review of Ecology and Systematics* 18**, 489-522.

Bacilieri R, Ducousso A, Petit RJ, Kremer A (1996) Mating system and asymmetric hybridization in a mixed stand of European oaks. ***Evolution* 50**, 900-908.

Backert S, Lurz R, Börner T (1996) Electron microscopic investigation of mitochondrial DNA from *Chenopodium album* (L.) ***Current Genetics* 29**, 427-436.

Backert S, Lurz R, Oyarzabal OA, Börner T (1997) High content, size and distribution of single-stranded DNA in the mitochondria of *Chenopodium album* (L.). ***Plant Molecular Biology* 33**, 1037-1050.

Baker HG (1959) The contribution of autecological and genecological studies to our knowledge of the past migration of plants. ***American Naturalist* 93**, 255-271.

Balfourier F, Imbert C, Charmet G (2000) Evidence for phylogeographic structure in *Lolium* species related to the spread of agriculture in Europe: a cpDNA study. ***Theoretical and Applied Genetics* 101**, 131-138.

Banks JA, Birky, CW Jr (1985) Chloroplast DNA diversity is low in a wild plant, *Lupinus texensis*. ***Proceedings of the National Academy of Sciences, USA* 82**, 6950-6954.

Barton NH, Whitlock MC (1997) The evolution of metapopulations. In: ***Metapopulation Dynamics: Ecology, Genetics, and Evolution*** (eds. Hanski I, Gilpin M), pp. 183-210. Academic Press, New York.

Baur E (1909) Das Wesen und die Erblichkeitsverhältnisse der 'Varietates albomarginatae hort.' von *Pelargonium zonale*. ***Zeitschrift für induktive Abstammungs- und Vererbungslehre* 1**, 330-335.

Bedbrook JR, Bogorad L (1976) Endonuclease recognition sites mapped on *Zea mays* chloroplast DNA. ***Proceedings of the National Academy of Sciences, USA* 7**, 4309-4313.

Belahbib N, Pemonge M-H, Ouassou A, Sbay H, Kremer A, Petit RJ (2001) Frequent cytoplasmic exchanges between oak species that are not closely related: *Quercus suber* and *Q. ilex* in Morocco. ***Molecular Ecology*** **10**, 2003-2012.

Belhassen E, Atlan A, Couvet D, Gouyon P-H, Quétier F (1993) Mitochondrial genome of *Thymus vulgaris* L. (Labiate) is highly polymorphic between and among natural populations. ***Heredity*** **71**, 462-472.

Belliard G, Pelletier G, Vedel F, Quétier F (1978) Morphological characteristics and chloroplast DNA distribution in different cytoplasmic parasexual hybrids of *Nicotiana tabacum*. ***Molecular and General Genetics*** **165**, 231-237.

Belliard G, Vedel F, Pelletier G (1979) Mitochondrial recombination in cytoplasmic hybrids of *Nicotiana tabacum* by protoplast fusion. ***Nature*** **281**, 401-403

Bendich AJ (1987) Why do chloroplasts and mitochondria contain so many copies of their genome? ***BioEssays*** **6**, 279-281.

Bendich AJ, Gauriloff LP (1984) Morphometric analysis of Cucurbit mitochondria: the relationship between chondriome volume and DNA content. ***Protoplasma*** **119**, 1-7.

Bendich AJ, Smith SB (1990) Moving pictures and pulsed-field electrophoresis show linear DNA molecules from chloroplasts and mitochondria. ***Current Genetics*** **17**, 421-425.

Bengtsson BO, Andersson H (1997) The population genetics of plant mitochondrial plasmids. ***Journal of Theoretical Biology*** **188**, 163-176.

Bensasson D, Zhang D-X, Hartl DL, Hewitt GM (2001) Mitochondrial pseudogenes: evolution's misplaced witnesses. ***Trends in Ecology and Evolution*** **16**, 314-321.

Birky, CW Jr (1983) Relaxed cellular controls and organelle heredity. ***Science*** **222**, 468-475.

Birky, CW Jr (1988) Evolution and variation in plant chloroplast and mitochondrial genomes. In: ***Plant Evolutionary Biology*** (eds. Gottlieb LB, Jain SK), pp. 23-53. Chapman and Hall, New York.

Birky, CW Jr (1991) Evolution and population genetics of organelle genes: mechanisms and models. In: ***Evolution at the Molecular Level*** (eds. Selander RK, Clark AG, Whittham TS), pp. 112-134. Sinauer Associates Inc., Sunderland.

Birky, CW Jr (1995) Uni-parental inheritance of mitochondrial and chloroplast genes: mechanisms and evolution. ***Proceedings of the National Academy of Sciences, USA*** **92**, 11331-11338.

Birky, CW Jr (2001) The inheritance of genes in mitochondria and chloroplasts: laws, mechanisms and models. ***Annual Review of Genetics*** **35**, 125-148.

Birky, CW Jr, Fuerst P, Maruyamma T (1989) Organelle gene diversity under migration, mutation, and drift: equilibrium expectations approach to equilibrium, effects of heteroplasmic cells, and comparison to nuclear genes. ***Genetics*** **103**, 513-527.

Blanchard JL, Schmidt GW (1995) Pervasive migration of organellar DNA to the nucleus in plants. ***Journal of Molecular Evolution*** **41**, 397-406.

Blanchard JL, Schmidt GW (1996) Mitochondrial DNA migration events in yeast and humans: integration by a common end-joining mechanism and alternative perspectives on nucleotide substitution pattern. ***Journal of Molecular Evolution*** **13**, 537-548.

Bogorad L (2001) Emergence of plant molecular biology viewed through the portal of chloroplast research. ***Current Science*** **80**, 153-160.

Bonnett HT, Glimelius K (1990) Cybrids of *Nicotiana tabacum* and *Petunia hybrida* have an intergeneric mixture of chloroplasts from *P. hybrida* and mitochondria identical or similar to *N. tabacum*. ***Theoretical and Applied Genetics*** **79**, 550-555.

Bossema I (1979) Jays and oaks: an eco-ethological study of a symbiosis. ***Behaviour* 70**, 1-117.

Bousquet J, Strauss SH, Doerksen AH, Price RA (1992) Extensive variation in evolutionary rate of *rbc*L gene sequences among seed plants. ***Proceedings of the National Academy of Sciences, USA* 89**, 7844-7848.

Bowe LM, dePamphilis CW (1996) Effects of RNA editing and gene processing on phylogenetic reconstruction. ***Molecular Biology and Evolution* 13**, 1159-1166.

Bradshaw AD (1972) Some of the evolutionary consequences of being a plant. ***Evolutionary Biology* 5**, 25-47.

Burban C, Petit RJ, Carcreff E, Jactel H (1999) Rangewide variation of the maritime pine bast scale *Matsucoccus feytaudii* Duc. (Homoptera: Matsucoccidae) in relation to the genetic structure of its host. ***Molecular Ecology* 8**, 1593-1602.

Cain ML, Damman H, Muir A (1998) Seed dispersal and the Holocene migration of woodland herbs. ***Ecological Monographs* 68**, 325-347.

Cain ML, Milligan BG, Strand AE (2000) Long-distance seed dispersal in plant populations. ***American Journal of Botany* 87**, 1217-1227.

Cardi T, Bastia T, Monti L, Earle ED (1999) Organelle DNA and male fertility variation in *Solanum* spp. and interspecific somatic hybrids. ***Theoretical and Applied Genetics* 99**, 819-828.

Caron H, Dumas S, Marque G, Messier C, Bandou E, Petit RJ, Kremer A (2000) Spatial and temporal distribution of chloroplast DNA polymorphism in a tropical tree species. ***Molecular Ecology* 9**, 1089-1098.

Ceplitis A (2001) The importance of sexual and asexual reproduction in the recent evolution of *Allium vineale*. ***Evolution* 55**, 1581-1591.

Chambers JC, MacMahon JA (1994) A day in the life of a seed: movements and fates of seeds and their implications for natural and managed systems. ***Annual Review of Ecology and Systematics* 25**, 263-292.

Chan H, Wildman SG (1972) Chloroplast DNA codes for the primary structure of the large subunit of Fraction I protein. ***Biochemichal and Biophysical Acta* 277**, 677-680.

Chat J, Chalak L, Petit RJ (1999) Strict paternal inheritance of chloroplast DNA and maternal inheritance of mitochondrial DNA in interspecific crosses of kiwifruit. ***Theoretical and Applied Genetics* 99**, 314-322.

Chat J, Decroocq S, Decroocq V, Petit RJ (2002) A case of chloroplast heteroplasmy in kiwifruit (*Actinidia deliciosa*) that is not transmitted during sexual reproduction. ***Journal of Heredity* 93**, 293-300.

Chat J, Decroocq S, Petit RJ (2003) A one-step organelle capture: gynogenetic kiwifruits with paternal chloroplasts. ***Proceedings of the Royal Society of London, B* 270**, 783-789.

Chesnoy L (1987) L'origine des organites du cytoplasme embryonnaire chez les Gymnospermes. ***Bulletin de la Société Botanique de France* 134**, 51-56.

Chiang TY, Schaal BA, Peng CI (1998) Universal primers for amplification and sequencing of a noncoding spacer between the atpB and rbcL genes of chloroplast DNA. ***Botanical Bulletin of Academia Sinica* 39**, 245-250.

Chiba Y (1951) Cytochemical studies on chloroplasts. I. Cytologic demonstration of nucleic acids in chloroplasts. ***Cytologia (Tokyo)* 16**, 259-264.

Chiu WL, Sears BB (1985) Recombination between chloroplast DNAs does not occur in sexual crosses of *Oenothera*. ***Molecular and General Genetics* 198**, 525-528.

Chiu WL, Sears BB (1993) Plastome-genome interactions affect plastid transmission in *Oenothera*. ***Genetics* 133**, 989-997.

Cho Y, Qiu Y-L, Kuhlman P, Palmer JD (1998) Explosive invasion of plant mitochondria by a group I intron. ***Proceedings of the National Academy of Sciences, USA* 95**, 14244-14249.

Clark JS, Macklin E, Wood L (1998a) Stages and spatial scales of recruitment limitation in southern Appalachian forests. ***Ecological Monographs* 68**, 213-235.

Clark JS, Fastie C, Hurtt G, Jackson ST, Johnson C, King GA, Lewis M, Lynch J, Pacala S, Prentice C, Schupp EW, Webb III T, Wyckoff P (1998b) Reid's paradox of rapid plant migration. ***BioScience* 48**, 13-24.

Clegg MT (1987) Preface, plant molecular evolution. ***American Naturalist* 130**, S1-S5.

Clegg MT, Brown AHD, Whitfeld PR (1984) Chloroplast DNA diversity in wild and cultivated barley: implications for genetic conservation. ***Genetical Research Cambridge* 43**, 339-343.

Clegg MT, Gaut BS, Learn GH, Morton BR (1994) Rates and patterns of chloroplast DNA evolution. ***Proceedings of the National Academy of Sciences, USA* 91**, 6795-6801.

Comes HP, Abbott RJ (2001). Molecular phylogeography, reticulation and lineage sorting in the Mediterranean species complex of *Senecio* sect. *Senecio* (Asteraceae). ***Evolution* 55**, 1943-1962.

Comps B, Gömöry D, Letouzey J, Thiébaut B, Petit RJ (2001) Diverging trends between heterozygosity and allelic richness during postglacial colonization in the European beech. ***Genetics* 157**, 389-397.

Connett MB (1987) Mechanisms of maternal inheritance of plastids and mitochondria: developmental and ultrastructural evidence. ***Plant Molecular Biology Reporter* 4**, 193-205.

Correns C (1906) Die Vererbung der Geschlechtsformen bei den gynodioecischen Pflanzen. ***Berichte der Deutschen Botanischen Gesellschaft* 24**, 459-474.

Correns C (1908) Die Rolle der männliche Keimzellen bei der Geschlechtsbestimmung der gynodioecischen Pflanzen. ***Berichte der Deutschen Botanischen Gesellschaft* 26**, 686-701.

Correns C (1909) Vererbungsversuche mit blass (gelb) grünen und buntblättrigen Sippen bei *Mirabilis jalapa*, *Urtica pilulifera* und *Lunaria annua*. ***Zeitschrift für induktive Abstammungs- und Vererbungslehre* 1**, 291-329.

Corriveau JL, Coleman AW (1988) Rapid screening method to detect potential bi-parental inheritance of plastid DNA and results for over 220 angiosperm species. ***American Journal of Botany* 75**, 1443-1458.

Covello PS, Gray MW (1989) RNA editing in plant mitochondria. ***Nature* 341**, 662-666.

Cruzan MB, Templeton AR (2000) Paleoecology and coalescence: phylogeographic analysis of hypotheses from the fossil record. ***Trends in Ecology and Evolution* 15**, 491-496.

Cruzan MB, Arnold ML, Carney SE, Wollenberg KR (1993) CpDNA inheritance in interspecific crosses and evolutionary inference in Louisiana irises. ***American Journal of Botany* 80**, 344-350.

Cummins JE (1998) Chloroplast-transgenic plants are not a gene flow panacea. ***Nature Biotechnology* 16**, 401.

Cwynar LC, MacDonald GM (1987) Geographical variation of lodgpole pine in relation to population history. ***American Naturalist* 129**, 463-469.

Dally AM, Second G (1990) Chloroplast DNA diversity in wild and cultivated species of rice (Genus Oryza, section Oryza): cladistic-mutation and genetic-distance analysis. ***Theoretical and Applied Genetics* 80**, 209-222.

Daniell H, Datta R, Varma S, Gray S, Lee S (1998) Containment of herbicide resistance through genetic engineering of the chloroplast genome. ***Nature Biotechnology* 16**, 345-348.

Darwin C (1859) ***On the Origin of Species by Means of Natural Selection***. John Murray, London.

Davis MB, Shaw RG (2001) Range shifts and adaptive responses to quaternary climate change. ***Science* 292**, 673-679.

de Haan AA, Mateman AC, Van Dijk PJ, Van Damme JMM (1997) New CMS types in *Plantago lanceolata* and their relatedness. ***Theoretical and Applied Genetics* 94**, 539-548.

Deguilloux M-F, Pemonge M-H, Petit RJ (2002) Novel perspectives in wood certification and forensics: dry wood as a source of DNA. ***Proceedings of the Royal Society of London, B* 269**, 1039-1046.

Demesure B, Sodzi N, Petit RJ (1995) A set of universal primers for amplification of polymorphic non-coding regions of mitochondrial and chloroplast DNA in plants. ***Moleular Ecology* 4**, 129-131.

dePamphilis CW, Palmer JD (1990) Loss of photosynthetic and chlororespiratory genes from the plastid genome of a nonphotosynthetic plant. ***Nature* 348**, 337-339.

Derepas A, Dulieu H (1992) Inheritance of the capacity to transfer plastids by the pollen parent in *Petunia hybrida* Hort. ***Journal of Heredity* 83**, 6-10.

Desplanque B, Viard F, Bernard J, Forcioli D, Saumitou-Laprade P, Cuguen J, Van Dijk H (2000) The linkage disequilibrium between chloroplast DNA and mitochondrial DNA haplotypes in *Beta vulgaris ssp. maritima* (L.): the usefulness of both genomes for population genetic studies. ***Molecular Ecology* 9**, 141-154.

Doebley J (1990) Molecular evidence and the evolution of maize. ***Economic Botany* 44**, 6-27.

Donaldson PA, Bevis EE, Pandeya RS, Gleddie SC (1994) Random chloroplast segregation and frequent mtDNA rearrangements in fertile somatic hybrids between *Nicotiana tabacum* L. and *N. glutinosa* L. ***Theoretical and Applied Genetics* 87**, 900-908.

Dong J, Wagner DB (1994) Paternally inherited chloroplast polymorphism in Pinus: estimation of diversity and population subdivision, and tests of disequilibrium with a maternally inherited mitochondrial polymorphism. ***Genetics* 136**, 1187-1194.

Downie SR, Palmer JD (1992) Use of chloroplast DNA rearrangements in reconstructing plant phylogeny. In: ***Molecular Systematics of Plants*** (eds. Soltis PS, Soltis DE, Doyle JJ), pp. 14-35. Chapman and Hall, New York, London.

Dowton M, Campbell NJH (2001) Intramitochondrial recombination – is it why some mitochondrial genes sleep around? ***Trends in Ecology and Evolution* 16**, 269-271.

Doyle JJ, Morgante M, Tingey SV, Powell W (1998) Size homoplasy in chloroplast microsatellites of wild perennial relatives of soybean (*Glycine* subgenus *Glycine*). ***Molecular Biology and Evolution* 15**, 215-218.

Draper CK, Hays JB (2000) Replication of chloroplast, mitochondrial and nuclear DNA during growth of unirradiated and UVB-irradiated *Arabidopsis* leaves. ***The Plant Journal* 23**, 255-265.

Dujon B, Belfort M, Butow RA, Jacq C, Lemieux C, Perlman PS Vogt VM (1989) Mobile introns: definition of terms and recommended nomenclature. ***Gene* 82**, 115-118.

Dulieu H, Derepas A, Cornu A (1990) Le rôle du pollen dans la transmission des chloroplastes et des mitochondries. Etude d'un cas particulier chez *Petunia*. ***Bulletin de la Société Botanique de France*** **137**, 49-56.

Duminil J, Pemonge M-H, Petit RJ (2002) A set of 35 consensus primer pairs amplifying genes and introns of plant mitochondrial DNA. ***Molecular Ecology Notes*** **2**, 428-430.

Dumolin S, Demesure B, Petit RJ (1995) Inheritance of chloroplast and mitochondrial genomes in pedunculate oak investigated with an efficient PCR method. ***Theoretical and Applied Genetics*** **91**, 1253-1256.

Dumolin-Lapègue S, Pemonge M-H, Petit RJ (1997a) An enlarged set of consensus primers for the study of organelle DNA in plants. ***Molecular Ecology*** **6**, 393-397.

Dumolin-Lapègue S, Demesure B, Fineschi S, Le Corre V, Petit RJ (1997b) Phylogeographic structure of white oaks throughout the European continent. ***Genetics*** **146**, 1475-1487.

Dumolin-Lapègue S, Pemonge M-H, Petit RJ (1998) Association between chloroplast and mitochondrial lineages in oaks. ***Molecular Biology and Evolution*** **15**, 1321-1331.

Dumolin-Lapègue S, Pemonge M-H, Gielly L, Taberlet P, Petit RJ (1999a) Amplification of oak DNA from ancient and modern wood. ***Molecular Ecology*** **8**, 2137-2140.

Dumolin-Lapègue S, Kremer A, Petit RJ (1999b) Are chloroplast and mitochondrial DNA species-independant in oaks? ***Evolution*** **53**, 1406-1413.

Dynesius, M, Jansson R (2000) Evolutionary consequences of changes in species' geographical distributions driven by Milankovitch climate oscillations. ***Proceedings of the National Academy of Sciences, USA*** **97**, 9115-9120.

Eigel L, Koop H-U (1992) Transfer of defined numbers of chloroplasts into albinos protoplasts by subprotoplast/protoplast microfusion: chloroplasts can be 'cloned', by using suitable plastome combinations or selective pressure. ***Molecular and General Genetics*** **233**, 479-482.

El Mousadik A, Petit RJ (1996) Chloroplast DNA phylogeography of the argan tree of Morocco. ***Molecular Ecology*** **5**, 547-555.

Ellstrand NC (1992) Gene flow by pollen: implications for plant conservation genetics. ***Oikos*** **63**, 77-86.

Ennos RA (1994) Estimating the relative rates of pollen and seed migration among plant populations. ***Heredity*** **72**, 250-259.

Erickson L, Kemble R (1990) Paternal inheritance of mitochondria in rapeseed (*Brassica napus*). ***Molecular and General Genetics*** **222**, 135-139.

Erickson L, Kemble R, Swanson E (1989) The Brassica mitochondrial plasmid can be sexually transmitted: pollen transfer of a cytoplasmic genetic element. ***Molecular and General Genetics*** **222**, 135-139.

Eyre-Walker A, Gaut BS (1997) Correlated rates of synonymous site evolution across plant genomes. ***Molecular Biology and Evolution*** **14**, 455-460.

Fairbanks DJ, Rytting B (2001) Mendelian controversies: a botanical and historical review. ***American Journal of Botany*** **88**, 737-752.

Fauré S, Noyer J-L, Carreel F, Horry J-P, Bakry F, Lanaud C (1994) Maternal inheritance of chloroplast genome and paternal inheritance of mitochondrial genome in bananas (*Musa acuminata*). ***Current Genetics*** **25**, 265-269.

Fejes E, Engler D, Maliga P (1990) Extensive homologous chloroplast DNA recombination in the pt14 *Nicotiana* somatic hybrid. ***Theoretical and Applied Genetics*** **79**, 28-32.

Feliner GN, Aguilar JF, Rossello JA (2002) Reticulation or divergence: the origin of a rare serpentine endemic assessed with chloroplast, nuclear and RAPD markers. ***Plant Systematics and Evolution* 231**, 19-38.

Fenner M (2000) ***Seeds – The Ecology of Regeneration in Plant Ecology.*** 2nd ed. CAB International, Wallingford, UK.

Fineschi S, Taurchini D, Villani F, Vendramin GG (2000) Chloroplast DNA polymorphism reveals little geographical structure in *Castanea sativa* Mill. (Fagaceae) throughout southern European countries. ***Molecular Ecology* 9**, 1495-1503.

Fitter JT, Rose RJ (1993) Investigation of chloroplast DNA heteroplasmy in *Medicago sativa* L. using cultured tissue. ***Theoretical and Applied Genetics* 86**, 65-70.

Fitter JT, Thomas MR, Rose RJ, Steele-Scott N (1996) Heteroplasmy of the chloroplast genome of *Medicago sativa* L. cv 'Regen S' confirmed by sequence analysis. ***Theoretical and Applied Genetics* 93**, 685-690.

Forcioli D, Saumitou-Laprade P, Valero M, Vernet P, Cuguen J (1998) Distribution of chloroplast DNA diversity within and among populations in gynodioecious *Beta vulgaris* ssp. *maritima* (Chenopodiaceae). ***Molecular Ecology* 7**, 1193-1204

Forsthoefel NR, Bohnert HJ, Smith SE (1992) Discordant inheritance of mitochondrial and plastid DNA in diverse alfalfa genotypes. ***Journal of Heredity* 83**, 342-345.

Freeman JS, Jackson HD, Steane DA, McKinnon GE, Dutkowski GW, Potts BM, Vaillancourt RE (2001) Chloroplast DNA phylogeography of *Eucalyptus globulus*. ***Australian Journal of Botany* 49**, 585-596.

Frey JE (1998) Genetic flexibility of plant chloroplasts. ***Nature* 398**, 115-116.

Fröst S, Vaivars L, Carlbom C (1970) Reciprocal extrachromosomal inheritance in rye (*Secale cereale* L.). ***Hereditas* 65**, 251-260.

Gaut BS, Clegg MT (1993) Molecular evolution of the *Adh*1 locus in the genus *Zea*. ***Proceedings of the National Academy of Sciences, USA* 90**, 5095-5099.

Gaut BS, Muse SV, Clark WD, Clegg MT (1992) Relative rates of nucleotide substitution at the *rbc*L locus of monocotyledonous plants. ***Journal of Molecular Evolution* 35**, 292-303.

Gemmel NJ, Allendorf FW (2001) Mitochondrial mutations may decrease population viability. ***Trends in Ecology and Evolution* 16**, 115-117.

Gepts P (n.d.) The evolution of crop plants. http://agronomy.ucdavis.edu/gepts/pb143/pb143.htm.

Gielly L, Taberlet P (1994) The use of chloroplast DNA to resolve plant phylogenies: noncoding *versus rbc*L sequences. ***Molecular Biology and Evolution* 11**, 769-777.

Gillham NW (1994) ***Organelle Genes and Genomes.*** Oxford University Press, Oxford, New York.

Glimelius K, Chen K, Bonnett HT (1981) Somatic hybridization in *Nicotiana*: segregation of organellar traits among hybrid and cybrid plants. ***Planta* 153**, 504-510.

Golenberg EM, Giannasi DE, Clegg MT, Smiley CJ, Durbin ML, Henderson D, Zurawski G (1990) Chloroplast DNA sequence from a Miocene Magnolia species. ***Nature* 341**, 660-662.

Golenberg EM, Clegg MT, Durbin ML, Doebley J, Ma DP (1993) Evolution of a noncoding region of the chloroplast genome. ***Molecular Phylogenetics and Evolution* 2**, 52-64.

Goloubinoff P, Pääbo S, Wilson AC (1993) Evolution of maize inferred from sequence diversity of an *Adh2* gene segment from archaeological specimens. ***Proceedings of the National Academy of Sciences, USA* 90**, 1997-2001.

Goulding SE, Olmstead RG, Morden CW, Wolfe KH (1996) Ebb and flow of the chloroplast inverted repeat. ***Molecular and General Genetics* 252**, 195-206.

Govindaraju DR, Wagner DB, Smith GP, Dancik BP (1988) Chloroplast DNA variation within individual trees of a *Pinus banksiana-Pinus contorta* sympatric region. ***Canadian Journal of Forest Research* 18**, 1347-1350.

Govindaraju DR, Dancik BP, Wagner DB (1989) Novel chloroplast DNA polymorphism in a sympatric region of two pines. ***Journal of Evolutionary Biology* 2**, 49-59.

Gray MW (1996) RNA editing in plant organelles: a fertile field. ***Proceedings of the National Academy of Sciences, USA* 93**, 8157-8159.

Green DF, Johnson EA (1995) Secondary dispersal of tree seeds on snow. ***Journal of Ecology* 85**, 329-340.

Griffin KL, Anderson OR, Gastrich MD, Lewis JD, Lin G, Schuster W, Seemann J, Tissue DT, Turnbull MH, Whitehead D (2001) Plant growth in elevated CO_2 alters mitochondrial number and chloroplast fine structure. ***Proceedings of the National Academy of Sciences, USA* 98**, 2473-2478.

Grivet D (2002) ***Phylogéographie et evolution moléculaire comparée d'arbres forestiers à l'aide des marqueurs chloroplastiques***. Thesis, University of Nancy I.

Grivet D, Petit RJ (2002a) Phylogeography of the common ivy (*Hedera sp.*) in Europe: genetic differentiation through space and time. ***Molecular Ecology* 11**, 1351-1362.

Grivet D, Petit RJ (2002b) Chloroplast DNA phylogeography of the hornbeam in Europe: evidence for a bottleneck at the outset of postglacial colonization. ***Conservation Genetics* 4**, 47-56.

Grivet D, Jeandroz S, Favre J-M (1999) Nad1 b/c intron polymorphism reveals maternal inheritance of the mitochondrial genome in *Picea abies*. ***Theoretical and Applied Genetics* 99**, 346-349.

Grivet D, Heinze B, Vendramin GG, Petit RJ (2001) Genome walking with consensus primers: application to the large single copy region of chloroplast DNA. ***Molecular Ecology Notes* 1**, 345-349.

Grun P (1976) ***Cytoplasmic Genetics and Evolution***. Columbia University Press, New York.

Gualberto J, Lamattina L, Bonnard G, Weil JH, Grienenberger JM (1989) RNA editing in wheat mitochondria results in the conservation of protein sequences. ***Nature* 341**, 660-662.

Hagemann R (2000) Erwin Baur or Carl Correns: who really created the theory of plastid inheritance? ***Journal of Heredity* 91**, 435-440.

Hagemann R, Schröder M-B (1989) The cytological basis of the plastid inheritance in angiosperms. ***Protoplasma* 152**, 57-64.

Hamilton MB (1999) Four primer pairs for the amplification of chloroplast intergenic regions with intraspecific variation. ***Molecular Ecology* 8**, 521-523.

Hampe A, Bairlein F (2000) Modified dispersal-related traits in disjunct populations of bird-dispersed *Frangula alnus* (Rhamnaceae): a result of its Quaternary distribution shifts? ***Ecography* 23**, 603-613.

Hanski I (1999) Metapopulation dynamics. ***Nature* 396**, 41-49.

Hanson MR (1991) Plant mitochondrial mutations and male sterility. ***Annual Review of Genetics* 25**, 461-486.

Hanson MR, Conde MF (1985) Functioning and variation of cytoplasmic genomes: lessons from cytoplasmic-nuclear interactions affecting male fertility in plants. ***International Review of Cytology*** **94**, 213-267.

Hanson MR, Folkerts O (1992) Structure and function of the higher plant mitochondrial genome. ***International Review of Cytology*** **141**, 129-172.

Harper JL (1977) ***Population Biology of Plants***. Academic Press, London/New York/Toronto Sydney/San Francisco.

Harris SA, Ingram R (1991) Chloroplast DNA and biosystematics: the effects of intraspecific diversity and plastid transmission. ***Taxon*** **40**, 393-412.

Havey MJ (1997) Predominant paternal transmission of the mitochondrial genome in cucumber. ***Journal of Heredity*** **88**, 232-235.

Havey MJ, McCreight JD, Rhodes B, Taurick G (1998) Differential transmission of the *Cucumis* organellar genomes. ***Theoretical and Applied Genetics*** 97, 122-128.

Hedgpeth J, Goodman HM, Boyer HW (1972) DNA nucleotide sequence restricted by the R1 endonuclease. ***Proceedings of the National Academy of Sciences, USA*** **69**, 3448-3452.

Heinze B (2001) A data base for PCR primers in the chloroplast genome. Poster presented at Plant and Animal Genome X, January 12-16, 2002, San Diego. http://fbva.forvie.ac.at/200/1859.html.

Hermann M, Bock R (1999) Transfer of plastid RNA-editing activity to novel sites suggests a critical role for spacing in editing-site recognition. ***Proceedings of the National Academy of Sciences, USA*** **96**, 4856-4861

Higgins SI, Richardson DM (1999) Predicting plant migration rates in a changing world: the role of long-distance dispersal. ***American Naturalist*** **153**, 464-475.

Hipkins VD, Marshall KA, Neale DB, Rottmann WH, Strauss SH (1995) A mutation hotspot in the chloroplast genome of a conifer (Douglas-fir: *Pseudotsuga*) is caused by a variability in the number of direct repeats derived from a partially duplicated tRNA gene. ***Current Genetics*** **27**, 572-579.

Honda SI, Hongladarom T, Wildman SG (1964) Characteristic movements of organelles in streaming cytoplasm of plant cells. In: ***Primitive Motile Systems in Cell Biology*** (eds. Allen RD, Kamiya N), pp. 485-502. Academic Press, New York.

Hu X-S, Ennos RA (1997) On estimation of the ratio of pollen to seed flow among plant populations. ***Heredity*** **79**, 541-552.

Hu ZM, Hu SY, Zhong ZJ (1996) Paternal inheritance of plastid DNA in genus *Pharbitis*. ***Acta Botanica Sinica*** **38**, 253-256.

Huang S, Chiang YC, Schaal BA, Chou CH, Chiang TY (2001) Organelle DNA phylogeography of *Cycas taitungensis*, a relict species in Taiwan. ***Molecular Ecology*** **10**, 2669-2681.

Hupfer H, Swiatek M, Hornung S, Herrmann RG, Maier RM, Chiu WL, Sears B (2000) Complete nucleotide sequence of the *Oenothera elata* plastid chromosome, representing plastome I of the five distinguishable Euoenothera plastomes. ***Molecular and General Genetics*** **263**, 581-585.

Husband BC, Barrett SCH (1996) A metapopulation perspective in plant population biology. ***Journal of Ecology*** **84**, 461-469.

Ibarra-Manríquez G, Martínez Ramos M, Oyama K (2001) Seedling functional types in a lowland rain forest in Mexico. ***American Journal of Botany*** **88**, 1801-1812.

Imbert E (2002) Ecological consequences and ontogeny of seed heteromorphism. ***Perspectives in Plant Ecology, Evolution and Systematics*** **5**, 13-36.

Izhar S, Schlicter M, Swartzberg D (1983) Sorting out of cytoplasmic elements in somatic hybrids of petunia and prevalence of the heteroplasmon through several meiotic cycles. ***Molecular and General Genetics*** **190**, 468-474.

Johnson LB, Palmer JD (1989) Heteroplasmy of chloroplast DNA in *Medicago*. ***Plant Molecular Biology*** **12**, 3-11.

Jordano P, Schupp EW (2000) Seed disperser effectiveness: the quantity component and patterns of seed rain for *Prunus mahaleb*. ***Ecological Monographs*** **70**, 591-615.

Kanazawa A, Tsutsumi N, Hirai A (1994) Reversible changes in the composition of the population of mtDNAs during dedifferentiation and regeneration in tobacco. ***Genetics*** **138**, 865-870.

Kanno A, Watanabe N, Nakamura I, Hirai A (1993) Variation in chloroplast DNA from rice (*Oryza sativa*): differences between deletions mediated by short direct-repeat sequences within a single species. ***Theoretical and Applied Genetics*** **86**, 579-584

Kempken F (1995) Horizontal transfer of a mitochondrial plasmid. ***Molecular and General Genetics*** **248**, 89-94.

King RA, Ferris C (2002) A variable minisatellite sequence in the chloroplast genome of *Sorbus* L. (Rosaceae: Maloideae). ***Genome*** **45**, 570-576.

Kirk JTO (1971) Chloroplast structure and biogenesis. ***Annual Review of Biochemistry*** **40**, 161-196.

Kirk JTO, Tilney-Bassett RAE (1978) ***The Plastids: Their Chemistry, Structure, Growth and Inheritance***. Elsevier/North-Holland Biomedical Press, New York.

Kobayashi K, Yoshikawa J, Suzuki M (2000) DNA identification of *Picea species* of the Last Glacial Age in northern Japan. ***Japanese Journal of Historical Botany*** **8**, 67-80.

Köhler RH, Cao J., Zipfel WR, Webb WW, Hanson MR (1997) Exchange of protein molecules through connections between higher plant chloroplasts. ***Science*** **276**, 2039-2042.

Kollmann J (2000) Dispersal of fleshy-fruited species: a matter of spatial scale? ***Perspectives in Plant Ecology, Evolution and Systematics*** **3**, 29-51.

Kolodner R, Tewari KK (1972) Circular mitochondrial DNA (70 x 10^6 daltons) from pea leaves. ***The FASEB Journal*** **31**, 876.

Kondo T, Tsumura Y, Kawahara T, Okamura M (1998) Inheritance of chloroplast and mitochondrial DNA in interspecific hybrids of *Chamaecyparis* spp. ***Breeding Science*** **78**, 177-179

König AO, Ziegenhagen B, van Dam BC, Csaikl UM, Coart E, Degen B, Burg K, De Vries SMG, Petit RJ (2002) Chloroplast DNA variation of oaks in western Central Europe and genetic consequences of human influences. ***Forest Ecology and Management*** **156**, 147-166.

Kremer A, Petit RJ, Zanetto A, Fougère V, Ducousso A, Wagner DB, Chauvin C (1991) Nuclear and organelle diversity in *Quercus robur* and *Q. petraea*. In: ***Genetic Variation in European Populations of Forest Trees***. (eds. Müller-Starck G, Ziehe M), pp. 141-166. Sauerländer's Verlag, Frankfurt am Main.

Kuroiwa T (1991) The replication, differentiation, and inheritance of plastids with emphasis on the concept of organelle nuclei. ***International Review of Cytology*** **128**, 1-62.

Lampa GK, Bendich AJ (1979) Changes in chloroplast DNA levels during development of pea (*Pisum sativum*). ***Plant Physiology*** **64**, 126-130.

Lampa GK, Bendich AJ (1984) Changes in mitochondrial DNA levels during development of pea (*Pisum sativum*). ***Planta* 162**, 463-468.

Lansman RA, Avise JC, Huettel MD (1983) Critical experimental test of the possibility of 'paternal leakage' of mitochondrial DNA. ***Proceedings of the National Academy of Sciences, USA* 80**, 1969-1971.

Laporte V, Cuguen J, Couvet D (2000) Effective population sizes for cytoplasmic and nuclear genes in a gynodioecious species: the role of the sex determination system. ***Genetics* 154**, 447-458.

Larkin PJ, Scowcroft WR (1981) Somaclonal variation: a novel source of variability from cell cultures for plant improvement. ***Theoretical and Applied Genetics* 60**, 197-214.

Laroche J, Li P, Maggia L, Bousquet J (1997) Molecular evolution of angiosperm mitochondrial introns and exons. ***Proceedings of the National Academy of Sciences, USA* 94**, 5722-5727.

Latta RG, Linhart YB, Fleck D, Elliot M (1998) Direct and indirect estimates of seed vs pollen movement within a population of ponderosa pine. ***Evolution* 52**, 61-67.

Lavin M, Mathews S, Hughes C (1991) Chloroplast DNA variation in *Gliricidia sepium* (Leguminosae): intraspecific phylogeny and tokogeny. ***American Journal of Botany* 78**, 1576-1585.

Lax AR, Vaughn KC, Duke SO, Endrizzi JE (1987) Structural and physiological studies of a plastome cotton mutant with slow sorting out. ***Journal of Heredity* 78**, 147-152

Le Corre V, Machon N, Petit RJ, Kremer A (1997a) Colonization with long-distance seed dispersal and distribution of maternally inherited diversity in forest trees: a simulation study. ***Genetical Research, Cambridge* 69**, 117-125.

Le Corre V, Dumolin-Lapègue S, Kremer A (1997b) Genetic variability at enzyme and RAPD loci in sessile oak *Quercus petraea* (Matt.) Liebl.: the role of history and geography. ***Molecular Ecology* 6**, 519-529.

Lee DJ, Blake TK, Smith SE (1988) Bi-parental inheritance of chloroplast DNA and the existence of heteroplasmic cells in alfalfa. ***Theoretical and Applied Genetics* 76**, 545-549.

Leon P, Arroyo A, Mackenzie S (1998) Nuclear control of plastid and mitochondrial development in higher plants. ***Annual Review of Plant Physiology and Plant Molecular Biology* 49**, 453-480.

Lev-Yadun S (2000) Why are underground flowering and fruiting more common in Israel than anywhere else in the world? ***Current Science* 79**, 289.

Levin DA (1979) The nature of plant species. ***Science* 204**, 381-384.

Levin DA, Kerster HW (1974) Gene flow in seed plants. ***Evolutionary Biology* 7**, 139-220.

Levings CS, Pring DR (1976) Restriction endonuclease analysis of mitochondrial DNA from normal and Texas cytoplasmic male-sterile maize. ***Science* 193**, 158-160.

Levy F, Neal CL (1999) Spatial and temporal genetic structure in chloroplast and allozyme markers in *Phacelia dubia* implicate genetic drift. ***Heredity* 82**, 422-431.

Liepelt S, Kuhlenkamp V, Anzidei M, Vendramin GG, Ziegenhagen B (2001) Pitfalls in determining size homoplasy of microsatellite loci. ***Molecular Ecology Notes* 1**, 332-335.

Liepelt S, Bialozyt R, Ziegenhagen B (2002) Wind-dispersed pollen mediates postglacial gene flow among refugia. ***Proceedings of the National Academy of Sciences, USA* 99**, 14590-14594.

Lin X, Kaul S, Rounsley S, Shea TP, Benito MI, Town CD, Fujii CY, Mason T, Bowman CL, Barnstead M, Feldblyum TV, Buell CR, Ketchum KA, Lee J, Ronning CM, Koo HL, Moffat KS, Cronin LA, Shen M, Pai G, Van Aken S, Umayam L, Tallon LJ, Gill JE, Adams MD, Carrera AJ, Creasy TH, Goodman HM, Somerville CR, Copenhaver GP, Preuss D, Nierman WC, White O, Eisen JA, Salzberg SL, Fraser CM, Venter JC (1999) Sequence and analysis of chromosome 2 of the plant *Arabidopsis thaliana*. ***Nature* 402**, 761-768.

Lineberger RD (n.d.) Origin, development, and propagation of chimeras. http://aggie-horticulture.tamu.edu/tisscult/chimeras/chimeralec/chimeras.html.

Lonsdale DM (1989) The plant mitochondrial genome. ***Biochemistry of Plants* 15**, 229-295.

Loridon K, Saumitou-Laprade P (2002) Detection of gene-anchored amplification polymorphism (GAAP) in the vicinity of plant mitochondrial genes. ***Molecular Genetics and Genomics* 267**, 329-337.

Lu SY, Peng CI, Cheng YP, Hong KH, Chiang TY (2001) Chloroplast DNA phylogeography of *Cunninghamia konishii* (Cupressaceae), an endemic conifer of Taiwan. ***Genome* 44**, 797-807.

Luo H, Van Coppenolle B, Seguin M, Boutry M (1995) Mitochondrial DNA polymorphism and phylogenetic relationships in *Hevea brasiliensis*. ***Molecular Breeding* 1**, 51-63

Lynch M, Blanchard JL (1998) Deleterious mutation accumulation in organelle genomes. ***Genetica* 102/103**, 29-39.

Lynch M, Bürger R, Butcher D, Gabriel W (1993) The mutational meltdown in asexual populations. ***Journal of Heredity* 84**, 339-344.

MacDonald GM (1993) Fossil pollen analysis and the reconstruction of plant invasions. ***Advances in Ecological Research* 24**, 67-110.

Mackenzie S, He S, Lyznik A (1994) The elusive plant mitochondrion as a genetic system. ***Plant Physiology* 105**, 775-780.

Manicacci D, Couvet D, Belhassen E, Gouyon P-H, Atlan A (1996) Founder effects and sex ratio in the gynodioecious *Thymus vulgaris* L. ***Molecular Ecology* 5**, 63-72.

Martînez-Zapater JM, Gil P, Capel J, Somerville CR (1992) Mutations at the *Arabdopsis* Chm locus promote rearrangements of the mitochondrial genome. ***Plant Cell* 4**, 889-899.

Mason RJ, Holsinger KE, Jansen RK (1994) Bi-parental inheritance of the chloroplast genome in *Coreopsis* (Asteraceae). ***Journal of Heredity* 85**, 171-173.

Masoud SA, Johnson LB, Sorensen EL (1990) High transmission of paternal plastid DNA in alfalfa plants demonstrated by restriction fragment polymorphic analysis. ***Theoretical and Applied Genetics* 79**, 49-55.

Maynard Smith J, Szathmáry E (1995) ***The Major Transitions in Evolution***. Oxford University Press, Oxford.

McCauley DE (1995) The use of chloroplast DNA polymorphism in studies of gene flow in plants. ***Trends in Ecology and Evolution* 10**, 198-202.

McCauley DE (1997) The relative contributions of seed and pollen movement to the local genetic structure of *Silene alba*. ***Journal of Heredity* 88**, 257-263.

McCauley DE (1998) The genetic structure of a gynodioecious plant: nuclear and cytoplasmic genes. ***Evolution* 52**, 255-260.

McCauley DE, Richards CM, Emery SE, Smith RA, McGlothlin JW (2001) The interaction of genetic and demographic processes in plant metapopulations: a case study of *Silene alba*. In: ***Integrating Ecology and Evolution in a Spatial Context*** (eds. Silvertown J, Antonovics J), pp. 177-196. Blackwell Science, Oxford.

McKendrick SL, Leake JR, Taylor DL, Read DJ (2000). Symbiotic germination and development of myco-heterotrophic plants in nature: ontogeny of *Corallorhiza trifida* Châtel and characterisation of its mycorrhizal fungi. ***New Phytologist*** **145**, 523-537.

Medgyesy P, Fejes E, Maliga P (1985) Interspecific chloroplast recombination in a *Nicotiana* somatic hybrid. ***Proceedings of the National Academy of Sciences, USA*** **82**, 6960-6964.

Medgyesy P, Pay A, Marton L (1986) Transmission of paternal chloroplasts in *Nicotiana*. ***Molecular and General Genetics*** **204**, 195-198.

Mejnartowicz M (1991) Inheritance of chloroplast DNA in *Populus*. ***Theoretical and Applied Genetics*** **82**, 477-480.

Mendel G (1866) Versuche über Pflanzen-Hybriden. ***Verhandlungen des naturforschenden Vereines, Abhandlungen, Brünn*** **4**, 3-47.

Michaelis P (1935) Entwicklungsgeschichtlich genetische Untersuchungen an *Epilobium*. III. Zur Frage der Übertragung von Pollenschlauchplasma in die Eizelle und ihre Bedeutung für die Plasmavererbung. ***Planta*** **23**, 486-500.

Michaelis P (1949) Über Abänderung des plasmatischen Erbgutes. ***Zeitschrift für induktive Abstammungs- und Vererbungslehre*** **86**, 101-112.

Michaelis P (1967) The investigation of plasmon segregation by the pattern-analysis. ***Nucleus*** **10**, 1-14.

Milligan BG (1992) Is organelle DNA strictly maternally inherited? Power analysis of a binomial distribution. ***American Journal of Botany*** **79**, 1325-1328.

Mitton B (1994) Molecular approaches to population biology. ***Annual Review of Ecology and Systematics*** **25**, 45-69.

Möbius M (1920) Über die Grösse der Chloroplasten. ***Berichte der Deutschen Botanischen Gesellschaft*** **38**, 224-232.

Mogensen HL (1988) Exclusion of male mitochondria and plastids during syngamy in barley as a basis for maternal inheritance. ***Proceedings of the National Academy of Sciences, USA*** **85**, 2594-2597.

Mogensen HL (1996) The hows and whys of cytoplasmic inheritance in seed plants. ***American Journal of Botany*** **83** 383-404.

Morand ME, Vedel F, Raquin C, Frascaria-Lacoste N (2001) A rapid and efficient method for the isolation of mitochondrial DNA from angiosperm tree species. Application to the restriction fragment length polymorphism distinction between European and American ashes. ***Molecular Ecology*** **1**, 67-70.

Mori SA, Brown JL (1994) Report on wind dispersal in a lowland moist forest in central French Guiana. ***Brittonia*** **46**, 105-125.

Morton BR, Clegg MT (1993) A chloroplast DNA mutational hotspot and gene conversion in a noncoding region near *rbcL* in the grass family (Poaceae). ***Current Genetics*** **24**, 357-365.

Muller HJ (1964) The relation of recombination to mutational advance. ***Mutation Research*** **1**, 2-9.

Muller M-H, Prosperi JM, Santoni S, Ronfort J (2001) How mitochondrial DNA diversity can help to understand the dynamics of wild-cultivated complexes. The case of *Medicago sativa* in Spain. ***Molecular Ecology*** **10**, 2753-2763.

Muloko-Ntoutoume N, Petit RJ, White L, Abernethy K (2000) Chloroplast DNA variation in a rainforest tree (*Aucoumea klaineana*, Burseraceae) in Gabon. ***Molecular Ecology*** **9**, 359-364.

Nagata, H, Saito C, Sakai A, Kuroiwa H, Kuroiwa T (1999) The selective increase or decrease of organellar DNA in generative cells just after pollen mitosis one controls cytoplasmic inheritance. ***Planta*** **209**, 53-65.

Neale DB, Sederoff RR (1989) Paternal inheritance of chloroplast DNA and maternal inheritance of mitochondrial DNA in loblolly pine. ***Theoretical and Applied Genetics*** **77**, 212-216.

Neale DB, Saghai-Maroof MA, Allard RW, Zhang Q, Jorgensen RA (1986) Chloroplast DNA diversity in populations of wild and cultivated barley. ***Genetics*** **130**, 1105-1110.

Neale DB, Marshall KA, Sederoff RS (1989) Chloroplast and mitochondrial DNA are paternally inherited in *Sequoia sempervirens* D. Don Endl. ***Proceedings of the National Academy of Sciences, USA*** **86**, 9347-9349.

Neale DB, Marshall KA, Harry DE (1991) Inheritance of chloroplast and mitochondrial DNA in incense-cedar (*Calocedrus decurrens*). ***Canadian Journal of Forest Research*** **21**, 717-720.

Newton KJ (1988) Plant mitochondrial genomes: organization, expression and variation. ***Annual Review of Plant Physiology and Plant Molecular Biology*** **39**, 503-532.

Oddou-Muratorio S, Petit RJ, Le Guerroué B, Guesnet D, Demesure B (2001) Pollen-*versus* seed-mediated gene flow in a scattered forest tree species. ***Evolution*** **55**, 1123-1135.

Olson MS, McCauley DE (2000) Linkage disequilibrium and phylogenetic congruence between chloroplast and mitochondrial haplotypes in *Silene vulgaris*. ***Proceedings of the Royal Society of London, B*** **267**, 1801-1808.

Olson MS, McCauley DE (2002) Mitochondrial DNA diversity, population structure, and gender association in the gynodioecious plant *Silene vulgaris*. ***Evolution*** **56**, 253-262.

Orive ME, Asmussen MA (2000) The effects of pollen and seed migration on nuclear-dicytoplasmic systems. II. A new method for estimating plant gene flow from joint nuclear-cytoplasmic data. ***Genetics*** **155**, 833-854.

Ouborg NJ, Piquot Y, Van Groenendael JM (1999) Population genetics, molecular markers and the study of dispersal in plants. ***Journal of Ecology*** **87**, 551-568.

Owens JN, Catalano GL, Morris ML, Aitken-Christie J (1995) The reproductive biology of Kauri (*Agathis australis*) II. Male gametes, fertilization and cytoplasmic inheritance. ***International Journal of Plant Sciences*** **156**, 404-416.

Palmer JD (1985) Comparative organization of chloroplast genomes. ***Annual Review of Genetics*** **19**, 325-354.

Palmer JD (1987) Chloroplast DNA evolution and biosystematic uses of chloroplast DNA variation. ***American Naturalist*** **130**, S6-S29.

Palmer JD (1990) Contrasting modes and tempos of genome evolution in land plant organelles. ***Trends in Genetics*** **6**, 115-120.

Palmer JD (1992) Comparison of chloroplast and mitochondrial genome evolution in plants. In: ***Cell Organelles*** (ed. Hermann R), pp. 99-133. Springer, Vienna.

Palmer JD, Shields CR, Cohen DB, Orton TJ (1983) An unusual mitochondrial DNA plasmid in the genus *Brassica*. ***Nature*** **301**, 725-728.

Palmer JD, Adams KL, Cho Y, Parkinson CL, Qiu Y-L, Song K (2000) Dynamic evolution of plant mitochondrial genomes: mobile genes and introns and highly variable mutation rates. *Proceedings of the National Academy of Sciences, USA* **97**, 6960-6966.

Petit RJ (1992) ***Polymorphisme de l'ADN chloroplastique dans un complexe d'espèces: les chênes blancs européens.*** Thesis, Paris XI-Orsay.

Petit RJ, Grivet D (2002) Optimal randomization strategies when testing the existence of a phylogeographic structure. *Genetics* **161**, 469-471.

Petit RJ, Kremer A, Wagner DB (1993a) Geographic structure of chloroplast DNA polymorphisms in European oaks. *Theoretical and Applied Genetics* **87**, 122-128.

Petit RJ, Kremer A, Wagner DB (1993b) Finite island model for organelle and nuclear genes in plants. *Heredity* **71**, 630-641.

Petit RJ, Pineau E, Demesure B, Bacilieri R, Ducousso A, Kremer A (1997) Chloroplast DNA footprints of postglacial recolonisation by oaks. *Proceedings of the National Academy of Sciences, USA* **94**, 9996-10001.

Petit RJ, Bialozyt R, Brewer S, Cheddadi R, Comps B (2001) From spatial patterns of genetic diversity to postglacial migration processes in forest trees. In: *Integrating Ecology and Evolution in a Spatial Context* (eds. Silvertown J, Antonovics J), pp. 295-318. Blackwell Science, Oxford.

Petit RJ, Latouche-Hallé C, Pemonge M-H, Kremer A (2002a) Chloroplast DNA variation of oaks in France and the influence of forest fragmentation on genetic diversity. *Forest Ecology and Management* **156**, 115-129.

Petit RJ, Csaikl UM, Bordács S, Burg K, Coart E, Cottrell J, van Dam B, Deans JD, Dumolin-Lapègue S, Fineschi S, Finkeldey R, Gillies A, Glaz I, Goicoechea PG, Jensen JS, König AO, Lowe AJ, Madsen SF, Mátyás G, Munro RC, Olalde M, Pemonge M, Popescu F, Slade D, Tabbener H, Tauchini D, Vries SGM, Ziegenhagen B, Kremer A (2002b) Chloroplast DNA variation in European white oaks: phylogeography and patterns of diversity based on data from over 2600 populations. *Forest Ecology and Management* **156**, 5-26.

Petit RJ, Brewer S, Bordács S, Burg K, Cheddadi R, Coart E, Cottrell J, Czaikl UM, van Dam B, Deans JD, Espinel S, Fineschi S, Finkeldey R, Glaz I, Goicoechea PG, Jensen JS, König AO, Lowe AJ, Madsen SF, Mátyás G, Munro RC, Popescu F, Slade D, Tabbener H, deVries SGM, Ziegenhagen B, de Beaulieu J-L, Kremer A (2002c) Identification of refugia and post-glacial colonisation routes of European white oaks based on chloroplast and fossil pollen evidence. *Forest Ecology and Management* **156**, 49-74.

Petit RJ, Duminil J, Fineschi S, Hampe A, Salvini D, Vendramin GG (2005) Comparative organization of chloroplast, mitochondrial and nuclear diversity in plant populations. *Molecular Ecology* **14**, 689-701.

Pichot C, El-Maâtaoui M, Raddi S, Raddi P (2001) Surrogate mother for endangered *Cupressus*. *Nature* **412**, 39.

Pitelka LF, the plant migration workshop group (1997) Plant migration and climate change. *American Scientist* **85**, 464-473.

Pons O, Petit RJ (1995) Estimation, variance and optimal sampling of gene diversity. I. Haploid locus. *Theoretical and Applied Genetics* **90**, 462-470.

Potts BM, Reid JB (1988) Hybridization as a dispersal mechanism. ***Evolution*** **42**, 1245-1255.

Powell W, Morgante M, McDevitt R, Vendramin GG, Rafalski JA (1995) Polymorphic simple sequence repeat regions in chloroplast genomes: applications to the population genetics of pines. *Proceedings of the National Academy of Scienes, USA* **92**, 7759-7763.

Provan J, Soranzo N, Wilson NJ, Goldstein DB, Powell W (1999) A low mutation rate for chloroplast microsatellites. ***Genetics* 153**, 943-947.

Provan J, Powell W, Hollingsworth PM (2001) Chloroplast microsatellites: new tools for studies in plant ecology and evolution. ***Trends in Ecology and Evolution* 16**, 142-147.

Pyke A (1999) Plastid division and development. ***The Plant Cell* 11**, 549-556.

Quétier F, Vedel F (1977) Heterogeneous populations of mitochondrial DNA molecules in higher plants. ***Nature* 268**, 365-368.

Race HL, Herrmann RG, Martin W (1999) Why organelles retained genomes? ***Trends in Genetics* 15**, 364-370.

Radetzky R (1990) Analysis of mitochondrial DNA and its inheritance in *Populus*. ***Current Genetics* 18**, 429-434.

Rajora OP, Dancik BP (1992) Chloroplast DNA inheritance in *Populus*. ***Theoretical and Applied Genetics* 84**, 280-285.

Rajora OP, Mahon JD (1994) Inheritance of mitochondrial DNA in lentil (*Lens culinaris* Medik.). ***Theoretical and Applied Genetics* 89**, 206-210.

Rajora OP, Mahon JD (1995) Paternal plastid DNA can be inherited in lentil. ***Theoretical and Applied Genetics* 90**, 607-610.

Rajora OP, Barrett JW, Dancik BP, Strobeck C (1992) Maternal transmission of mitochondrial DNA in interspecific hybrids of *Populus*. ***Current Genetics* 22**, 141-145.

Rand DM (2001) The units of selection on mitochondrial DNA. ***Annual Review of Ecology and Systematics* 32**, 415-448.

Reboud X, Zeyl C (1994) Organelle inheritance in plants. ***Heredity* 72**, 132-140.

Reid C (1899) ***The Origin of the British Flora***. Dulau, London.

Richards AJ (1997) ***Plant Breeding Systems***, 2nd edn. Stanley Thorne, Cheltenham.

Richards C (2000) Inbreeding depression and genetic rescue in a plant metapopulation. ***American Naturalist* 155**, 383-394.

Rieseberg LH, Soltis DE (1991) Phylogenetic consequences of cytoplasmic gene flow in plants. ***Evolutionary Trends in Plants* 5**, 65-84.

Rieseberg LH, Van Fossen C, Arias D, Carter RL (1994) Cytoplasmic male sterility in sunflower: origin, inheritance, and frequency in natural populations. ***Journal of Heredity* 85**, 233-238.

Röbbelen G (1966) Chloroplastdifferenzierung nach geninduzierter Plastommutation bei *Arabidopsis thaliana* (L.) Heynh. ***Zeitschrift für Pflanzenphysiologie* 55**, 387-403.

Röhr H, Kües U, Stahl U (1998) Organelle DNA of plants and fungi: inheritance and recombination. ***Progress in Botany* 60**, 39-87.

Rollo F, Venanzi FM, Amici A (1994) DNA and RNA from ancient plant seeds. In: ***Ancient DNA*** (eds. Hermann B, Hummel S), pp 218-236. Springer-Verlag, New York.

Rose RJ, Thomas MR, Fitter JT (1990) The transfer of cytoplasmic and nuclear genomes by somatic hybridisation. ***Australian Journal of Plant Physiology* 17**, 303-321.

Ross M (1990) Sexual asymmetry in hermaphroditic plants. ***Trends in Ecology and Evolution* 5**, 43-47.

Roussell DL, Thompson DL, Pallardy SG, Miles D, Newton KJ (1991) Chloroplast structure and function is altered in the NCS2 maize mitochondrial mutant. ***Plant Physiology* 96**, 232-238.

Rousset F (1997) Genetic differentiation and estimation of gene flow from *F*-statistics under isolation by distance. ***Genetics* 145**, 1219-1228.

Ruiz-Pesini E, Lapena AC, Diez-Sanchez C, Perez-Martos A, Montoya J, Alvarez E, Diaz M, Urries A, Montoro L, Lopez-Perez MJ, Enriquez JA (2000) Human mtDNA haplogroups associated with high or reduced spermatozoa motility. ***American Journal of Human Genetics* 67**, 682-696.

Russell SD (1987) Quantitative cytology of the egg and central cell of *Plumbago zeylanica* and its impact on cytoplasmic inheritance patterns. ***Theoretical and Applied Genetics* 74**, 693-699.

Russell SD (1991) Isolation and characterization of sperm cells in flowering plants. ***Annual Review of Plant Molecular Biology* 42**, 189-204.

Saghai Maroof MA, Zhang Q, Neale DB, Allard RW (1992) Associations between nuclear loci and chloroplast DNA genotypes in wild barley. ***Genetics* 131**, 225-231.

Saltonstall K (2002) Cryptic invasion by a non-native genotype of the common reed, *Phragmites australis*, into North America. ***Proceedings of the National Academy of Sciences, USA* 99**, 2445-2449.

Salvini D, Anzidei M, Fineschi S, Malvolti ME, Taurchini D, Vendramin GG (2001) Low genetic differentiation among Italian populations of *Populus tremula* L. estimated using chloroplast PCR-RFLP and microsatellite markers. ***Forest Genetics* 8**, 81-87.

Sapp J (1987) ***Beyond the Gene. Cytoplasmic Inheritance and the Struggle for Authority in Genetics.*** Oxford University Press, Oxford.

Sauer JD (1988) ***Plant Migration: The Dynamics of Geographic Patterning in Seed Plant Species.*** University of California Press, Berkley.

Saumitou-Laprade P, Pannenbecker G, Maggouta F, Jean R, Michaelis G (1989) A linear 10.4 kb plasmid in the mitochondria of *Beta maritima*. ***Current Genetics* 16**, 181-186.

Saumitou-Laprade P, Rouwendal GJA, Cuguen J, Krens FA, Michaelis G (1993) Different CMS sources found in *Beta vulgaris* ssp *maritima*: mitochondrial variability in wild populations revealed by a rapid screening procedure. ***Theoretical and Applied Genetics* 85**, 529-535.

Schaal BA, Leverich WJ, Rogstad SH (1991) A comparison of methods for assessing genetic variation in plant conservation biology. In: ***Genetics and Conservation of Rare Plants*** (eds. Falk DA, Holsinger KE), pp. 123-134. Oxford University Press, Oxford, New York.

Schaal BA, Hayworth DA, Olsen KM, Rauscher JT, Smith WA (1998) Phylogeographic studies in plants: problems and prospects. ***Molecular Ecology* 7**, 465-474.

Schoen DJ, Lloyd DG (1984) The selection of cleistogamy and heteromorphic diaspores. ***Botanical Journal of the Linnean Society* 23**, 303-322.

Schötz F (1975) Untersuchungen über die Plastidenkonkurrenz bei *Oenothera. V.* ***Biologisches Zentralblatt* 94**, 17-26.

Schuster W, Brennicke A (1987) Plastid, nuclear and reverse transcriptase sequences in the mitochondrial genome of *Oenothera*: is genetic information transferred between organelles via RNA? ***The EMBO Journal* 6**, 2857-2863.

Scowcroft WR (1979) Nucleotide polymorphism in chloroplast DNA of *Nicotiana debneyi*. ***Theoretical and Applied Genetics* 55**, 133-137.

Sears BB (1980) The elimination of plastids during spermatogenesis and fertilization in the plant kingdom. ***Plasmid* 4**, 233-255.

Sederoff RR Levings CS, Timothy DH, Hu WWL (1981) Evolution of DNA sequence organization in mitochondrial genomes of *Zea*. ***Proceedings of the National Academy of Sciences, USA*** **78**, 5953-5957.

Selosse MA, Albert B, Godelle P (2001) Reducing the genome size of organelles favours gene transfer to the nucleus. ***Trends in Ecology and Evolution*** **16**, 135-141.

Sharbel TF, Mitchell-Olds T (2001) Recurrent polyploid origins and chloroplast phylogeography in the *Arabis holboellii* complex (Brassicaceae). ***Heredity*** **87**, 59-68.

Shiina T, Hayashi K, Ishii N, Morikawa K, Toyoshima Y (2000) Chloroplast tubules visualized in transplastomic plants expressing green fluorescent protein. ***Plant Cell Physiology*** **41**, 367-371.

Shinozaki K, Ohme M, Tanaka M (1986) The complete nucleotide sequence of the tobacco chloroplast genome: its gene organization and expression. ***The EMBO Journal*** **5**, 2043-2049.

Shore JS, Triassi M (1998) Paternally biased cpDNA inheritance in *Turnera ulmifolia* (Turneraceae). ***American Journal of Botany*** **85**, 328-332.

Sidow A, Wilson AC, Pääbo S (1991) Bacterial DNA in Clarkia fossils. ***Philosophical Transactions of the Royal Society of London, B*** **333**, 429-433.

Silvertown J, Charlesworth D (2001) ***Introduction to Plant Population Biology***. Blackwell Science, Oxford.

Small ID, Isaac PG, Leaver CJ (1987) Stoichiometric differences in DNA molecules containing the *atpA* gene suggest mechanisms for the generation of mitochondrial genome diversity in maize. ***The EMBO Journal*** **6**, 865-869.

Small ID, Suffolk R, Leaver CJ (1989) Evolution of plant mitochondrial genomes via substoichiometric intermediates. ***Cell*** **58**, 69-76.

Smith SE (1989) Bi-parental inheritance of organelles and its implications in crop improvement. ***Plant Breeding Reviews*** **6**, 361-393.

Soliman K, Fedak G, Allard RW (1987) Inheritance of organelle DNA in barley and *Hordeum* x *Secale* intergeneric hybrids. ***Genome*** **29**, 867-872.

Soltis DE, Mayer MS, Soltis PS, Edgerton M (1991) Chloroplast-DNA variation in *Tellima grandiflora* (Saxifragaceae) ***American Journal of Botany*** **78**, 1379-1390.

Soltis DE, Soltis PS, Milligan BG (1992) Intraspecific chloroplast DNA variation: systematic and phylogenetic implications. In: ***Molecular Systematics of Plants*** (eds. Soltis PS, Soltis DE, Doyle JJ), pp. 117-150. Chapman and Hall, New York, London.

Soltis DE, Gitzendanner MA, Strenge DD, Soltis PS (1997) Chloroplast DNA intraspecific phylogeography of plants from the Pacific Northwest. ***Plant Systematics and Evolution*** **206**, 353-373.

Song KM, Osborn TC and Williams PH (1988) Brassica taxonomy based on nuclear restriction fragment length polymorphisms (RFLPs) 1. Genome evolution of diploid and amphidiploid species. ***Theoretical and Applied Genetics*** **75**, 784-794.

Song K, Lu P, Tang K and Osborn TC (1995) Rapid genome change in synthetic polyploids of *Brassica* and its implications for polyploid evolution. ***Proceedings of the National Academy of Sciences, USA*** **92**, 7719-7723.

Soranzo N, Provan J, Powell W (1999) An example of microsatellite length variation in the mitochondrial genome of conifers. ***Genome*** **42**, 158-161.

Sork VL, Nason J, Campbell DR, Fernandez JF (1999) Landscape approaches to historical and contemporary gene flow in plants. ***Trends in Ecology and Evolution*** **14**, 219-224.

Southern EM (1975) Detection of specific sequences among DNA fragments separated by gel electrophoresis. ***Journal of Molecular Biology* 98**, 503-517.

Sperisen C, Büchler U, Gugerli F, Matyas G, Geburek T, Vendramin GG (2001) Tandem repeats in plant mitochondrial genomes: application to the analysis of population differentiation in the conifer Norway spruce. ***Molecular Ecology* 10**, 257-263.

Squirrell J, Hollingsworth PM, Bateman RM, Dickson JH, Light MHS, MacConaill M, Tebbitt MC (2001) Partitioning and diversity of nuclear and organelle markers in native and introduced populations of *Epipactis helleborine* (Orchidaceae). ***American Journal of Botany* 88**, 1409-1418.

Stebbins GL (1950) ***Variation and Evolution in Plants***. Columbia University Press, New York.

Stehlik I, Schneller JJ, Bachman K (2001) Resistance or emigration: response of the high-alpine plant *Eritrichium nanum* (L.) Gaudin to the ice age within the Central Alps. ***Molecular Ecology* 10**, 357-370.

Steinborn R, Linke B, Nothnagel T, Börner T (1995) Inheritance of chloroplast and mitochondrial DNA in alloplasmic forms of the genus *Daucus*. ***Theoretical and Applied Genetics* 91**, 632-638.

Stubbe W (1960) Untersuchungen zur genetischen Analyse des Plastoms von *Oenothera*. ***Zeitschrift für Botanik* 48**, 191-218.

Suyama Y, Kawamuro K, Kinoshita I, Yoshimura K, Tsumura Y, Takahara H (1996) DNA sequence from a fossil pollen of *Abies* spp. from Pleistocene peat. ***Genes and Genetic Systems* 71**, 145-149.

Szalanski AL, Steinauer G, Bischof R, Petersen J (2001) Origin and conservation genetics of the threatened Ute ladies'-tresses, *Spiranthes diluvialis* (Orchidaceae). ***American Journal of Botany* 88**, 177-180.

Szmidt AE, Lu M-Z, Wang X-R (2001) Effects of RNA editing on the coxI evolution and phylogeny reconstruction. ***Euphytica* 118**, 9-18.

Taberlet P, Gielly L, Bouvet J (1991) Universal primers for amplification of three non-coding regions of chloroplast DNA. ***Plant Molecular Biology* 17**, 1105-1109.

Taberlet P, Fumagalli L, Wust-Saucy A-G, Cosson J-F (1998) Comparative phylogeography and postglacial colonization routes in Europe. ***Molecular Ecology* 7**, 453-464.

Tarayre M, Saumitou-Laprade P, Cuguen J, Couvet D, Thompson JD (1997) The spatial genetic structure of cytoplasmic (cpDNA) and nuclear (allozyme) markers within and among populations of the gynodioecious *Thymus vulgaris* (Labiatae) in southern France. ***American Journal of Botany* 84**, 1675-1684.

Taylor DR, Zeyl C, Cooke E (2002) Conflicting levels of selection in the accumulation of mitochondrial defects in *Saccharomyces cerevisiae*. ***Proceedings of the National Academy of Sciences, USA* 99**, 3690-3694.

Thanh ND, Medgyesy P (1989) Limited chloroplast gene transfer via recombination overcomes plastome-genome incompatibility between *Nicotiana tabacum* and *Solanum tuberosum*. ***Plant Molecular Biology* 12**, 87-93.

The Arabidopsis Genome Initiative (2000) Analysis of the genome sequence of the flowering plant *Arabidopsis thaliana*. ***Nature* 408**, 796-815

Thompson DS (1961) Studies on the inheritance of male-sterility in carrot, *Daucus carota* L. var. *sativa*. ***Proceedings of the American Society of Horticultural Sciences* 78**, 332-338.

Tilney-Bassett RAE (1970) Genetics and plastid physiology in *Pelargonium*. III. Effect of cultivar and plastids on fertilization and embryo survival. ***Heredity* 20**, 451-466.

Tilney-Bassett RAE (1975) Genetics of variegated plants. In: ***Genetics and Biogenesis of Mitochondria and Chloroplasts*** (eds. Birky, CW Jr, Perlman PS, Byers TJ), pp. 268-308. Ohio State University Press, Columbus.

Tremblay NO, Schoen DJ (1999) Molecular phylogeography of *Dryas integrifolia*: glacial refugia and postglacial recolonization. ***Molecular Ecology*** **8**, 1187-1198.

Tsumura Y, Suyama Y (1998) Differentiation of mitochondrial DNA polymorphisms in populations of five Japanese *Abies* species. ***Evolution*** **52**, 1031-1042.

Tyler T (2000) Detecting migration routes and barriers by examining the distribution of species in an apomictic species complex. ***Journal of Biogeography*** **27**, 979-988.

Unseld M, Marienfeld JR, Brandt P, Brennicke A (1997). The mitochondrial genome of *Arabidopsis thaliana* contains 57 genes in 366,924 nucleotides. ***Nature Genetics*** **15**, 57-61.

Vaillancourt RE, Jackson HD (2000) A chloroplast DNA hypervariable region in eucalypts. ***Theoretical and Applied Genetics*** **101**, 473-477.

van der Pijl L (1972) ***Principles of Dispersal in Higher Plants.*** Springer, Berlin.

Vaughn JC, Mason MT, Sper-Whitis GL, Kuhlman P, Palmer JD (1995) Fungal origin by horizontal transfer of a plant mitochondrial group I intron in the chimeric *coxI* gene of *Peperomia*. ***Journal of Molecular Evolution*** **1**, 563-572.

Vaughn KC (1981) Plastid fusion as an agent to arrest sorting out. ***Current Genetics*** **3**, 243-245.

Vedel F, Quétier F, Bayen M (1976). Specific cleavage of chloroplast DNA from higher plants by *EcoRI* restriction nuclease. ***Nature*** **263**, 440-442.

Vedel F, Quétier F, Dosba F, Doussinault G (1978). Study of wheat phylogeny by *EcoRI* analysis of chloroplastic and mitochondrial DNAs. ***Plant Science Letters*** **13**, 97-102.

Vendramin GG, Lelli L, Rossi P, Morgante M (1996) A set of primers for the amplification of 20 chloroplast microsatellites in Pinaceae. ***Molecular Ecology*** **5**, 595-598.

Vendramin GG, Anzidei M, Madaghiele A, Sperisen C, Bucci G (2000) Chloroplast microsatellites analysis reveals the presence of population subdivision in Norway spruce. ***Genome*** **43**, 68-78.

Verhoeven HA, Blass J (1988) Direct cell to cell transfer of mitochondria by microinjection. In: ***Progress in Plant Protoplast Research*** (eds. Puite KJ, Dons JJM, Huizing HJ, Kool AJ, Koorneef AJ, Krens FA), pp 299-300. Kluwer Academic, Dordrecht, Boston, London.

Vielle-Calzada JP, Crane CF, Stelly DM (1996) Apomixis: the asexual revolution. ***Science*** **274**, 1322-1323.

Wade MJ, McKnight ML, Shaffer HB (1994) The effects of kin-structured colonization on nuclear and cytoplasmic genetic diversity. ***Evolution*** **48**, 1114-1120.

Wagner DB, Furnier GR, Saghai-Maroof MA, Williams SM, Dancik BP (1987) Chloroplast DNA polymorphisms in lodgepole and jack pines and their hybrids. ***Proceedings of the National Academy of Sciences, USA*** **84**, 2097-2100.

Wagner DB, Govindaraju DR, Yeatman CW, Pitel JA (1989) Paternal chloroplast DNA inheritance in a diallel cross of jack pine (*Pinus banksiana* Lamb.). ***Journal of Heredity*** **80**, 483-485.

Wagner DB, Dong J, Carlson MR, Yanchuk AD (1991a) Paternal leakage of mitochondrial DNA in *Pinus*. ***Theoretical and Applied Genetics*** **82**, 510-514.

Wagner DB, Sun Z-X, Govindaraju DR, Dancik BP (1991b) Spatial patterns of chloroplast DNA and cone morphology variation within populations of a *Pinus banksiana-Pinus contorta* sympatric region. ***American Naturalist* 138**, 156-170.

Wakeley J (1996) The excess of transitions among nucleotide substitutions: new methods of estimating transition bias underscore its significance. ***Trends in Ecology and Evolution* 11**, 158-163.

Walsh JB (1993) The effects of different levels of genetic exchange on organelle evolution. ***Journal of Heredity* 84**, 415-418.

Wang BC, Smith TB (2002) Closing the seed dispersal loop. ***Trends in Ecology and Evolution* 17**, 379-385.

Wang G-Z, Miyashita NT, Tsunewaki K (1997) Plasmon analyses of *Triticum* (wheat) and *Aegilops*: PCR-single-strand conformational polymorphism (PCR-SSCP) analyses of organellar DNAs. ***Proceedings of the National Academy of Sciences, USA* 94**, 14570-14577.

Wang X-R, Szmidt AE, Lu M-Z (1996) Genetic evidence for the presence of cytoplasmic DNA in pollen and megagametophytes and maternal inheritance of mitochondrial DNA in *Pinus*. ***Forest Genetics* 3**, 37-44.

Weising K, Gardner RG (1999) A set of conserved PCR primers for the analysis of single sequence repeat polymorphisms in chloroplast genomes of dicotyledonous angiosperms. ***Genome* 42**, 9-19.

Wenny DG, Levey DJ (1998) Directed seed dispersal by bellbirds in a tropical cloud forest. ***Proceedings of the National Academy of Sciences, USA* 95**, 6204-6207.

Werren JH, Beukeboom L (1998) Sex determination, sex ratios and genetic conflict. ***Annual Review of Ecology and Systematics* 29**, 233-261.

Whatley JM (1982) Ultrastructure of plastid inheritance: green algae to angiosperms. ***Biological Review* 57**, 527-569.

White EE (1990) Chloroplast DNA in *Pinus monticola*. 1. Physical map. ***Theoretical and Applied Genetics* 79**, 119-124.

Whitlock MC, McCauley DE (1999) Indirect measures of gene flow and migration: F_{ST} doesn't equal 1/(4*Nm*+1). ***Heredity* 82**, 117-125.

Whittle C-A, Johnston MO (2002) Male-driven evolution of mitochondrial and chloroplastidial DNA sequences in Plants. ***Molecular Biology and Evolution* 19**, 938-949.

Widmer A, Baltisberger M (1999) Molecular evidence for allopolyploid speciation and a single origin of the narrow endemic *Draba ladina* (Brassicaceae). ***American Journal of Botany* 86**, 1282-1289.

Wilkinson DM (1997) Plant colonization: are wind dispersed seeds really dispersed by birds at larger spatial and temporal scales? ***Journal of Biogeography* 24**, 61-65.

Wittzell H (1999) Chloroplast DNA variation and reticulate evolution in sexual and apomictic sections of dandelions. ***Molecular Ecology* 8**, 2023-2035.

Wolfe KH, Li W-H, Sharp PM (1987) Rates of nucleotide substitution vary greatly among plant mitochondrial, chloroplast, and nuclear DNAs. ***Proceedings of the National Academy of Sciences, USA* 84**, 9054-9058.

Wolfson R, Higgins KG, Sears BB (1991) Evidence for replication slippage in the evolution of *Oenothera* chloroplast DNA. ***Molecular Biology and Evolution* 8**, 709-720.

Wright S (1968-1978) ***Evolution and the Genetics of Populations***, 4 vols. University of Chicago Press, Chicago.

Wu J, Krutovskii KV, Strauss SH (1998) Abundant mitochondrial genome diversity, population differentiation and convergent evolution in pines. ***Genetics* 150**, 1605-1614.

Yang TW, Yang YA, Xiong Z (2000) Paternal inheritance of chloroplast DNA in interspecific hybrids in the genus *Larrea* (Zygophyllaceae). ***American Journal of Botany* 87**, 1452-1458.

Yu HS, Russell SD (1994) Quantitative differences in sperm cells and organelles of tobacco (*Nicotiana tabacum* L.) grown under differing environmental conditions. ***Theoretical and Applied Genetics* 89**, 814-817.

Zhu T, Mogensen HL, Smith SE (1993) Quantitative, three-dimensional analysis of alfalfa egg cells in two genotypes: implications for bi-parental plastid inheritance. ***Planta* 190**, 143-150.

Zohary D, Hopf M (2000) ***Domestication of Plants in the Old World.*** 3rd edition. Oxford University Press, Oxford.

Zurawski G, Clegg MT (1987) Evolution of higher plant chloroplast DNA-encoded genes: implications for structure-function and phylogenetic studies. ***Annual Review of Plant Physiology and Plant Molecular Biology* 38**, 391-418.

Zurawski G, Clegg MT, Brown AHD (1984) The nature of nucleotide sequence divergence between barley and maize chloroplast DNA. ***Genetics* 106**, 735-749.

Part II

Broad phylogeographical studies

Chapter 3

Phylogeography of South European mammals

Ettore Randi
E-mail: met0217@iperbole.bo.it

Abstract
Current distributions and the structure of natural biomes in Europe have been largely affected by the Quaternary glaciations, which were prevalent for ca. 80% of the past two milion years. The evolutionary consequences of the Pleistocene climatic changes, the location of refugia in southern Eurasia, and their roles as generators of biodiversity for the entire continent, are still debated. Paleoecological and molecular findings suggested that both Mediterranean and non-Mediterranean Eurasian refuges likely hosted temperate species during the last glaciations. Additionally, the occurrence of cryptic northern refugia was recently postulated. The biogeography, taxonomy and conservation biology of southern European mammals will take advantage of the information stemming from phylogeographic research. In this paper I will draft the main climatic events that determined the paleoecology of southern Europe during the Late Pleistocene, I will review the assumptions and consequences of alternative phylogeographic models, and discuss some case studies of mammalian species.

Keywords: phylogeography, southern Europe, Pleistocene, refugia, hedgehogs, wood mouse, brown bear, *Mus*, chamois, Italian hare, wolves

1. Quaternary climatic changes and the distribution of biodiversity in southern Europe

Pleistocene climatic changes have deeply affected the patterns of biodiversity in the Palaearctic Region (Hewitt 2000). The repeated cycles of climatic cooling and warming, which were particularly intense during the last 700 kyr (kyr = thousand years), determined the expansion or the contraction of the Arctic and Alpine ice caps. At the Last Glacial Maximum (LGM), 20-14 kyr before present (BP), the average soil temperatures were 10-20°C lower than present, permafrost extended to southern France and Germany, and huge ice caps covered the Pyrenees and the Alps. The last deglaciation phase in Europe began ca. 14 kyr BP, but the climatic amelioration was interrupted by a steep reversal, the Younger Dryas (ca. 10.5 kyr BP), a cold and dry stage that pro-

S. Weiss and N. Ferrand (eds.), Phylogeography of Southern European Refugia, 101–126.

moted a southward spread of tundra and the retreat of temperate forests. The expansion of temperate forests was possible only at the end of the Younger Dryas, and by 6 kyr BP the vegetation across most of Europe was similar to the present (Dawson 1996).

The distributions of mammalian species that are associated with temperate and Mediterranean habitats have likely followed the distributions of forests during the last glacial-interglacial cycle (Hewitt 1996). Thus, the distributions of many animal species have been severely restricted to small refugia during the LGM and the Younger Dryas. The Iberian, Italian and Balkan peninsulas have been identified as the three main glacial refuge areas during the Pleistocene in Europe. The expansion of populations and their distribution ranges were definitely possible in the Holocene after the end of the Younger Dryas.

The traditional view (based on analyses of terrestrial proxies in Europe) argues that there have been four main Pleistocene cold stages in the Northern Hemisphere, starting ca. 600 kyr BP, that is the Günz (600-500 kyr BP), Mindel (450-350 kyr BP), Riss (200-135 kyr BP) and Würm (120-18 kyr BP) glaciations (Figure 1). These four glaciations were named after rivers in southern Germany where specific geological observations were made. Glacial and interglacial stages in North America, or locally in the European Alps, British Isles and northern Europe were named differently. Also the reference dates of glacial stages are approximate, and they differ in the various regions of the Northern Hemisphere. Moreover, each stage was punctuated by many global or local reversals, and climatic anomalies were not uncommon. For example, during the last Ice Age there were ca. 24 interstadials, rapid climatic switches, when average temperatures changed by 10-12°C in just 5-10 years. Intervals of warming and cooling are called, respectively, interstadials and stadials. Deep-sea drilling cores indicate that the Quaternary was a period of continuous climate change, and that there were many more than the four glaciations that have been described from the terrestrial record in Europe. Therefore, paleoclimatic frameworks should be not generalized, and local paleoclimatic conditions should be reconstructed using specific data sets (Dawson 1996).

2. Phylogeography

The Pleistocene climatic changes had manifold consequences on landscape structure in southern Europe. At LGM the distributions of plant and animal species were strongly conditioned by the huge Scandinavian and Alpine ice caps, the extension of continental permafrost and tundra, the lowering of the sea level (the Mediterranean was 120 m lower than present) and the presence of land bridges in the Mediterranean (Figure 2). Landscape changes and the

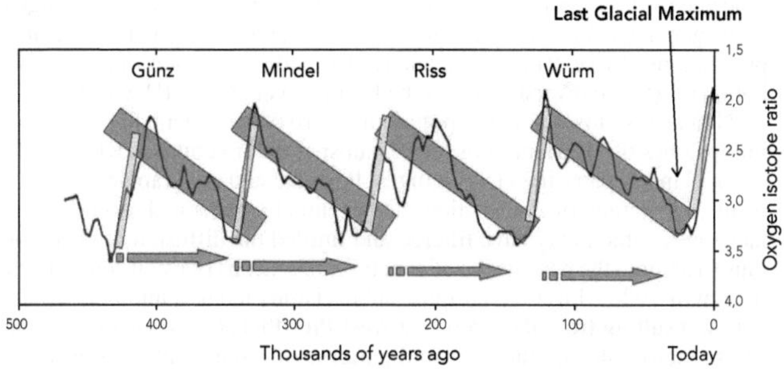

***Figure 1**. The four main glacial maxima of the Pleistocene in Europe, shown in relation to the changing oxygen-isotope ratio (d ^{18}O), a proxy for climatic change drawn from ocean sediment cores.*

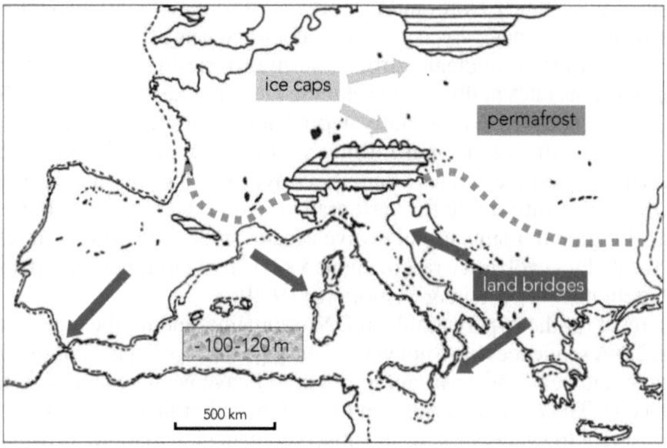

***Figure 2**. The main landscape changes in southern Europe at LGM. Glaciers covered most of northern Europe, the Pyrenees and the Alps. Permafrost existed over most of the European continent. The Mediterranean was 100-120 m lower than present resulting in land bridges, such as those connecting the Iberian Peninsula and Africa, Corsica and Sardinia, Sicily and the Italian Peninsula, and the Italian and Balkan peninsulas.*

repeated cycles of population contraction and expansion left genetic signals in the genomes of plant and animal organisms and in the genetic structure of populations (Avise *et al.* 1987), and led to a structured (non-random) distribution of genetic diversity across the European continent (Hewitt 1996).

Nowadays closely related populations are often genetically subdivided into patches of vicariant allospecies, subspecies or ecotypes, which differentiated in allopatry in glacial refuges. Postglacial demographic expansions resulted in parapatric distributions and eventually generated secondary contact zones, which may have filtered and limited the diffusion of genotypes, thus sustaining the persistence of genetic subdivisions (tension zones; Barton & Hewitt 1985). Phylogeographic subdivisions led to a number of suture zones, resulting from the main east/west Plio-Pleistocene vicariant events. These factors shaped the manifold patterns of biodiversity in Europe and around the Mediterranean, which, during the last few thousand years were (and still are) deeply affected, and in part destroyed by human activities. Describing the phylogeographic patterns and inferring past population dynamics is essential to develop a sound framework for conservation biology in Europe (Randi 2003).

Phylogeography makes use of molecular and geographical data to infer the role of historical factors in the distribution of current patterns of biodiversity (Avise 2000). Non-random patterns of geographical and phylogenetic distributions of genetic diversity can be identified using the appropriate genetic markers (mtDNA, nuclear introns and microsatellites). It is important to realize that the rates of divergence of these DNA markers are different. Nuclear DNA sequences (nDNA) usually evolve at rates lower or much lower than 1% sequence divergence per million years (myr); protein-coding mtDNA genes usually evolve at rates around 2%/myr (which is the 'standard' average mtDNA substitution rate for vertebrates), while the hypervariable mtDNA control-region domains might evolve at rates of 20-200%/myr (Hare 2001). Microsatellites evolve at a rate of about 5×10^{-4} mutations per locus per gamete in vertebrates (Goldstein & Schlötterer 1999).

Except for the hypervariable mtDNA domains, and for the microsatellites, most DNA sequences have not yet had enough time to diverge during the Quaternary, and very few new haplotypes could have been generated during the Holocene. Thus, mutation rates at mitochondrial or nuclear DNA sequences suggest that the generation of the observed genetic variability often predates late Pleistocene paleobiogeographical events. The kind of phylogeographic structure that have been described in many natural populations, has likely been produced by DNA lineage sorting in populations with fluctuating effective sizes (N_e). Past population dynamics strongly affected the distribution of genetic diversity, and, therefore, the coalescent theory (Rosenberg & Nordborg 2002) is the most suited framework for Pleistocene phylogeography. The differences in distribution of DNA markers observed among populations are likely due to the sorting of mutations that were generated in the past (Avise *et al.* 1984).

Thus, theory and practice of phylogeographic inference at the species and population levels are different, although logically connected. Biological species are usually reciprocally monophyletic for polymorphic DNA markers, and just one or a few gene trees should faithfully mirror the true species tree. In contrast, intraspecific populations are usually poly- or paraphyletic, and trees from several unlinked genes are necessary to describe the variability of the coalescent and thus approach the species tree (Avise *et al.* 1984; Edwards & Beerli 2000). However, it is not simple to find single-copy nuclear sequences evolving at an informative rate (2-20%/myr). Even introns often evolve 5-10 times slower than mtDNA sequences, and that is why most of the phylogeographic studies published so far are based on mtDNA data. Microsatellites have much higher mutation rates, but their stepwise mutation mechanism (Kimura & Ohta 1978) may produce allele homoplasy and complicate the reconstruction of allelic genealogies (Garza *et al.* 1995; Nauta & Weissing 1996; Guillaume *et al.* 2001). However, novel population genetic approaches (Bayesian coalescent modeling of population structure and model-based inference on past population dynamics) promise to be very useful in phylogeography (Cornuet & Luikart 1996; Kuhner *et al.* 1998; Wilson & Balding 1998; Beaumont 1999; Estoup & Cornuet 1999; Pritchard *et al.* 2000; Balding *et al.* 2001; Garza & Williamson 2001).

3. Refugia, colonization routes and the phylogeographic structure of mammalian species in southern Europe

A general framework of southern European phylogeography relies on the idea that Quaternary glacial/interglacial cycles affected the distributions of plant and animal communities and species, which contracted into southern refugia and expanded recolonizing deglaciated regions. Cycles of population isolation/expansion left detectable genetic signatures, leading to a structured distribution of genetic diversity in natural populations across the European Continent. However, in practice, just a few species have been studied in detail, and many aspects of European phylogeography are still poorly known.

A debated question concerns the number and location of refuge areas in Europe (Bilton *et al.* 1998; Willis & Whittaker 2000; Stewart & Lister 2001). Three major glacial refuge areas have been identified, in southwestern Iberia, southern Italy and in the southern Balkans (plus Greece and Turkey), plus a number of putative, but still not yet identified, cryptic northern refuges (Figure 3). According to Hewitt's model (1996, 2000), current biodiversity in central and western Europe originated mainly from the three major Mediterranean glacial refuge areas (southwestern Iberia, southern Italy, southern Balkans, plus northern Turkey and the Caspian Sea shores). These glacial refuges correspond to areas with temperate broadleaf forest and Mediterranean forest biomes at LGM. Postglacial recolonization was sustained mainly by leading edge expansion or long distance dispersal of small propagules. Repeated founding events from the same source population resulted in a loss of heterozygosity in newly founded northern populations. In this perspective later migrants would have contributed little to the composition of the new populations. Postglacial recolonization routes were constrained (in species-specific ways) by four mountain range barriers, the Pyrenees, Alps, Balkans and the Carpathians.

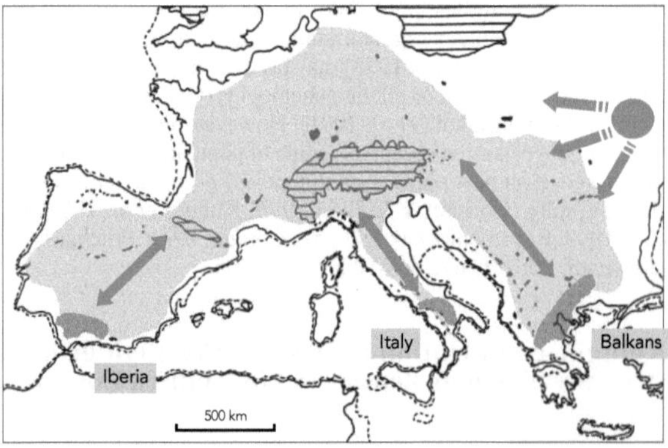

Figure 3. *Locations of the three major glacial refuge areas that have been identified in south-western Iberia, southern Italy and in the southern Balkans (plus Greece and Turkey). Other putative non-Mediterranean (top right), or cryptic northern refuges have been hypothesized.*

As an alternative to Hewitt's view, Bilton *et al.* (1998) suggested that Mediterranean refugia are 'hotspots' of endemism, but they were not the main sources of postglacial recolonization in central and northern Europe. Bilton *et al.* suggested a role for non-Mediterranean refugia, such as the southern slopes of the Carpathian Mountains, the hills of Crimea, the southwestern Ural Mountains, and the northern slopes of the Altai Mountains. These areas were probably temperate and not so arid as central and northern Europe and Asia during the last glaciation. Tree cover could have developed and temperate mammals could have survived at LGM on the slopes of mountain and sheltered river valleys. Consequently, central and northern Europe could have been colonized mainly by populations originating from non-Mediterranean refugia, because the Pyrenees, the Alps and the Balkans were also barriers to dispersal. Populations isolated in the Mediterranean peninsulas underwent genetic differentiation and eventually speciated, but they did not contribute to repopulate areas in central Europe. This model is supported by the high number of mammalian endemics in Iberia (*Sorex granarius*, *Galemys pyrenaicus*, *Talpa occidentalis*, *Microtus cabrerae*, *M. lusitanicus*), in Italy (*Sorex samniticus*, *Talpa romana*, *Microtus savii*) and in the Balkans (*Talpa stankovici*, *Dinaromys bogdanovi*, *Microtus felteni*, *M. thomasi*) (Amori *et al.* 1996).

A recent paper by Stewart & Lister (2001) further suggests that cryptic northern refugia could have also existed. During the late glacial stages, treeless steppe and tundra covered wide regions in Europe, Asia and North America. Temperate vegetation and woodlands would have survived only beyond the permafrost boundary. However, palynological and fossil findings indicate that temperate refugia might have also existed in sheltered areas in northern Europe during the LGM. Thus, small populations of temperate trees, mammals (e.g. *Cervus elaphus* in Devon, UK), and mammalian communities could have survived in northern Europe during glacials These populations could have contributed to the recolonization of central Europe, starting from the cryptic northern refugia.

These frameworks allow delineating a number of phylogeographic scenarios, which are more complex than simple models of linear range population expansion/contraction, and that may lead to alternative phylogeographic and population genetic predictions. Following, I will draft four phylogeographic models, with examples, which have been described in some mammalian species distributed in southern Europe.

3.1 Southern refuges

The classical and probably most frequently described phylogeographic patterns can be explained assuming north-south cyclical periods of distribution range and population expansion and contraction (Hewitt 1996, 2000). These cycles were synchronous with glacial/interglacial stages, and led to population differentiation in southern refuges, and postglacial recolonization following a variety of dispersal routes (Figure 4). This model predicts the preservation of the maximum genetic diversity in ancestral refuge populations. In fact, population contraction does not necessarily lead to loss of diversity, except for cases of very strong bottlenecks. Genetic diversity is predicted to decrease with the increasing of geographic distance from Pleistocene refuges, particularly if recolonization was realized by propagules with small effective population size (the leading-edge model). These dynamics are expected to produce two peculiar patterns of phylogeographic differentiation, that is 1) patches of species, subspecies or populations that diverge from east to west across Europe, and 2) a northward decline in within population genetic diversity. This model is clearly exemplified by the phylogeography of *Erinaceus* (Santucci *et al.* 1998; Seddon *et al.* 2001). There are two species of hedgehogs, the Western hedgehog, *Erinaceus europaeus*, and the Eastern hedgehog, *E. concolor*. These species can be subdivided into three phylogeographic units, one distributed in northeast and central Europe, one in central and northern Europe, and the third endemic to Sicily. The northernmost populations have lower mtDNA haplotype diversity than southern populations (Seddon *et al.* 2001).

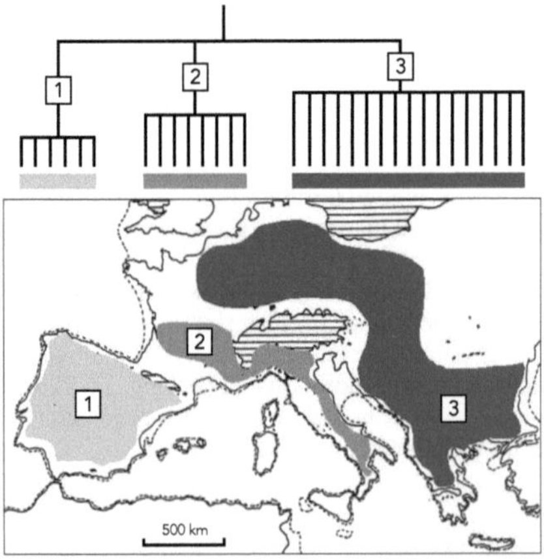

Figure 4. *The southern refuge phylogeographic model. Postglacial colonization of populations that differentiated in southern refugia is expected to produce genetic trees showing significant phylogenetic structure and concordant phylogeographic distributions of the main evolutionary lineages (3.1, 3.2 and 3.3).*

Phylogeography of the hedgehogs: *Erinaceus europaeus*, *Erinaceus concolor*

The hedgehogs include a widespread western Palaearctic group of populations, subdivided in two species with confused subspecific taxonomy, which are present in the European fossil record from the Pliocene. Allozymes (Filippucci & Simson 1996), mtDNA (Santucci *et al.* 1998; Seddon *et al.* 2001) and nuclear intron sequences (Seddon *et al.* 2001) suggested that *E. europaeus* and *E. concolor* split at ca. 9% mtDNA divergence (corresponding to 3.2-4.5

myr BP; using a rate of 11.06 – 12.60 x 10^{-9} substitutions/site/year). The main genetic split leading to speciation, has been attributed to a Pliocene cold period at ca. 6 myr BP. *E. concolor* further split into two clades showing 4% mtDNA divergence, corresponding to 1.7-2.2 myr. *E. europaeus* split into three clades at 4% mtDNA divergence, corresponding to 1.7-2.2 myr. The phylogeographic structure of *Erinaceus* indicates a major historical east/west genome subdivision, which led to speciation in the Pliocene.

The mtDNA data show signals of a postglacial northward population expansion starting from three refugia: Iberia, Italy and the Balkans. Apparently, the Pyrenees and the Alps were not barriers to dispersal for *Erinaceus*. The colonization routes used by hedgehogs look similar to those of the oak (*Quercus*). In fact, *E. europaeus* is associated with deciduous woodlands, which support its invertebrate prey base. The molecular data also indicate a deep divergence of the Sicilian mtDNA haplotypes, perhaps due to a long-term isolation of hedgehog populations in Sicily and southern Calabria from peninsular Italy, which could have been caused by flooded lowlands in central Calabria during the Pleistocene. The phylogeography of the hedgehog supports a prediction of Hewitt's model, that is a rapid population expansion leading to serial bottlenecks with progressive loss of gene diversity. In fact, a single mitotype was found in England, Scotland and Ireland (clade E2; see Seddon *et al.* 2001), and a single mitotype was also found in Estonia (clade C1; Seddon *et al.* 2001). Mutation rates and divergence of mtDNA sequences suggested that clade diversity was not generated in the refugia during the last glacial (100 kyr BP), but was rather preexistent. Therefore, the geographical distribution of mtDNA haplotypes is due to lineage sorting, and demographic considerations are relevant to explain the observations.

Nuclear DNA analyses (nucleotide sequences from two introns) showed sharp species distinction and confirmed the basal split shown by the mtDNA data. Intron sequences from *concolor* split in two clades, corresponding to the two mtDNA clades, but they failed to identify the three mtDNA clades that were found in *europaeus*. The haploid mtDNA has an effective population size (*Ne*) equal to 1/4 that of nDNA, and it is more sensitive to bottlenecks (Birky *et al.* 1989). It is possible that bottlenecks were stronger in *concolor* than in *europaeus*, and that slowly evolving intron sequences did not sort in *europaeus*. Alternative explanations could include male-biased dispersal, but there is no evidence to support sex-biased dispersal patterns in *Erinaceus*.

The parapatric distributions of *E. europaeus* and *E. concolor* overlap in central Europe (along a north-south line from the Baltic to the eastern Alps), in areas where other species show contact or hybrid zones: the house mouse (*Mus domesticus* and *M. musculus*), Bombina toads (*Bombina variegata* and *B. bombina*), crows (*Corvus corone* and *C. cornix*), grass snakes (*Natrix natrix*), oaks (*Quercus robur* group), and shrews (*Sorex araneus* group). Most of these secondary contacts are tension zones (Barton & Hewitt 1985).

Phylogeography of the wood mouse *Apodemus sylvaticus*

The wood mouse is a widespread Western Palaearctic forest species that is present in the Pleistocene fossil record. The taxonomy of the Mediterranean populations is confused. Four subspecies are recognized: *A. s. sylvaticus*, distributed in central and northern Europe (west of the Pyrenees and north of the Alps); the Mediterranean *A. s. dichrurus*, distributed in southern Spain, France, Italy, Sicily, Sardinia, Corsica; *A. s. callipides* from the Cantabrics and Pyrenees; and *A. s. milleri*, endemic to north and central Italy. There are some insular populations of probable anthropogenic origin (Filippucci *et al.* 2002).

Allozyme and mtDNA data allowed splitting the species into three main clades: 1) a northwestern group, corresponding to *A. s. sylvaticus* and ranging from the Pyrenees to Scandinavia through central Europe (these data negate the presence of distinct '*dichrurus*' populations in Mediterranean southern Spain, France, and of '*callipides*' in the Cantabrics and Pyrenees); 2) an Italian-Balkanic group, corresponding to *A. s. milleri* and including wood mice of anthropogenic origin in the Tyrrhenian islands, Sardinia and Corsica; and 3) a distinct Sicilian clade corresponding to *A. s. dichrurus* (Filippucci *et al.* 2002). The phylogeography of *Apodemus sylvaticus* suggests a postglacial colonization of central and northern Europe from an Iberian refuge. The Alps were a biogeographic barrier to the postglacial dispersal of the Italian populations. Some wood mouse populations of Tyrrhenian islands are of anthropogenic origin, and the Sicilian wood mouse population is deeply divergent, likely due to ancient isolation. There were probably two glacial refugia for the wood mouse, one in Iberia, and another in Italy-Balkans. Trans-Adriatic populations could have been connected since the end of the LGM via the late Quaternary, north-Adriatic land bridge (Filippucci *et al.* 2003).

The Alps apparently blocked the dispersal of other mammalian species in Italy, such as *Talpa caeca* and *Pitymys savii*, and were apparently a barrier to postglacial colonization in the brown bear as well, whereas the Pyrenees were not.

Phylogeography of the brown bear *Ursus arctos*

The brown bear is a Pleistocene species, which probably originated in central Asia and dispersed eastward into northeastern Asia, crossed Beringia during the Wisconsin, and reached North America ca. 50-70 kyr BP. It dispersed also westward into western Europe during the late Pleistocene, when it was sympatric with *U. spelaeus*, a cold-adapted species, which went extinct after the LGM. The brown bear was formerly widespread, but, during the last few centuries strongly declined and was eradicated almost everywhere in Europe, mainly as a consequence of direct human persecution and habitat loss. Nowadays bears survive in fragmented populations restricted to mountain ranges in the Cantabrian, Pyrenees and in Italy, apart from the more or less contiguous range in eastern and southeastern Europe. In the Italian Apennines there are probably less than 60 bears, but the exact number is uncertain, and in the Alps there are 15 bears, which are sustained by an ongoing reintroduction project. Brown bears from Slovenia are currently expanding their range and dispersing towards the eastern Italian Alps.

The phylogeography of the brown bear has been described in a number of papers (Randi *et al.* 1994; Taberlet & Bouvet 1994; Kohn *et al.* 1995). In Europe there are two main mtDNA lineages: the eastern and the western mtDNA clades, which differ by ca. 7%. The western lineage includes two clades, corresponding to populations that likely originated in the Iberian and Balkan refuges. The estimated mtDNA divergence time is about 800 kyr, which is older than the postulated times of glacial population isolation. Therefore, it is likely that current brown bear phylogeographic structure is due to stochastic lineage sorting of pre-existing haplotypes. The current phylogeographic structure allows the inference of two main postglacial colonization routes in Europe: 1) brown bears from the Iberian refuge dispersed to central Europe and reached southern Scandinavia; and 2) brown bears from southeastern Europe dispersed northward and colonized northeastern Scandinavia. Therefore, the Alps and the Balkans were barriers to the dispersal of brown bears. Brown bear colonization from a southern and a northern route generated a population contact zone in central Scandinavia, which can be identified by mtDNA genotyping. Apparently the northern and southern population do not mix, but the Scandinavian brown bear contact zone is maintained mainly by limited female dispersal. Male dispersal produced moderate gene flow across the contact zone, which has been identified by microsatellite markers (Waits *et al.* 2000).

Small populations of brown bear in Europe are mostly monomorphic at their mtDNA. Kohn *et al.* (1995) found that 23 of 28 distinct mtDNA haplotypes were fixed in only one locality. Isolation and long-lasting demographic declines likely reduced the genetic diversity of Italian brown bears, which showed a mtDNA haplotype closely related to that of Croatian brown bears, and reduced diversity at microsatellite markers. Genetic and demographic data suggest distinct conservation strategies for the Alpine and Apennine brown bear populations. The Apennine brown bears should be managed as a single unit, avoiding any restocking and supporting the current trends of population expansion towards other neighboring areas. The Alpine brown bears are geographically and genetically contiguous to the Croatian (plus other Balkan) populations. Alpine and contiguous populations of brown bears should be managed as a conservation unit, by supportive reintroduction in the Alps, and by supporting the natural population expansion from Slovenia and Croatia. Reintroduced brown bears in the Alps and population fragments in the Apennines are currently monitored by non-invasive genetics, using DNA obtained from scats or by hair trapping.

3.2 Eastern colonization waves

A second group of phylogeographic patterns likely originates from differential cycles of population dispersal and colonization from eastern Europe (or Eurasia) into western Europe. In this model the colonization of central and western Europe is entirely due to the expansion of eastern populations during subsequent interglacials. For example, a first east-to-west dispersal phase during mid-Pleistocene has been followed by a second dispersal phase after the LGM (Figure 5). Populations or species from the first dispersal phase survived in southern refuges (Iberia, Italy, Balkans), where they differentiated becoming adapted to southern climates and habitats. However, these populations did not have chances to expand and recolonize central Europe, because they were blocked by a second dispersal wave of eastern populations, which followed the eastern advance of mixed-deciduous forest, or that colonized mountain habitats (Figure 5). This process led to the prediction that taxa currently distributed in 'Mediterranean' refuge areas are phylogenetically more closely related to each other than to the postglacial dispersing taxa, and that within population genetic diversity would eventually decline westward. This model is exemplified by the phylogeography of the two extant species of chamois. *Rupicapra pyrenaica*, which originated from a first colonization during the Riss glaciation, and show a disjunct distribution in the Pyrenees and in the central Italian Apennines. *Rupicapra rupicapra*, which evolved in mountain areas in central Asia, colonized the mountain ranges in central Europe at the end of the Würm glaciation (Masini & Lovari 1988).

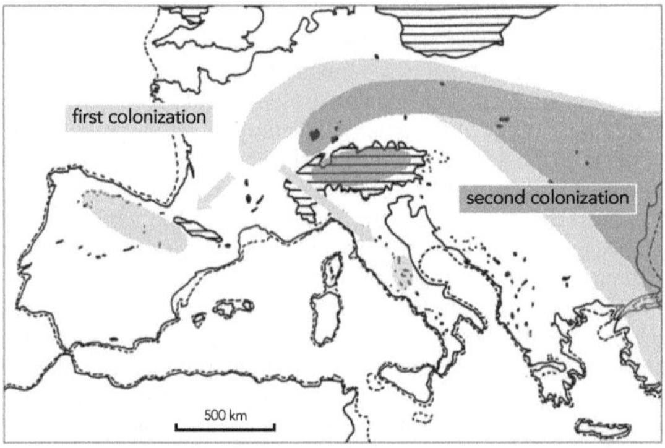

Figure 5. *The eastern colonization phylogeographic model. Colonization of central and western Europe was entirely due to the expansion of eastern populations during subsequent interglacials.*

Phylogeography of *Mus musculus* and *Mus domesticus*

The subgenus *Mus* is phylogenetically split into two clades: 1) the 'aboriginal' species, that is the wild mice populations living in natural habitats; and 2) the 'commensal' species living in close association with humans and villages (Sage *et al.* 1993). The aboriginal species include *M. spretus* (distributed in North Africa, Spain and France; sympatric with *M. domesticus*), *M. macedonicus* (Bulgaria, Greece, Cyprus; sympatric with *M. domesticus*) and *M. spicilegus* (Ukraine, Rumania, Bulgaria to eastern Austria; sympatric with *M. musculus*). The commensal species complex includes *M. domesticus* (naturally distributed in western Europe, North Africa, Near East, and translocated world-wide), *M. musculus* (in eastern and northern Europe, and northern Asia), *M. castaneus* (in southeastern Asia), and *M. bactrianus* (south of the Himalayas). The taxonomic ranks of these forms are debated: species rank was supported by Sage *et al.* (1993), while a subspecies rank was supported by Boursot *et al.* (1993). Calibrations of mtDNA sequence divergence suggest that *Mus* speciated during the Pliocene-Pleistocene, and that there were at least two east-west waves of speciation. The aboriginal mice speciated early (in the Early/Middle Pleistocene), while the commensal mice speciated during the last ca. 900 kyr. *M. musculus* dispersed into western Europe recently, probably in consequence to commensalism. Populations from central Asia (*musculus*) are very polymorphic (at allozyme loci and at the mtDNA, showing lineages that are also present in populations of *bactrianus* and *castaneus*). These populations are much more polymorphic than the peripheral populations, and they are phylogenetically at the center of a star phylogeny. These data suggest that northern India is the center of origin of *M. musculus*. Ancestral *M. musculus* radiated to the Middle East and the Caspian Sea where populations started to differentiate in *M. m. musculus* and *M. m. domesticus*, and expanded towards the periphery of their ranges. *M. musculus* dispersed into western Europe recently due to commensalism and dispersal by humans (passive transport and ecological dependence). The fossil record suggests that there was a northern route (*musculus* expanded into central and northern Europe ca. 4-2.8 kyr BP), and a southern route (*domesticus* expanded in the Middle East and western Mediterranean regions ca. 12-8 kyr BP). When these populations met they formed a secondary north-south contact zone across central Europe, running from Jutland to the Black Sea. European populations of *domesticus* have no clear mtDNA phylogeographic structure, due to high current gene flow or ancestral shared polymorphisms. The observed mtDNA clades originated before the period of mice colonization of Europe. Populations in the Mediterranean islands have no reduced genetic variability, suggesting multiple colonization events and recent rapid expansion. In contrast, *M. spretus* shows declining diversity from North Africa to Spain to France suggesting natural colonization of Europe through Gibraltar and the Pyrenees.

The Apennine chamois *Rupicapra pyrenaica ornata*

Genus *Rupicapra* evolved during the Middle Pleistocene (ca. 500 kyr BP), and it is present in the fossil record of central Europe in association with arctic species, like *Rangifer* (reindeer) and *Preovibos* (a muskox). *Rupicapra* is a survivor of a Plio-Pleistocene evolutionary radiation of the subfamily Caprinae, which originated in southeastern Asia, probably in the Miocene (more than 6 myr BP), and dispersed widely in Asia, then towards North America and western Europe in the Plio-Pleistocene. The Rupicaprini originated during the Miocene in Asia and dispersed during the Late Miocene-early Pliocene and Pleistocene. *Rupicapra* is a mountain-dwelling genus, which probably originated in some mountain areas west of the Himalayas, and dispersed westward during mid Pleistocene cold periods. There are two species of *Rupicapra*: *R. pyrenaica* (with 2-3 subspecies) and *R. rupicapra* (with 6-7 subspecies). The pattern of speciation and subspeciation suggests two different waves of colonization in western Europe: 1) a first colonization of *pyrenaica* populations during the Riss glaciation (the only chamois Riss fossil is *pyrenaica*, a skull from the Pyrenees); 2) a second colonization (during the Würm glaciation) of *rupicapra* populations, which evolved and differentiated in mountain areas in the Caucasus and that were more adapted to cold climates (Figure 6). A number of Würm fossils indicate the presence of both *pyrenaica* and *rupicapra* in the Pyrenees, the Alps and the Apennines (Masini & Lovari 1988).

Speciation and phylogeography of the Italian hare, *Lepus corsicanus*

Lepus corsicanus, the Italian hare, was described as a new species in 1898 by de Winton, who used specimens that were collected in Corsica (de Winton 1898). Later it was shown that the Corsican population originated from an anthropogenic introduction of hares from Italy before the 16th century (Vigne 1992). The Italian hare was soon considered just a subspecies of the brown hare *L. europaeus*, and it was supposed to be extinct due to overhunting and subsequent massive releases of non-indigenous brown hares. The morphology of museum samples was studied by Palacios (1966), who revalidated the species rank for the Italian hare. However, Palacios used museum samples collected mainly in the 1800s. Therefore, until 1966 it was unknown if the Italian hare survived in the wild or was definitely extinct. A few samples were collected in 1974-75 in Calabria, and thereafter in Sicily in 1996, and samples available for morphologic and molecular analyses were collected (Lo Valvo *et al.* 1997).

Nucleotide sequences of the mtDNA control region and cytochrome *b* indicate that *L. corsicanus* is a phylogenetically distinct species, which can be identified by concordant morphologic and mtDNA traits (Pierpaoli *et al.* 1999; Figure 7). The Italian hare seems to be reproductively isolated and apparently does not hybridize with sympatric brown hares. Phylogenetic analyses sug-

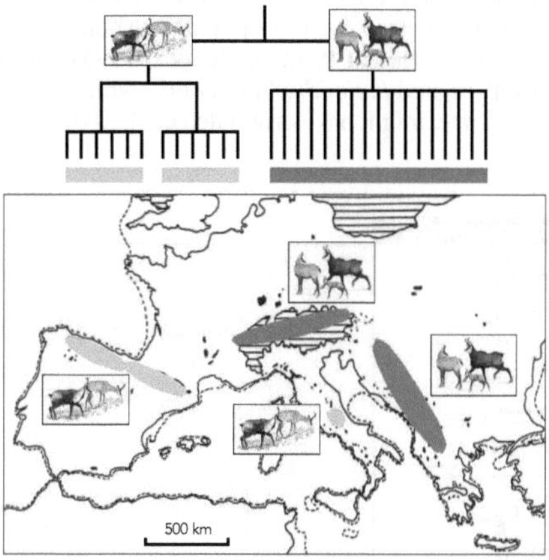

Figure 6. *Phylogeography of the chamois.* Rupicapra pyrenaica *is distributed in Spain, France, and in the Italian Apennines;* R. rupicapra *is distributed in the Italian Alps and in several locations in the Balkans and southeastern Europe.*

gest that Italian and brown hares are not closely related sister taxa, but belong to two distinct evolutionary lineages, one including *L. corsicanus*, the Alpine hare *L. timidus*, and two species endemic to Iberia, *L. granatensis* and *L. castroviejoi*; the other one including *L. europaeus*, and the African species *L. capensis*, *L. habessinicus* and *L. starcki* (Figure 7). To obtain an estimate of interspecific divergence times in *Lepus* we have applied the standard calibration of mtDNA cytochrome *b* divergence rate of 2-4%/myr (Avise *et al.* 1998; Santucci *et al.* 1998). The basal split between the two main phylogenetic lineages was at ca. 12% sequence divergence, corresponding to 6-3 mya. The lower estimate of 3 mya is more concordant with the available fossil record, which documents the first appearance of *Lepus* in the Villafranchian, about 2.5 myr BP.

Most of the studied African and European species of hare showed pairwise mtDNA divergences of 10-13%. Therefore, these species could have originated during the last 3 myr, i.e. in the second half of Pliocene and at the Plio/Pleistocene boundary, a period that was marked by great climatic and paleogeographic changes, culminating with the beginning of major glacial cycles about 2.5 mya (deMenocal 1995). It is plausible that Plio/Pleistocene climate changes fostered allopatric isolation in glacial refuge areas located in southern parts of the Iberian, Italian and Balkan peninsulas, and mediated speciation through repeated cycles of population contraction and expansion in Europe and Africa (Hewitt 1996; deMenocal 1995). This calibration suggests that closely related *L. timidus/corsicanus*, which split at 3.3% mtDNA sequence divergence, might have originated ca. 820 000 years ago.

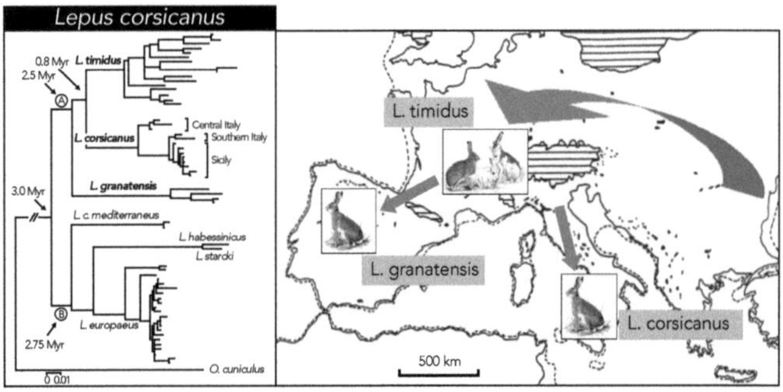

Figure 7*. A mitochondrial DNA phylogeny of European species of* Lepus. *Initial colonization of ancestral populations of hares from eastern Europe gave rise to* L. timidus, *the endemic Iberian species* L. granatensis *and* L. castroviejoi, *and* L. corsicanus *in southern Italy.*

In this perspective we speculate that the species of hares related to *L. corsicanus*, represent relictual taxa originating from a common ancestor during an early phase of speciation and dispersal of *Lepus* in western Europe (Figure 7). These species could have been confined to marginal distributions as a consequence of habitat changes and comparatively recent dispersal of a strong competitor, *L. europaeus*, in western Europe (Figure 8). *L. granatensis* and *L. corsicanus* have localized distributions in Spain, southern Italy and Sicily; that is, in areas that have been indicated as glacial refuges for plant and animal communities. The mountain hare has a circumpolar distribution and occurs in isolated populations in Ireland, Scotland and in the Alps. However, the fossil record documents a much more widespread distribution of the mountain hare during the last glacial period throughout most of central Europe and Britain (Lopez Martinez 1980). The range of mountain hares might have become restricted to northern latitudes or isolated mountain ranges in response to interspecific competition with other lagomorphs, and notably the brown hare. The Italian hare probably differentiated in isolated refuges in southern Italy during the last glaciation. The Italian hare is an endangered endemic species and urgently needs conservation efforts to ensure its survival.

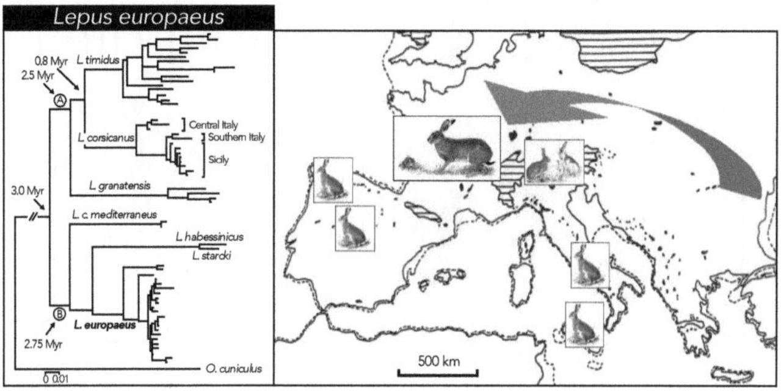

Figure 8. *A secondary colonization of hare populations from the east, and diffusion of* L. europaeus *in Europe.*

3.3 Population shifts

A third kind of phylogeographic pattern can be explained assuming whole north-to-south population range shifts. During interglacials, populations moved northwards from southern refuges, following climate ameliorations and in parallel with forest spreading (Figure 9A). Populations in former refuges went extinct because climates became too warm and/or habitats too dry (i.e. habitat changed from deciduous forest to Mediterranean maquis)

In this scenario, postglacial colonization and population expansion did not lead to bottlenecks, and northern populations might show greater gene diversity than southern populations. Southern populations might be relictual, or originated by secondary colonization waves into suboptimal habitats, and therefore are expected to be the least polymorphic. There will be no phylogeographic pattern, because populations disappeared from glacial refuge areas, and the current distribution ranges are completely disconnected from past Quaternary distributions (Figure 9B). For example, the northern populations are ancestral, while the southern populations are recent. This model is exemplified by the phylogeography of the noctule bat, *Nyctalus noctula* (Petit *et al.* 1999).

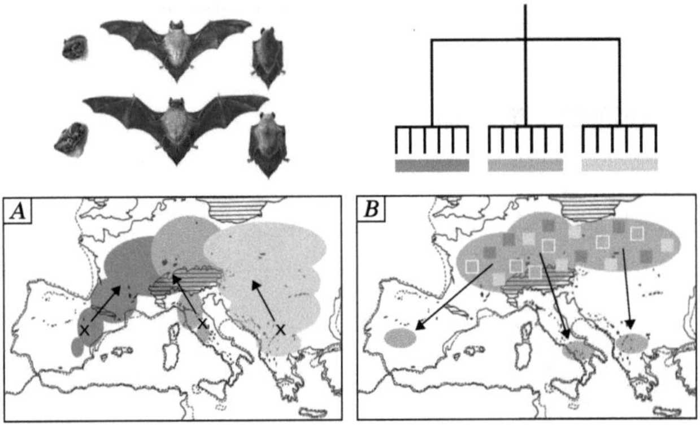

Figure 9. Population shifts. Expanding populations from southern refuges admixed in central Europe. The refugial populations went extinct. Consequently, the genetic trees might show phylogenetic structure, but the genotypes admix in the postglacial populations and there is no phylogeographic structure.

3.4 No global phylogeographic structure

Finally there are species that do not show any apparent large-scale phylogeographic pattern (Figure 10). Extensive historical or current dispersal and gene flow could have generated these patterns. This model is exemplified by the apparent lack of phylogeographic structuring among wolf populations worldwide (Wayne *et al.* 1992; Vilà *et al.* 1999). However, certain local populations of species without large-scale phylogeographic structure can show genetic distinction due to isolation in particular refuge areas. This is the case with the wolf population surviving in peninsular Italy.

Population genetics of wolves *Canis lupus*

The wolf is a Pleistocene Eurasian species, which originated ca. 700 kyr BP. Presently, wolf populations live in a wide variety of habitats: desert (Arabia, Near East), Mediterranean shrublands (Greece, Italy), broadleaf forests (Apennines and Alps), coniferous forests (Siberia), and frozen tundra (Arctic islands, Ellesmere) (Mech 1970). Wolf populations in Europe strongly declined in the 18th and 19th centuries, due mainly to direct human persecution (Delibes 1990). Wolves were eradicated from all central and northern European countries after the Second World War. The demographic decline continued till the end of the 1960s, when isolated wolf populations survived in Iberia, Italy, Finland and Greece (Zimen & Boitani 1975; Breitenmoser 1998). During the last 30 years wolves recovered naturally, showing positive demographic trends and episodes of recolonization in Italy, France, Switzerland, Germany, Sweden and Norway (Randi *et al.* 2000; Valière *et al.* 2003).

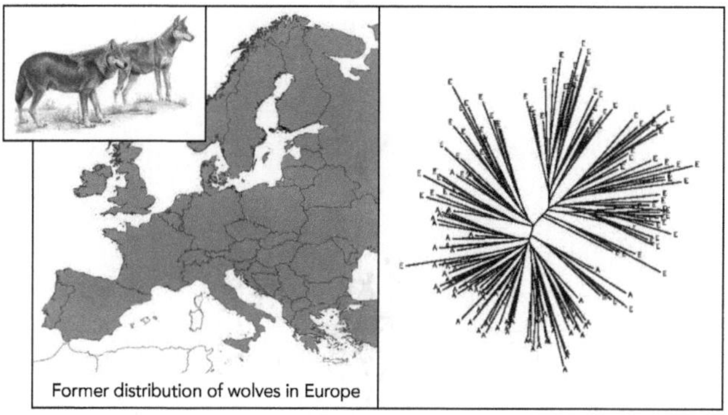

Figure 10. *No phylogeographic structure and a star phylogeny, as produced by recent colonization and widespread gene flow throughout Europe.*

Wolf populations show wide morphological diversity and are subdivided in 10-12 controversial subspecies, but they have no clear phylogeographic structure in America or in Eurasia. Mitochondrial DNA sequencing (Vilà *et al.* 1999; Randi *et al.* 2000) showed that many wolf populations have unique mtDNA haplotypes, but both haplotype and population trees showed no clear geographic structure (Figure 10). Nested clade analyses suggested that cladograms were not randomized with respect to geography, but there were obviously no strong associations between haplogroups and their geographic locations. Some closely related haplotypes were sampled in the same locations, but others were sampled in very distant places. In the phylogenetic trees or networks, Asian haplotypes were usually internal, while European and American ones were terminal. Terminal, recent haplotypes have restricted geographical distributions, thus suggesting recent restrictions to gene flow. Italian wolves show a unique mtDNA CR haplotype, which is monomorphic in the population and different from any other mtDNA haplotype identified so far in the other wolf populations and dog breeds world-wide (Randi *et al.* 2000). Assuming a very low effective population size, that is a ratio of effective to observed population size (N_e/N_0) of about 0.2:0.3, the observed mtDNA monomorphism might have been produced by random drift in the declining and isolated Italian wolf population during the last 100-150 years. However, recent microsatellite analyses (Randi & Lucchini 2002; Lucchini *et al.* 2002) indicate that nuclear gene diversity is not low in the Italian wolves, but that their allele frequencies are sharply different from dogs and from any other wolf populations in Europe (Figure 11).

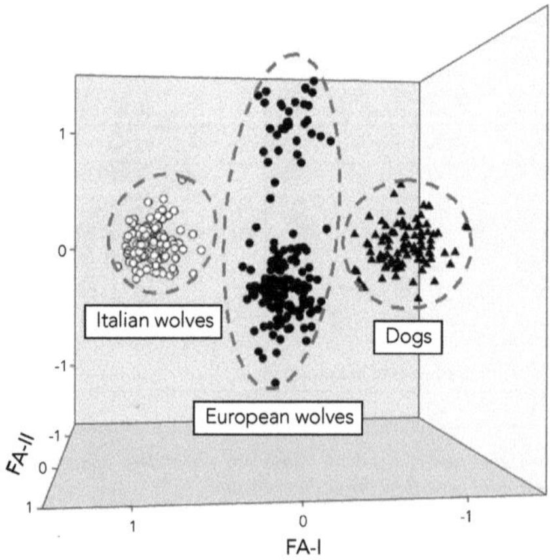

***Figure 11**. Factorial component plotting of multi-locus microsatellite genotypes showing genetic distinction among wolves (*Canis lupus*) and dogs.*

Results of a Bayesian coalescent model indicate that wolves in Italy underwent a 100- to 1000-fold population contraction over the past 4700 to 23 800 years. The inferred population decline was stronger and more protracted in peninsular Italy than elsewhere in Europe, suggesting that wolves have apparently been genetically isolated for thousand of generations south to the Alps. Ice caps covering the Alps at the LGM, and the wide expansion of the Po River, which cut the alluvial plains throughout all the Holocene, might have provided effective geographic barriers to wolf dispersal (Figure 12). More recently, the admixture of Alpine and Apennine wolf populations could have been prevented by deforestation, which was already widespread in the 15th century in northern Italy. Results of this study suggest that, despite the high potential rates of dispersal and gene flow, local wolf populations may not mix for long periods of time (Carmichael *et al.* 2001).

The Italian wolf population is currently undergoing rapid expansion after a protracted period of demographic decline, which lasted for centuries. The Italian wolves are expanding in parts of their historical range in the Apennines, and they are recolonizing the western Alps. The genetic integrity of wolf populations might be threatened by crossbreeding with feral or free-ranging dogs, which are both widespread and common, particularly in central and southern Italy.

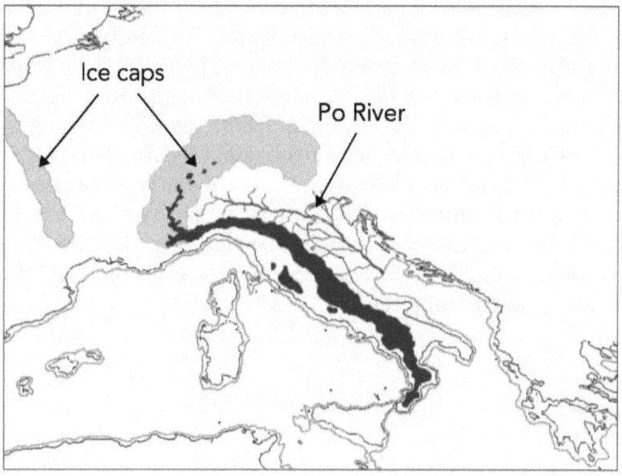

Figure 12*. Distribution range (dark areas) of wolves in Italy and in the Alps. The Pyrenean and Alpine glaciers (gray areas) and the Po River are indicated.*

Conclusions

Information on phylogeographic structuring and genetic diversity in some European mammal populations is available, although much research work still remains to be done. The discovery of the patterns of geographic structuring in a growing number of species would allow, by comparative phylogeography (Taberlet *et al.* 1998; Arbogast & Kenagy 2001), the identification of geographic regions sharing common natural histories (Moritz 1994; Avise 2000). These regions could be identified as 'hotspots' of biodiversity, and should be particularly targeted for conservation projects. However, for the moment, the phylogeography of each mammalian species appears to be largely species-specific. Although some common patterns might emerge, each species seems to have reacted independently to the late Pleistocene climate and landscape changes. Phylogeographic reconstructions should, however, pay attention to the consequences of the stochastic processes that may affect the geographical distribution of genetic variability (Knowles 2000; Knowles & Maddison 2002). Stochastic processes in phylogeography are still severely underestimated, but they might significantly modify the conclusions that can be inferred from empirical data sets. Another critical issue that should be carefully evaluated is the necessary distinction between the gene and the species trees. We are still using haplotype divergence to infer population distinction and divergence times. However, coalescent theory makes clear that these connections are often not obvious, and should be carefully modeled by taking into account complex scenarios of historical changes in the genetically effective size and structure of the populations (Gaggiotti & Excoffier 2000; Wakeley 2000; Ray *et al.* 2003). Finally, both distribution ranges and recent demography of many species have been radically affected by anthropogenic factors during the last few millennia and, particularly during the last few centuries. The consequences of ancient natural climate and habitat changes, and/or of recent human actions should be carefully identified, in order to be able to provide reliable historical reconstructions of the natural histories of animal species and communities.

References

Amori G, Angelici FM, Prigioni C, Vigna Taglianti A (1996) The mammal fauna of Italy: a review. ***Hystrix*** **8**, 3-7.

Arbogast BS, Kenagy GJ (2001) Comparative phylogeography as an integrative approach to historical biogeography. ***Journal of Biogeography*** **28**, 819-825.

Avise JC (2000) ***Phylogeography***. Harvard University Press, Cambridge.

Avise JC, Neigel JE, Arnold J (1984) Demographic influences on mitochondrial DNA lineage survivorship in animal populations. ***Journal of Molecular Evolution*** **20**, 99-105.

Avise JC, Arnold J, Ball RM, Bermingham E, Lamb T, Neigel JE, Reeb CA, Saunders NC(1987) Intraspecific phylogeography: the mitochondrial DNA bridge between population genetics and systematics. ***Annual Review of Ecology and Systematics*** **18**, 489-522.

Avise JC, Walker DE, Johns GC (1998) Speciation durations and Pleistocene effects on vertebrate phylogeography. ***Proceedings of the Royal Society of London, B*** **265**, 1707-1712.

Balding DJ, Bishop M, Cannings C (eds.) (2001) ***Handbook of Statistical Genetics***. John Wiley & Sons, Chichester.

Barton NH, Hewitt GM (1985) Analysis of hybrid zones. ***Annual Review of Ecology and Systematics*** **16**, 113-148.

Beaumont MA (1999) Detecting population expansion and decline using microsatellites. ***Genetics*** **153**, 2013-2029.

Bilton DT, Mirol PM, Mascheretti S, Fredga K, Zima J, Searle JB (1998) Mediterranean Europe as an area of endemism for small mammals rather than a source of northwards post-glacial colonization. ***Proceedings of the Royal Society of London, B*** **265**, 1219-1226.

Birky CW Jr, Fuerst P, Maruyama T (1989) Organelle gene diversity under migration, mutation and drift: equilibrium expectations, approach to equilibrium, effects of heteroplasmic cells, and comparison to nuclear genes. ***Genetics*** **121**, 613-627.

Boursot P, Din W, Anand R, Darviche D, Dod B, Von Deimling F, Talwar GP, Bonhomme F (1993) Origin and radiation of the house mouse: mitochondrial DNA phylogeny. ***Journal of Evolutionary Biology*** **9**, 391-415.

Breitenmoser U (1998) Large predators in the Alps: the fall and rise of man's competitors. ***Biological Conservation*** **83**, 279-289.

Carmichael LE, Nagy JA, Larter NC, Strobeck C (2001) Prey specialization may influence patterns of gene flow in wolves of the Canadian Northwest. ***Molecular Ecology*** **10**, 2787-2798.

Cornuet JM, Luikart G (1996) Description and power analysis of two tests for detecting recent population bottlenecks from allele frequency data. ***Genetics*** **144**, 2001-2014.

Dawson AG (1996) ***Ice Age Earth: Late Quaternary Geology and Climate***. Routledge, New York.

de Winton WE (1898) On the hares of Western Europe and North Africa. ***Annual Magazine of Natural History, London*** **1**, 149-158.

Delibes M (1990) Status and conservation needs of the wolf in the Council of Europe member States. ***Nature and Environment Series*** **47**, 1-46.

deMenocal PB (1995) Plio-Pleistocene African climate. ***Science*** **270**, 53-59.

Edwards SV, Beerli P (2000) Gene divergence, population divergence, and the variance in coalescence time in phylogeographic studies. ***Evolution*** **54**, 1839-1854.

Estoup A, Cornuet JM (1999) Microsatellite evolution: inferences from population data. In: ***Microsatellites: Evolution and Applications*** (eds. Goldstein DB, Schlötterer C), pp. 49-65. Oxford University Press, Oxford, New York.

Filippucci MG, Simson S (1996) Allozyme variation and divergence in Erinaceidae (Mammalia: Insectivora). ***Israel Journal of Zoology*** **42**, 335-345.

Filippucci MG, Macholàn M, Michaux JR (2002) Genetic variation and evolution in the genus *Apodemus* (Muridae, Rodentia). ***Biological Journal of the Linnean Society*** **75**, 395-419.

Gaggiotti OE, Excoffier L (2000) A simple method of removing the effect of a bottleneck and unequal population sizes on pairwise genetic differences. ***Proceedings of the Royal Society of London, B*** **267**, 81-87.

Garza JC, Williamson EG (2001) Detection of reduction in population size using data from microsatellite loci. ***Molecular Ecology*** **10**, 305-318.

Garza JC, Slatkin M, Freimer N (1995) Microsatellite allele frequencies in humans and chimpanzees with implications for constraints on allele size. ***Molecular Biology and Evolution*** **12**, 594-603.

Goldstein DB, Schlötterer C (eds.) (1999) ***Microsatellites: Evolution and Applications***. Oxford University Press, Oxford, New York.

Guillaume Q, Ferrand N, Weiss S, Mougel F, Monnerot M (2001) Stationary distributions of microsatellite loci between divergent population groups of the European Rabbit (*Oryctolagus cuniculus*). ***Molecular Biology and Evolution*** **18**, 2169-2178.

Hare MP (2001) Prospects for nuclear gene phylogeography. ***Trends in Ecology and Evolution*** **16**, 700-706.

Hewitt GM (1996) Some genetic consequences of ice ages, and their role in divergence and speciation. ***Biological Journal of the Linnean Society*** **58**, 247-279.

Hewitt GM (2000) The genetic legacy of the Quaternary ice ages. ***Nature*** **405**, 907-913.

Kimura M, Ohta T (1978) Stepwise mutation model and distribution of allelic frequencies in a finite population. ***Proceedings of the National Academy of Sciences, USA*** 75, 2868–2872.

Knowles LL (2000) Tests of Pleistocene speciation in montane grasshoppers (genus *Melanopus*) from the Sky island of western North America. ***Evolution*** **54**, 1337-1348.

Knowles LL, Maddison WP (2002) Statistical phylogeography. ***Molecular Ecology*** **11**, 2623-2635.

Kohn M, Knauer F, Stoffella A, Schroeder W, Paabo S (1995) Conservation genetics of the European brown bear – A study using excremental PCR of nuclear and mitochondrial sequences. ***Molecular Ecology*** **4**, 95-103.

Kuhner MK, Yamato J, Felsenstein J (1998) Maximum likelihood estimation of population growth rates based on the coalescent. ***Genetics*** **149**, 429-434.

Lo Valvo M, Barera A, Seminara S (1997) Biometria e status della lepre appenninica (*Lepus corsicanus*, de Winton 1898) in Sicilia. ***Naturalista Siciliano*** **IV**, (XXI) 67-74.

Lopez Martinez N (1980) Les lagomorphes (*Mammalia*) du Pleistocene superieur de Jaurens. ***Nouvelles Archives de la Musee d'Histoire Naturelle de Lyon*** **18**, 5-16.

Lucchini V, Fabbri E, Marucco F, Ricci S, Boitani L, Randi E (2002) Non-invasive molecular tracking of colonizing wolf (*Canis lupus*) packs in the western Italian Alps. ***Molecular Ecology*** **11**, 857-868.

Masini F, Lovari S (1988) Systematics, phylogenetic relationships and dispersal of the chamois (*Rupicapra* spp.). ***Quaternary Research*** **30**, 339-349.

Mech D (1970) ***The Wolf: The Ecology and Behavior of an Endangered Species***. University of Minnesota Press, Minneapolis.

Moritz CC (1994) Defining 'evolutionary significant units' for conservation. ***Trends in Ecology and Evolution*** **9**, 373-375.

Nauta M, Weissing F (1996) Constraints on allele size at microsatellite loci: implication for genetic differentiation. ***Genetics*** **143**, 1021-1032.

Palacios F (1996) Systematics of the indigenous hares of Italy traditionally identified as *Lepus europaeus* Pallas, 1778 (Mammalia: Leporidae). ***Bonner zoologische Beiträge*** **56**, 59-91.

Petit E, Excoffier L, Mayer F (1999) No evidence of bottleneck in the postglacial recolonization of Europe by the noctule bat (*Nyctalus noctula*). ***Evolution*** **53**, 1247-1258.

Pierpaoli M, Riga F, Trocchi V, Randi E (1999) Species distinction and evolutionary relationships of the Italian hare (*Lepus corsicanus*) as described by mitochondrial DNA sequencing. ***Molecular Ecology*** **8**, 1805-1817.

Pritchard JK, Stephens M, Donnelly PJ (2000) Inference of population structure using multilocus genotype data. ***Genetics*** **155**, 945-959.

Randi E (2003) Conservation genetics of carnivores in Italy. ***Comptes Rendus Biologies* 326**, Supplement 1, S54-S60.

Randi E, Lucchini V (2002) Detecting rare introgression of domestic dog genes into wild wolf (*Canis lupus*) populations by Bayesian admixture analyses of microsatellite variation. ***Conservation Genetics* 3**, 31-45.

Randi E, Gentile L, Boscagli G, Huber D, Roth HU (1994) Mitochondrial DNA sequence divergence among some west European brown bear (*Ursus arctos*) populations. Lessons for conservation. ***Heredity* 73**, 480-489.

Randi E, Lucchini V, Christensen MF, Mucci N, Funk SM, Dolf G, Loeschcke F (2000) Mitochondrial DNA variability in Italian and east European wolves: detecting the consequences of small population size and hybridization. ***Conservation Biology* 14**, 464-473.

Ray N, Currat M, Excoffier L (2003) Intra-deme molecular diversity in spatially expanding populations. ***Molecular Biology and Evolution* 20**, 76-86.

Rosenberg NA, Nordborg M (2002) Genealogical trees, coalescent theory and the analysis of genetic polymorphisms. ***Nature Reviews, Genetics* 3**, 380-390.

Sage RD, Atchlet WR, Capanna E (1993) House mice as models in systematics. ***Systematic Biology* 42**, 523-561.

Santucci F, Emerson BC, Hewitt GM (1998) Mitochondrial DNA phylogeography of European hedgehogs. ***Molecular Ecology* 7**, 1163-1172.

Seddon JM, Santucci F, Reeve NJ, Hewitt GM (2001) DNA footprints of European hedgehogs, *Erinaceus europaeus* and *E. concolor*: Pleistocene refugia, postglacial expansion and colonization routes. ***Molecular Ecology* 10**, 2187-2198.

Stewart JR, Lister AM (2001) Cryptic northern refugia and the origins of modern biota. ***Trends in Ecology and Evolution* 16**, 608-612.

Taberlet P, Bouvet J (1994) Mitochondrial DNA polymorphism, phylogeography, and conservation genetics of the brown bear *Ursus arctos* in Europe. ***Proceedings of the Royal Society of London, B* 255**, 195-200.

Taberlet P, Fumagalli L, Wust-Saucy A-G, Cosson J-F (1998) Comparative phylogeography and postglacial colonization routes in Europe. ***Molecular Ecology* 7**, 453-464.

Valière N, Fumagalli L, Gielly L, Miquel C, Lequette B, Poulle M-L, Weber J-M, Arlettaz R, Taberlet P (2003) Long-distance wolf recolonization of France and Switzerland inferred from non-invasive genetic sampling over a period of 10 years. ***Animal Conservation* 6**, 83-92.

Vigne JD (1992) Zooarchaeology and the biogeographical history of the mammals of Corsica and Sardinia since the last ice age. ***Mammal Review* 2**, 87-89.

Vilà C, Amorim IR, Leonard JA, Posada D, Castroviejo J, Petrucci-Fonseca F, Crandall KA, Ellergren H, Wayne RK (1999) Mitochondrial DNA phylogeography and population history of the grey wolf *Canis lupus*. ***Molecular Ecology* 8**, 2089-2103.

Waits L, Taberlet P, Swenson JE, Sandegren F, Franzén R (2000) Nuclear DNA microsatellite analysis of genetic diversity and gene flow in the Scandinavia brown bear (*Ursus arctos*). ***Molecular Ecology* 9**, 421-431.

Wakeley J (2000) The effects of subdivision on the genetic divergence of populations and species. ***Evolution* 54**, 1092-1101.

Wayne RK, Lehman N, Allard MW, Honeycutt RL (1992) Mitochondrial DNA variability of the gray wolf: genetic consequences of population decline and habitat fragmentation. ***Conservation Biology* 6**, 559-569.

Willis KJ, Whittaker RJ (2000) The refugial debate. *Science* 287, 1406-1407.

Wilson IJ, Balding DJ (1998) Genealogical inference from microsatellite data. ***Genetics*** **150**, 499-510.

Zimen E, Boitani L (1975) Number and distribution of wolves in Italy. ***Zeitschrift für Säugetierkunde*** **40**, 102-112.

Chapter 4

Longitudinal patterns in species richness and genetic diversity in European oaks and oak gallwasps

Rachel J. Atkinson, Antonis Rokas and Graham N. Stone*
*E-mail: graham.stone@ed.ac.uk

Abstract
While latitudinal patterns of genetic diversity are well known for many taxa in Europe, there has been little analysis of longitudinal patterns across Pleistocene glacial refugia. Here we analyze longitudinal patterns in two aspects of diversity (species richness and intraspecific genetic diversity) for two trophically related groups of organisms – oaks (Fagaceae, genus *Quercus*) and their associated gallwasps (Hymenoptera: Cynipidae) – across four southern refugial regions (Asia Minor, the Balkans, Italy, and the Iberian Peninsula). Two major patterns emerge: a) both intraspecific genetic diversity and species richness are lower in Iberia than in eastern refugia; and b) though depauperate, Iberia contains substantial endemic diversity both in terms of intraspecific variation and species. Sequence data for two of three gallwasp case studies show that Iberian lineages diverged from central and eastern European lineages well before the Pleistocene. However, coalescence times for Iberian lineages are far more recent, suggesting that Iberian populations may have a recent history of genetic bottlenecks. We suggest that these patterns are the result of two processes: a) the formation of distinct refugial biotas towards the end of the Pliocene, and b) longitudinal variation in the magnitude of climatic fluctuations during the Pleistocene. Specifically, we suggest that the lower richness in Iberia is the result of a recent history of stronger fluctuations and much more intense aridity than areas further east. Data for equivalent analyses in other taxa exist. Comparison of patterns across taxa should allow the genesis and testing of hypotheses for the origin and maintenance of longitudinal patterns of biodiversity in the Western Palaearctic.

Keywords: phylogeography, longitude, genetic diversity, species, gallwasp, cynipid, oak, *Quercus*, paleoclimate, Pliocene, Pleistocene

Introduction

Western Palaearctic phylogeography is dominated by the study of latitudinal patterns, particularly those resulting from range expansion following the retreat of the ice sheets at the end of the Pleistocene (Hewitt 1996, 1999;

S. Weiss and N. Ferrand (eds.), Phylogeography of Southern European Refugia, 127–151.

Taberlet *et al.* 1998). Many taxa show decreases in species richness, and intraspecific genetic diversity, with increasing latitude (Pamilo & Savolainen 1999; reviewed in Stone *et al.* 2002a). The origins for postglacial range expansion in Europe are southern glacial refugia, and while their precise identification varies among taxa, these can broadly be identified as Iberia (Spain, Portugal and northwest Africa), Italy, the Balkans (the Carpathian basin south to Greece) and Asia Minor (Anatolia south to Israel and east to Iran). Although these refugia extend over 45 degrees of longitude, and incorporate far more biotic diversity than an equivalent range in latitude, patterns in diversity among refugia have been far less studied. Intraspecific genetic differences among populations from different refugia are known for many taxa, and are a crucial tool in tracing the origin of northerly range expansions (see Hewitt 1999 for a recent review). However, to date there has been little emphasis on the processes generating diversity among the southern refugia that are the origins for postglacial range expansion. The higher species richness and intraspecific genetic diversity at lower latitudes means that while latitudinal patterns may determine local richness, interrefugial processes are probably more significant for an understanding of the biodiversity of the Western Palaearctic as a region. Analysis of such differences can potentially provide information on a) the timescale and sequence of colonization of these refugia, and b) the demographic processes operating in each refugium (Stone *et al.* 2002a,b). These processes combine to generate the patterns in species richness and genetic diversity we see today. Here we examine patterns across refugia for oaks (Fagaceae, genus *Quercus*) and an associated group of obligate parasites – the oak gallwasps (Hymenoptera, Cynipidae, Cynipini). Oaks are a dominant feature of the flora of the Western Palaearctic, and both oaks and gallwasps have been the subjects of extensive population genetic work in the region (summarized below). We address the following general questions.What are the longitudinal patterns in species richness and intraspecific genetic diversity in these taxa? What, if anything, can these patterns tell us about the timescale of the colonization of one refuge from others, and about variation in the demographies of populations in different refugia?

Oaks in the western palaearctic

Western Palaearctic oaks are members of two sections within the oak subgenus *Quercus* – section *Quercus sensu stricto* (the white oaks: e.g. *Q. robur*, *Q. petraea*, *Q. pubescens*, *Q. pyrenaica*), and section *Cerris* (semi-evergreen oaks such as *Q. cerris* and *Q. suber*, and evergreen oaks such as *Q. trojana*, *Q. coccifera*, and *Q. ilex*). Section *Cerris* oaks are characteristic of southern, more arid environments, while much of northern Europe harbors only two white oak species – *Quercus petraea* and *Quercus robur*. Species richness in the region as a whole (discussed in more detail below) remains contentious, and is complicated by high levels of interspecific hybridization. The most recent major revision (Camus 1936-8, 1938-9) recognized 76 species, but a more recent review of the literature by Goværts & Frodin (1998) interpreted many of Camus' species as either varieties or subspecies, and recognized only twenty-nine species – 13 in the section *Cerris*, and 16 in the section *Quercus sensu stricto*. It is this classification that we follow here.

Oak gallwasp biology

In the Western Palaearctic, oak gallwasps (the Tribe Cynipini of the Cynipidae) are obligate gall-inducing parasites of oaks (Askew 1984; Stone *et al.* 2002c). Females lay their eggs into specific oak tissues, and the eggs and resulting larvae induce structurally complex, species-specific galls. The wasp undergoes its entire development within the gall, which provides both food and protection. As a group, gallwasps attack both of the oak sections found in the Western Palaearctic, though individual species have specific host associations (Cook *et al.* 2002). Oak gallwasps are cyclical parthenogens, alternating between one sexual and one asexual generation. For most species, these two generations develop in spring and summer of the same year. In all but two of the ten genera native to the Western Palaearctic, both of these generations induce galls on oaks within a single section of the subgenus *Quercus*. The two exceptions include species (termed host-alternators) with lifecycles involving obligate alternation between hosts in the sections *Cerris* and *Quercus sensu stricto*. In the genus *Andricus*, the host-alternators are a monophyletic, species-rich clade in which the sexual generation develops on section *Cerris* oaks, and the summer asexual generation galls develop on one or more species in section *Quercus sensu stricto* (Cook *et al.* 2002). All Western Palaearctic members of the genus *Callirhytis* show the reciprocal pattern of host alternation, – the sexual generation galls develop on oaks in section *Quercus sensu stricto*, and the asexual generation galls develop on hosts in the section *Cerris*. Oak gallwasps as a group are found throughout the range of the subgenus *Quercus* (Stone *et al.* 2002c).

Oaks and oak gallwasps are native to all four of the major southern glacial refugia in the Western Palaearctic. As a group, only the white oaks have naturally extended their distributions beyond these refugia (Ferris *et al.* 1993; Dumolin-Lapègue *et al.* 1997; Petit *et al.* 2002a). Oak gallwasps whose lifecycles also involve only these oaks are also thought to have colonized northern Europe thousands of years ago (Stone *et al.* 2002a,b). More recently, a wave of northern range expansion has been triggered by human planting of *Quercus cerris*, allowing eight species (seven host-alternators and one species that only attacks *Quercus cerris*) to spread from Italy and the Balkans as far north as Britain (Stone & Sunnucks 1993; Csóka *et al.* 1998; Atkinson 2000; Rokas 2001; Stone *et al.* 2001, 2002c). As demonstrated for many taxa elsewhere in the world (e.g. Rosenzweig 1995; Chown & Gaston 2000; Gaston 2000), both taxa show a decrease in intraspecific genetic diversity and species richness with increasing latitude (Stone & Sunnucks 1993; Dumolin-Lapègue *et al.* 1997; Stone *et al.* 2002a,b).

Longitudinal patterns in diversity

Our interest in longitudinal patterns of diversity in oaks and their associated gallwasps is stimulated by two preliminary observations. First, as described below, recognized refugia show differences in species composition, including both widespread and endemic elements. Second, population genetic analyses of several gallwasp species show a decline in genetic diversity from east to west in the refugia surrounding the Mediterranean Sea (Atkinson 2000). Here we address longitudinal patterns in more detail, examining patterns in two aspects of diversity – intraspecific genetic diversity, and species richness. These two aspects of diversity are expected to reflect historical processes throughout the region, associated with sequence and timescale of the colonization of the refugia and more recent demographic impacts characteristic of specific refugia. Before presenting data for oaks and gallwasps, we provide a brief background to what is known for each of these timescales.

The ancient colonization of refugia

The Western Palaearctic radiations of the oak sections *Cerris* and *Quercus sensu stricto* are both thought to have originated in Asia Minor, followed by westwards expansion into Europe (Manos *et al.* 1999). Although there is no direct fossil evidence for the origin of oak gallwasp radiations, current patterns of species richness suggest an eastern origin (perhaps in North America) (Stone *et al.* 2001), also followed by westwards range expansion. For both groups, this raises the issue of whether the sequence of refuge occupation in Europe can be reconstructed. As in studies of northwards range expansion, DNA sequence data can potentially allow inference of the order of refuge colonization, the number of lineages involved, and the minimum age of sequences sampled in a particular refuge (Stone *et al.* 2002a). We apply these approaches, as far as possible, to three gallwasp species below.

As has been shown for many taxa during the Holocene, uni-directional range expansion usually comprises a series of bottlenecking events resulting in a steady decrease in genetic diversity with distance from the origin (reviewed by Stone *et al.* 2002a,b). The same pattern has been demonstrated over a far shorter timescale for the recent invasion of northern Europe by several host-alternating oak gallwasps (Stone & Sunnucks 1993; Csóka *et al.* 1998). A westwards colonization of refugia could also have involved stepping stone processes, resulting in a decrease of genetic diversity from east to west. However, as discussed below, genetic diversity within refugia will also reflect refuge-specific variation in the demographic history of populations.

An additional consideration in the oak-gallwasp interaction is that range expansion by the parasite is contingent on range expansion by its host(s). A gallwasp can only expand its range if its host does the same (as in the case of recent invasions by gallwasps associated with *Quercus cerris*), or if it shifts from one host to another where the ranges of the two overlap. Analysis of phylogenetic patterns of host association in the gallwasp genus *Andricus* shows that host shifts among oak sections are probably extremely rare in gallwasps (Cook *et al.* 2002). Detailed analysis of one gallwasp species – *Andricus kollari* – has also shown that populations associated with refugially distinct host oak species only hybridize extremely rarely, resulting in genetically distinct host ecotypes. If this effect is widespread, it may represent a major barrier to genetic exchange among refugia.

Population history since colonization

Patterns of temporal and latitudinal variation in the Pleistocene climate of the Western Palaearctic have been extensively studied (see Hewitt 1999 for a recent review). Van Andel (2002) has proposed that there has also been extensive variation in climate with longitude, such that populations distributed across refugia have probably experienced very different paleoclimatic histories following colonization. In particular, the magnitude of climatic and habitat fluctuation is thought to have been greater in Iberia than in other refugia (Van Andel 2002). Habitat fragmentation reduces effective population sizes, greatly affecting both speciation rates and extinction probability. Taken simply at face value, this leads to three general predictions for the Iberian biota: we expect it to be distinct (due to regional speciation), with individual species showing low genetic diversity and extensive spatial structure.

Here we review patterns in the distribution of oak gallwasp and oak species across the refugia, considering both absolute species richness and the similarity with biotas in neighboring refugia. We also review patterns of genetic diversity in three gallwasp and one oak species with circum-Mediterranean distributions; the host-alternators *Andricus kollari* and *Andricus quercustozae*, the non-host-alternator *Biorhiza pallida* and the oak *Q. robur*.

Methods

Oak gallwasp species richness

Gallwasp faunas for Iberia and the rest of Europe are drawn from a recently revised literature (Melika *et al.* 2000; Nieves-Aldrey 2001 respectively). Species lists for Turkey and Asia Minor have been generated from a combination of our own unpublished survey data and the limited literature available (Chodjai 1980). We use this information to compare the faunal composition of each refuge, based on total species number and representation of the different genera. We also calculate pairwise comparisons of species similarity for the different refugia using Sørensen's similarity quotient (SQ):

$$SQ = \frac{2J}{(A + B)},$$

where J is the number of species shared by two samples. A is the number of species in sample A and B is the number of species in sample B. This similarity quotient is for pairwise comparison of qualitative data (i.e. species lists) and ranges from 0 (where there are no species in common) to 1 (where all species are shared by the two data sets) (Cole 1985; Dupre 2000).

Oak gallwasp genetic diversity

We use data from the case studies of three oak gallwasp species, *Andricus kollari*, *A. quercustozae* and *Biorhiza pallida*. The case studies are based on three types of genetic data: i) sequence data for a 433 bp fragment of the mitochondrial cytochrome *b* gene; ii) two nuclear sequences from the internal transcriber region (ITS); and iii) genetic diversity measures from polymorphic allozyme loci. Details of the methods, data and analytical techniques are given in full for *A. kollari* in Stone *et al.* (2001), for *B. pallida* in Rokas *et al.* (2001) and for *A. quercustozae* in Atkinson (2000) and Rokas (2001).

We analyze longitudinal patterns in two measures of genetic diversity for *A. kollari* and *A. quercustozae*: average expected heterozygosity, and mean number of alleles per locus (allelic diversity). Average expected heterozygosity is the heterozygosity calculated from observed gene frequencies, averaged over polymorphic loci. Both of these measures are commonly employed in analyses of latitudinal patterns in genetic diversity (Stone *et al.* 2002a), and the mean number of alleles per locus is particularly sensitive to population bottlenecks. The genetic makeup of refugial populations for all three species is illustrated using sequence-based phylogenies whose construction is described in detail elsewhere (Atkinson 2000; Rokas 2001; Stone *et al.* 2001). The diversity among sequences associated with a specific refuge is illustrated using maximum % genetic divergence for the mtDNA cytochrome *b* data. This measure is sensitive to the model of sequence evolution assumed, and our estimates are derived using the best-fit model of evolution identified for each dataset using likelihood ratio tests incorporated in the program Modeltest (Posada & Crandall 1998). These estimates can then be used to generate (very approximate) ages for the most recent common ancestor of sequences within each refuge, using the assumption of the molecular clock and a rate of 2.3% divergence per million years (Brower & de Salle 1994).

Oak species richness

Because of high levels of interspecific hybridization in oaks, the number of species in the Western Palaearctic remains unresolved. Here we use the recently revised flora proposed by Goværts & Frodin (1998), which recognizes twenty-nine species – 13 in the section *Cerris*, and 16 in the section *Quercus sensu stricto*. We have constructed species lists for each refuge based on distributional information from the Flora Europaea (Jalas & Suominen 1988), more detailed area-specific floras from Iberia (Castroviejo *et al.* 1986), the Balkans (Turrill 1929), Turkey (Yaltirik 1982), the former USSR (Konarov 1936), Iran (Browicz & Menitsky 1971), Iraq (Townsend & Guest 1980), Palestine (Zohary 1966) and three volumes of 'Ecosystems of the World' (Castri *et al.* 1981; Orington 1983; Röhrig & Ulrich 1991). We then use these lists in a comparison of species richness between refugia (using Sørensen's quotient, as described above), including identification of refuge-specific species.

White oak haplotype diversity

The high levels of hybridization between white oak species, and variation among refugia in the taxa able to hybridize, make it difficult to identify species- and refuge-specific patterns. While acknowledging these problems, here we analyze haplotype diversity data for the only oak so far sampled in depth in three refugia – *Q. robur*. The data are drawn from published Europe-wide surveys of genetic diversity in the white oaks (Petit *et al.* 2002b), with specific sources for Iberia (Olalde *et al.* 2002), the Balkans (Bordács *et al.* 2002), and Italy (Fineschi *et al.* 2002). The data comprise the frequency of each of 42 haplotypes described from white oaks across Europe. The haplotypes are classified by the length variants of 23 polymorphic restriction fragments, ranked from 1 to 6 according to the migration distance down the gel, with a further category for the presence of point mutations (Petit *et al.* 2002b). We compare the list of haplotypes present in each of the three refugia and use a multivariate cluster analysis to determine the similarity of these haplotypes. The cluster analysis algorithm used here involves a single linkage method and Manhattan distances, although identical groupings are returned with a wide diversity of methods. This approach is used here for the purposes of illustration only. We also present values of total mean genetic diversity (v_T) calculated by Petit *et al.* (2002b) using the software Hapermut. This measure is an estimation of the genetic diversity (the probability that two haplotypes selected at random within a refuge will differ) taking into account the genetic distance between the haplotypes.

Results

Oak gallwasp species richness

Across refugia, the lowest gallwasp species richness is found in Iberia (n=74), with higher richness in Asia Minor (n=82) and the Balkans (n=101). We emphasize that while the estimate for Iberia is extensively researched, the estimate for Asia Minor is certainly an underestimate, and so the relative paucity of Iberia is probably more pronounced than is currently suggested. The lower richness in Iberia is also evident at the level of gallwasp genera. Western Palaearctic oak gallwasps are currently divided into ten genera, all of which are represented in Asia Minor, the Balkans and Italy but three of which (*Aphelonyx, Chilaspis* and *Dryocosmus*) are absent from Iberia (Table 1) (Melika *et al.* 2000; Nieves-Aldrey 2001). These three genera all gall section *Cerris* oaks, specifically *Quercus cerris* and close relatives, but are unable to exploit *Quercus suber* or the evergreen members of this oak section found in Iberia.

Table 1. *The number of oak gallwasp species and oak species found in Asia Minor, the Balkans, Italy and Iberia. The number of species specific to a single refuge is given in parentheses. For the oak gallwasps the species numbers for each genus are also given.*

Genus	Asia Minor	Balkans	Italy	Iberia
Andricus	54(4)	69(6)	65(1)	41(7)
Aphelonyx	1	1	1	0
Biorhiza	1	1	1	1
Callirhytis	2	3	1	3
Chilaspis	3(1)	2	1	0
Cynips	6	7	7	6
Dryocosmus	1	2(1)	1	0
Neuroterus	11	13(2)	10	7
Plagiotrochus	1	1	4	11(8)
Trigonaspis	2	2	2	5(3)
Total gallwasp	82(5)	101(9)	93(1)	74(18)
Total *Quercus*	17(7)	12(1)	11(1)	12(6)

Although each refuge contains endemic species, the faunas of the Balkans, Asia Minor and Italy are very similar (as indicated by the high values of Sørensen's quotient SQ: Table 2). In contrast, even though the Iberian fauna is relatively species-poor, it contains more endemic species than any other refuge – a pattern reflected in the low values of SQ in pairwise comparisons between Iberia and the other refugial areas (Table 2). These species are restricted to three genera, two of which – *Trigonaspis* and *Plagiotrochus* – have very low species richness in all the other refugia. The third, the largest of the oak gallwasp genera (*Andricus*) has refuge-specific species throughout the Western Palaearctic (Table 1).

Table 2. Comparison of gallwasp species lists between pairs of refugia using Sørensen's quotient.

Refuge	Asia Minor	Balkans	Italy
Balkans	0.809		
Italy	0.811	0.876	
Iberia	0.564	0.571	0.635

Oak gallwasp genetic diversity

a) Patterns of genetic diversity

A. quercustozae and *A. kollari* both show a decrease in genetic diversity with longitude from Anatolia in the east to Iberia in the west. This applies whether we measure genetic diversity as the mean number of alleles per locus or average heterozygosity per locus (Figures 1A & B). For *A. quercustozae* heterozygosity decreases from 0.24-0.30 in Anatolia to 0.07-0.08 in Iberia, and the mean number of alleles per locus halves from 2.83 to 1.42. The pattern for *A. kollari* is similar, as heterozygosity decreases from 0.25-0.29 in Anatolia and the Balkans to 0.12-0.17 in Iberia, and the mean number of alleles per locus also halves from 2.9 to 1.5 over the same range.

Although the genetic diversity in Iberia is the lowest of any refuge there are more alleles (private alleles) in this region than in any other refuge, as well as very significant differences to other refugia in the frequencies of shared (presumed ancestral) alleles (Atkinson 2000; Stone *et al.* 2001). Iberia also harbors refuge-specific private alleles for the third case-study species *Biorhiza pallida*.

b) Phylogeography based analysis

Phylogenies for the three gallwasp species are shown in Figures 2-4. We comment on four patterns that emerge for all three species: first, all three species sequences sampled in Iberia represent discrete, monophyletic clades. In contrast, sequences from the Balkans and Italy do not constitute refuge-specific clades for any of the species. For *A. kollari* and *A. quercustozae* the Iberian sequences form a single clade, while for *Biorhiza pallida* Iberian sequences and those sampled from immediately adjacent parts of southwestern France fall into two clades (Figure 4). Phylogenies for all three species point to very limited exchange of haplotypes among refugia. There is a suggestion that sequences for Turkish *Andricus quercustozae* fall into two groups, perhaps representing discrete refugial lineages (see Discussion).

Second, sequence diversity within refugia varies (Table 3). For *A. kollari* and *A. quercustozae*, the maximum divergence of sequences within Iberia is much less than that observed for sequences from Italy and the Balkans. For these species, animals sampled in Iberia have a much more recent common ancestor than samples for other refugia. The opposite pattern is found for *B. pallida* (Table 3).

***Table 3**. Maximum percentage divergence of mtDNA haplotype sequence diversity for* A. kollari, A. quercustozae *and* B. pallida *between and within the Iberian and combined Italian/ Balkan refugia. Values in parentheses are estimated times for the divergence using the approximation of 2.3% divergence per million years (see Methods).*

	A. kollari	*A. quercustozae*	*B. pallida*
Within Italy/Balkan refuge	4.07 (1.77 mya)	2.0 (0.87 mya)	1.2 (0.52 mya)
Within Iberian refuge	1.87 (0.81 mya)	1.1 (0.48 mya)	5.0 (2.18 mya)
Between the refugia	10.4 (4.5 mya)	12.6 (5.48 mya)	7.9 (3.43 mya)

A

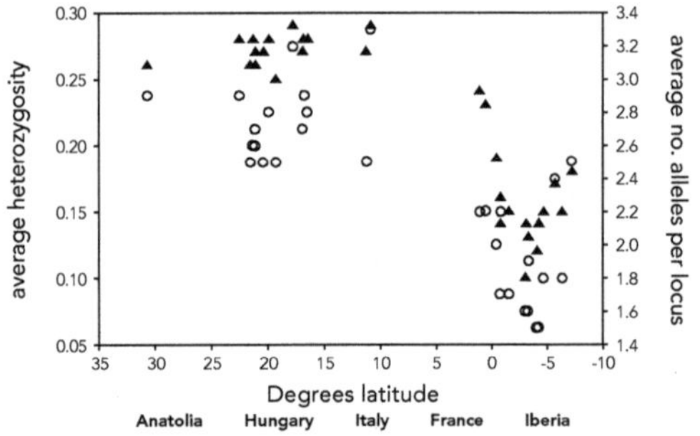

B

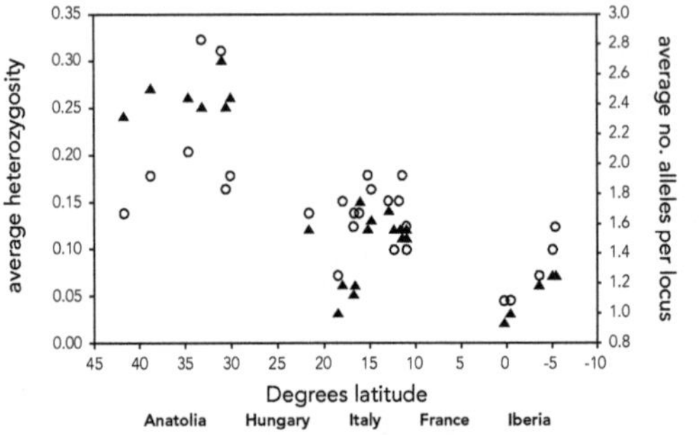

Figure 1**. The genetic diversity measured as average expected heterozygosity (right axis, triangles) and number of alleles per locus (left axis, circles) from polymorphic allozyme loci for **A) A. kollari ***and B)*** A. quercustozae *with degrees longitude east to west across the Mediterranean region.*

Third, in all species, the divergence in sequences between the common ancestor of all Iberian sequences and the common ancestor of all Italian/ Balkan sequences is substantial (Figures 2-4). For *A. kollari* and *A. quercustozae* this implies either a) that Iberian populations were founded relatively recently (compared to other refugia) from a very divergent ancestral haplotype or b) Iberian populations have been recently bottlenecked, such that current diversity represents only a small monophyletic fraction of the diversity present over time in this refuge.

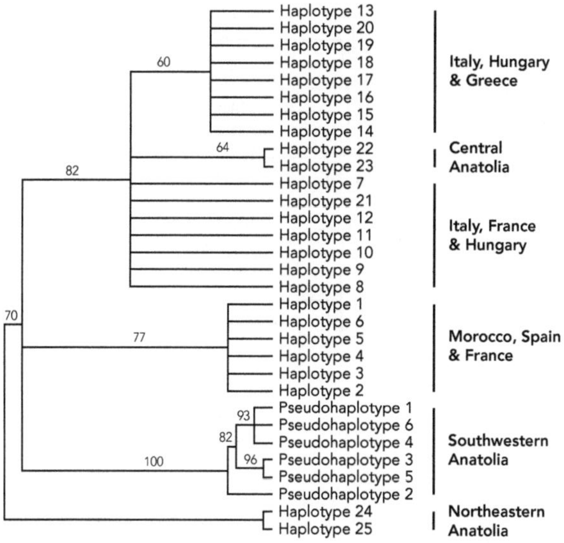

Figure 2. *A 50% majority-rule consensus tree from a maximum likelihood analysis for* A. quercustozae *using 47 individuals from 38 populations and data from a 433 bp fragment of the cytochrome* b *gene (from Atkinson 2000). Branches without a bootstrap value are supported by less than 50% of replicates.*

Fourth, in all three gallwasp species branches at the base of the tree are too poorly resolved to infer a colonization sequence. Identification of a refugial origin requires rooting the tree with an outgroup, and high bootstrap support for the basal nodes in the tree. For reasons that will not be discussed here, selection of an appropriate outgroup is problematic. However, even if the trees were rooted, bootstrap values for the basal nodes in *Andricus kollari* and *A. quercustozae* are too low (Figures 2 & 3) to support confident identification of which refugia harbor basal sequences, and which derived.

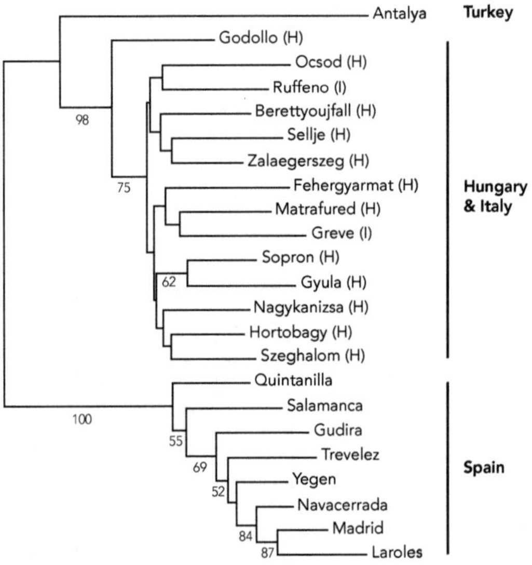

Figure 3. *Relationships, based on cytochrome* b *sequence data, among* A. kollari *populations from both native and recently-invaded regions. The lefthand cladogram is the strict consensus tree generated by maximum parsimony. Numbers on the cladogram are bootstrap percentages from 100 bootstraps, for maximum parsimony above the branch and maximum likelihood below. Branches without a bootstrap value are supported by less than 50% of replicates. Country codes are as follows: SP = Spain and SW FR = Southwest France, Italy (IT), Hungary (H) and Turkey (T). Samples labelled NW FR are from the invaded range of this species in northern France.*

Oak species richness

Two of the twenty-nine species identified by Goværts & Frodin (1998) are island endemics (*Q. sicula* from Sicily, and *Q. alnifolia* from Cyprus) and will not be considered further. Asia Minor is home to the highest number

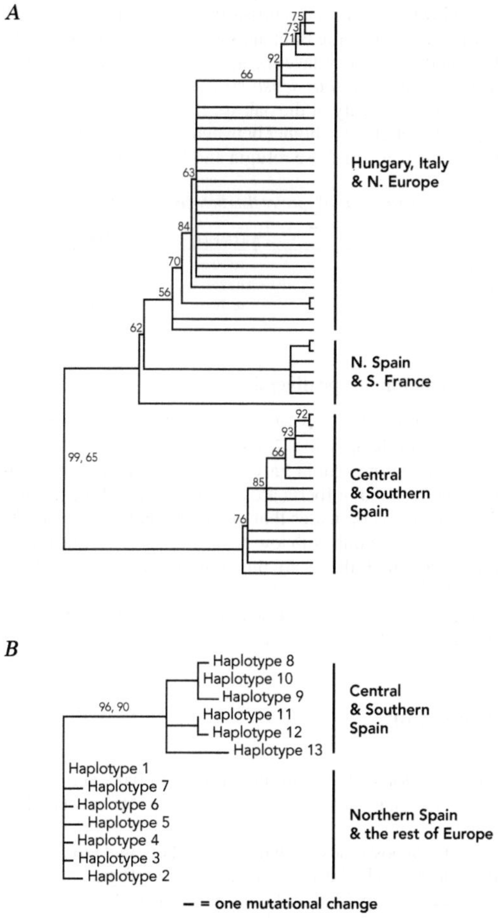

Figure 4. *A) A 50% majority-rule consensus tree for* B. pallida *for the ITS dataset using maximum parsimony and B) a consensus tree for the cytochrome* b *data set with bootstrap values for parsimony and maximum likelihood respectively (from Rokas* et al. *2001). Branches without a bootstrap value are supported by less than 50% of replicates.*

(n=17) of the remaining 27 species, while Iberia, Italy and the Balkans all have a lower and similar number of species (n=12, 11 and 12 respectively). However, although the Balkans and Italy only have one unique species each, Iberia and Asia Minor have six and seven respectively (Figure 2), representing the *Cerris* and *Quercus sensu stricto* sections equally. As for the oak gallwasps, the similarity of the oak species in the different refugia measured by Sørensen's quotient is higher between Asia Minor, the Balkans and Italy than it is between any of these refugia and Iberia (Table 4).

***Table 4**. Comparison of oak species lists between pairs of refugia using Sørensen's quotient.*

	Asia Minor	Balkans	Italy
Balkans	0.690		
Italy	0.571	0.783	
Iberia	0.276	0.417	0.435

White oak haplotype diversity

The haplotypic diversity of *Quercus robur* is higher in the Balkans than in either Italy or Iberia (with 17, 7 and 8 haplotypes respectively). However, while the haplotypes present in Italy are a subset of those found in the Balkans, all haplotypes found in Iberia are unique to this refuge (Figure 5). Multivariate cluster analysis indicates that, as for the gallwasps, the Iberian haplotypes form a distinct group with low genetic diversity, while the haplotypes present in Italy and the Balkans are much more divergent (Figure 6). This pattern is supported by comparison of values of mean genetic diversity (±1 standard deviation) within the Balkans and Iberia of 0.618 (0.055) and 0.158 (0.022) respectively (Petit *et al.* 2002b).

Discussion

Consideration of longitudinal patterns in oaks and gallwasps reveals four major effects, most of which involve contrasts between Iberia and the other refugia:

1) Iberia possesses a depauperate but distinct biota at both the species and the generic level relative to more easterly refugia.

2) Iberian populations of three widely distributed gallwasp species all possess haplotypes that are highly divergent from those sampled in other refugia.

3) For *Andricus kollari*, *A. quercustozae* and *Quercus robur*, Iberian populations show lower intraspecific genetic diversity than populations of the same species in more easterly refugia, and the two gallwasps show maximum diver-

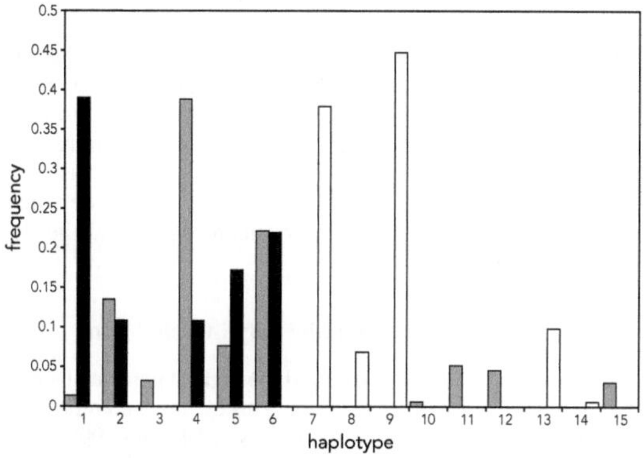

Figure 5. *The haplotypes and their frequencies for* Q. robur *collected in each refuge; Balkans n = 434 (grey bar), Italy n = 64 (dark bar), Iberia n = 161 (white bar). Data from Bordács* et al. *(2002); Fineschi* et al. *(2002); Olalde* et al. *(2002).*

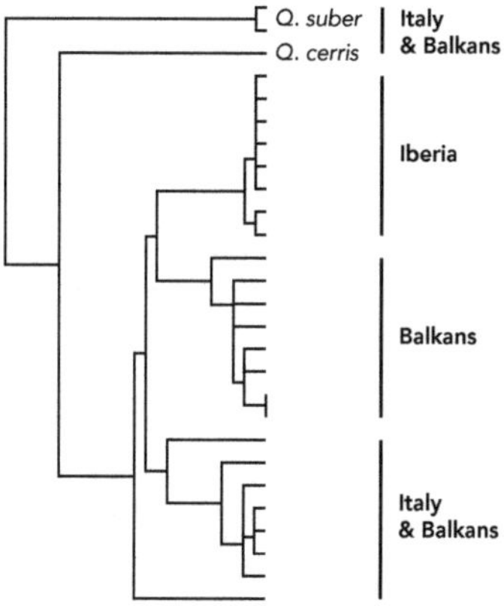

Figure 6. *A multivariate cluster analysis of* Q. robur *haplotypes based on RFLP data from Petit* et al. *(2002b) constructed using a single linkage method and Manhattan distance measure.*

sity in Asia Minor. This pattern is not universal, however (cf *Biorhiza pallida*).

4) For *Quercus robur*, Iberian populations appear to be derived from those further to the east. In all three gallwasp species, however, nodes at the base of the respective phylogenies are too poorly resolved to infer a colonization sequence. These observations can only be made in relation to patterns across the region as a whole, and before we discuss the unique nature of Iberian oak and gallwasp biota we will consider the gradient in species richness across the Mediterranean region and the deep separation of each refugium.

High species richness and genetic diversity in Asia Minor

The highest species richness for oaks and the highest genetic diversity for oak gallwasps are both found in Asia Minor. Turkey and Iran represent a major center of diversity for many taxa (e.g. Covas & Blondel 1998; Moghaddam *et al.* 2000), and may actually consist of at least two refugia (King & Ferris 1998; Kaya & Raynal 2001): a region with Anatolian and Mediterranean affinities in the west, and a region linked biologically to Iran and the Caucasus in the east. These two regions were represented for only one of the gallwasp species described here – *Andricus quercustozae* – and each contribute discrete lineages. The same pattern is seen in five other gallwasp species not discussed here (Atkinson 2000; Rokas 2001). The region's high diversity may thus be derived in part from its status as an original center of diversity for oaks and associated fauna, and also the compound nature of the refuge. Increases in our understanding of population structure and paleoclimates show that other regions may also represent multiple refugia – including Iberia (Gómez, this conference; Paulo *et al.* 2001). It may also be significant that many oak and gallwasp taxa are common to Turkey and the Balkan region, including species (such as *Quercus cerris*, *Q. robur* and *Q. petraea*) that are obligate hosts for large numbers of gallwasp species. This region may then constitute an eastern 'super-refuge', genetically distinct from Iberia and the western Mediterranean (Médail & Quézel 1997).

High genetic divergence between refugial areas

Using the general (and very approximate) assumption that 2.3% sequence divergence for the mitochondrial cytochrome *b* gene is equivalent to one million years, oak gallwasps across the Western Palaearctic shared a most recent common ancestor between 3.5-5.5 mya (we have no data on refugial divergence for the oaks). Refuge colonization thus substantially predates the Pleistocene Ice Ages, and probably occurred during the Pliocene epoch (5-1.8 mya). We also cannot estimate the order in which regions corresponding to later refugia were colonized. The Pliocene epoch was characterized by a warm, wet and relatively stable climate, with extensive forest cover across the Western Palaearctic. Around the northern Mediterranean this forest was similar to the evergreen sclerophyllous broad-leaved forest still found in eastern China, replaced in the drier south by more drought-tolerant woodland (Fanquette *et al.* 1999). The extensive forests at the beginning of the Pliocene probably allowed migration of species, and gene flow between populations, for oaks and gallwasps throughout much of the Mediterranean.

Through the Pliocene the climate gradually cooled due to the establishment of permanent ice in the Arctic and the development of the trade winds. In the Mediterranean region this resulted in a gradual increase in aridity, and by 2.8 mya the pattern of summer drought typical of today's Mediterranean climate had stabilized (Suc 1984). This led to the development of a more xerophytic Mediterranean flora in southern Spain, Italy and northern Africa (dominated by *Olea*, *Ceratonia* and evergreen *Quercus*). The first of the climatic fluctuations that were to dominate the Quaternary occurred about 2.4 mya, and there is evidence in the pollen record of spatial structuring in the vegetation due to an increased aridity of the area. This may have resulted in the fragmentation of the forest belt across the northern rim of the Mediterranean, which in turn would have prevented gene flow between isolated forest patches promoting the evolution of refuge-specific characteristics. This may have been the period in which cork oak, *Quercus suber*, became the dominant semi-deciduous section *Cerris* oak in Iberia, while *Q. cerris* and related species (*Q. libani*, *Q. ithaburensis*) remained widespread through Italy, the Balkans and Turkey. Intraspecific divergence into host-specific ecotypes in species such as *Andricus kollari* probably began in this period.

The similarity of the Italian and Balkan refugia

The apparent similarity of the oak and gallwasp fauna and flora in Italy and the Balkans suggests that there has been continuous or recent migration and gene flow between the two areas. The seasonality generated by climatic fluctuations in the late Pliocene appears to have been less severe in the eastern Mediterranean, and mixed deciduous and coniferous woodland are thought to have persisted in the Balkans, Israel and land surrounding the Black Sea (Thompson & Fleming 1996). This may well have allowed the persistence of larger populations, resulting in the retention throughout the region of many shared ancestral polymorphisms, and the spread of new ones. Gene flow in gallwasps and oaks could have occurred across land to the north of Italy, or across the Adriatic Sea. Today this is a distance of less than 70 km, and it is not clear what sort of a barrier this would represent to insects, such as gallwasps, that are probably extensively wind-dispersed (Stone & Sunnucks 1993). Data (not shown here) from polymorphic allozyme loci indicate significant differences in allele frequency between Italian and Balkan populations for *A. quercustozae*, and Italian *Andricus kollari* populations possess private alleles at low frequency (Stone *et al.* 2001). There are also significant refugial differences in the oak trees (Petit *et al.* 2002a). The low magnitude of differences between Italy and the Balkans suggests that either relatively high levels of genetic exchange prevented major differences from developing, or that extensive recent migration and gene flow have substantially masked (or homogenized) these differences. Subtler effects such as non-equilibrium allele frequencies or apparent linkage between loci that indicate incomplete mixing of the two source populations may help to distinguish between these alternatives (Atkinson, unpublished).

The uniqueness of Iberia

While Italy and the Balkans show similar diversity for both oaks and oak gallwasps, the Iberian gallwasp fauna is impoverished. For both oaks and gallwasps however, there are many unique Iberian elements, in the form of species and private allozyme alleles, unique sequences and haplotypes [a pattern also reported for oaks by Muir *et al.* (1999) using microsatellite markers]. We will consider these two aspects in turn.

i) The low gallwasp species richness in Iberia

The absence of *Q. cerris* from the Iberian Peninsula may have played a major part in this pattern. In the Western Palaearctic, species in seven gallwasp genera gall section *Cerris* oaks – host-alternators in the genera *Andricus* and *Callirhytis*, and species in the genera *Aphelonyx*, *Chilaspis*, *Dryocosmus*, *Neuroterus*, and *Plagiotrochus*, which only gall section *Cerris* oaks. Colonization of *Cerris* oaks from the oak section *Quercus sensu stricto* has occurred independently in many of these lineages, and (with the clear exception of *Plagiotrochus*) the ancestral section *Cerris* host is probably *Quercus cerris* (Stone *et al.* 2002c; Cook *et al.* 2002). This species is absent from Iberia, which is occupied by the closely related *Quercus suber*. Only a proportion of the gallwasps able to attack *Quercus cerris* have been able to shift hosts onto *Quercus suber* (Stone *et al.* 2001), suggesting that host specificity has limited immigration of species into the Iberian refuge. The genetic divergence among gallwasps exploiting section *Cerris* oaks is relatively ancient, suggesting that the distributions we see today predate the geographic separation of *Q. cerris* and *Q. suber*, rather than having evolved after the geographic separation of these oaks (Cook *et al.* 2002). The impact on species richness is dramatically illustrated for *Andricus* – the largest gallwasp genus in the Western Palaearctic with *ca.* 90 species. Of these, the majority belong to a single clade of host-alternators. As far as is known, all but one of these host-alternating species are able to exploit *Quercus cerris*, and of these a substantial proportion also have Iberian populations able to exploit *Q. suber*. In contrast, only one species – the Iberian endemic *Andricus picta* – is known, which exploits *Q. suber* but cannot (as far is known) exploit *Q. cerris*.

ii) The high degree of species endemism in Iberia

Endemic oak gallwasp faunas are a logical extension of the rarity of host shifts among oaks discussed above. Longitudinal variation in oak taxa among refugia is particularly marked in the oak section *Cerris*, and the vast majority of refuge-endemic gallwasps gall these oaks. Iberian endemics are members of three genera – *Plagiotrochus*, *Andricus*, and *Trigonaspis*, – and all of them are found on oak species either only or predominantly found in Iberia. The endemic *Trigonaspis* species induces galls on white oaks specific to the peninsula, the endemic *Plagiotrochus* species induce galls on *Q. coccifera* and *Q. ilex*, two species with distributions concentrated around the western end of the Mediterranean, and the endemic *Andricus* species utilize *Q. suber* for their spring generation. An explanation of the high level of endemism of oak trees is more difficult, perhaps reflecting the extent of isolation experienced by this refuge in time and space throughout the Pliocene and Pleistocene.

A puzzling feature in oak gallwasp distributions stems from the fact that *Quercus suber* is not entirely limited to Iberia, but is also found along the west coast of Italy. However, very few of the gallwasps (such as members of the genus *Plagiotrochus*) that exploit this host in Iberia are known from Italy. It is also currently unknown whether host-alternating species in Italy are able to exploit *Q. suber* as well as *Q. cerris*, perhaps existing as ecotypes as *Andricus kollari* does (Stone *et al.* 2001). The Italian populations of *Q. suber* are genetically very distinct from those in southwestern France and Spain (Lopes 2002), and it is possible that these differences are great enough to represent a barrier to host shifts by Iberian gallwasps that have coevolved with Iberian *Q. suber*.

iii) Endemic intraspecific variation in Iberian gallwasp populations

Two processes could potentially have been responsible for the generation of refuge-specific polymorphism in Iberian gallwasp populations. First, genetic drift and low levels of gene flow with other refugia may have resulted in the retention of different subsets of ancestral polymorphism in Iberia. Second, prolonged genetic isolation from other refugia may have allowed the spread and retention of new private variants. A feature of the endemic sequence variation in the gallwasps *A. kollari* and *A. quercustozae* is that it also has a far more recent common ancestor than is associated with the variation in other refugia. A qualitatively similar pattern is also found in *Q. robur*. We now consider the paleoclimatic processes that could have generated this.

In the Mediterranean region, Quaternary climatic cycling (from 1.8 mya to the present day) resulted in fluctuations between cold, arid glacial periods and warmer, wetter, interglacials. During the glacials most of Europe was covered in polar desert with deciduous forest (including *Quercus* species) restricted to sheltered mountainous areas in the western Balkans, Italy and southern Iberia (Van Andel & Tzedakis 1996). During the transitions to the interglacials these forests spread across southern Europe, but by the height of the interglacials the high temperatures in the Mediterranean led to a rapid replacement of this forest by a more open sclerophyllous Mediterranean forest. At these times the deciduous forest in refugial regions was again restricted to temperate microhabitats (Bennett *et al.* 1991), including higher altitudes (the mountain ranges in eastern Anatolia and central Italy) and higher latitudes (including northern sections of the Balkan Peninsula, Willis & Whittaker 2000) and along the northern coast of the Iberia and Anatolia.

While the whole of the Mediterranean region was affected by these climatic fluctuations, the severity of these conditions varied along the Mediterranean (Van Andel 2002). Van Andel & Tzedakis (1996) suggest that reduction in deciduous forest cover was much greater in Iberia during the glacials (when much of Iberia was covered in arid cold steppe) and interglacials (when the same region was too dry for forest cover) in comparison to refugia further east. Mediterranean forest was also restricted to the extreme south of the peninsula, as well as coastal areas of Italy, the Balkans, NW Africa and Anatolia.

Unlike Italy and the Balkans – where a pollen core from NW Greece indicates a continuous record of tree pollen for the last 430 k years (Tzedakis 1993) – there is little evidence for significant areas of deciduous forest in the whole of the Iberian Peninsula during the last glacial. The persistence and age of refuge-specific lineages shows, however, that pockets of suitable microhabitat must have existed and very low levels of pollen at mid-altitudes in the mountain ranges and within gorge systems that run across the country (Olalde *et al.* 2002) suggest that both environments may have provided local refuges. This inference is supported by the current distribution of oak haplotypes in Iberia, used by Olalde *et al.* (2002) to suggest that refugia may have been concentrated in the coastal areas with a few isolated pockets in the interior. Such severe habitat fragmentation is most likely to have resulted in low local population sizes, and restricted gene flow between populations. For host-alternators dependent on two oak taxa, such fragmentation would have resulted in local extinction from pockets lacking representatives from both oak sections. Bottlenecking events may well be reason for both the low diversity and recent common ancestry of Iberian gallwasp lineages relative to other refugia (Atkinson 2000; Stone *et al.* 2001). They are also consistent with the more dramatic spatial structuring of genetic variation of *Andricus kollari* in Iberia relative to more easterly refugia (Stone *et al.*, unpublished data).

B. pallida

Unlike *A. kollari* and *A. quercustozae*, the refugial genetic diversity for *B. pallida* is higher and older in Iberia than in refugia further east. Taken at face value, these data are compatible either with an Iberian origin for this gallwasp (though rooting with suitable outgroups would be needed to show this), or with higher rates of loss of genetic diversity in populations outside Iberia. However, it should be noted that *B. pallida* has been found to be infected with the endosymbiotic bacterium Wolbachia in all populations apart from those in central Iberia (Rokas *et al.* 2001). The low level of diversity at the cytochrome *b* gene and at allozyme loci in all but the Wolbachia-free populations are suggestive of some complex effect associated with this infection that may be masking any phylogeographical pattern. Thus it is difficult to determine whether the difference we have noted is due to differences in life-history characteristics, or due to the endosymbiont and we will not consider this species any further here.

Conclusions

By comparing data on oaks and oak gallwasps from Iberia, Italy, the Balkans and Asia Minor we have shown that there are longitudinal differences in species richness and genetic diversity between these refugia. We have discussed these in terms of ancient colonization of the refugia, and the history of refugia since colonization. Divergence times estimated from mtDNA sequences provide an indication of the relative timing of the patterns, and we have used these dates to construct a paleoecological history of the area to explain some of them.

The extensive phylogeographic literature on postglacial range expansion in Europe contains many other data sets similar to those used here and, as has been done for latitude (Taberlet *et al.* 1998; Hewitt 1999) a comparison of longitudinal patterns across different taxa may allow the identification of general trends. The degree to which concordances across taxa may be identified remains contentious (Taberlet 1998), and longitudinal trends may be even more challenging than those across latitude because of the greater age of the underlying processes. Nevertheless, longitudinal processes represent the raw material on which later latitudinal processes work, and as such are crucial in shaping the biodiversity of the Western Palaearctic. Furthermore, longitude rather than latitude is a more useful predictor in the detection of biodiversity hotspots (Reid 1998) in which conservation efforts may be concentrated.

Acknowledgements

We would like to thank Gyuri Csóka, Jose-Luis Nieves-Aldrey, Juli Pujade Villar and George Melika for their help in preparing oak gallwasp species lists. We also thank Joe Cann for discussion about the paleoclimate of the Mediterranean in the Pliocene and Pleistocene. RJA received a grant from the MacLagan trust (Royal Society of Edinburgh) to attend the conference in Vairão. GNS was supported by a NERC EDGE program grant (Gst/03/2035) and NERC grant (GR12847), a Small Ecological Research Grant from the British Ecological Society, and a Royal Society Research grant.

References

Askew RR (1984) ***The Biology of Gallwasps***. Oxford & IBH, New Delhi.

Atkinson RJ (2000) ***The Genetic Analysis of Natural History, Reproductive Strategy and Population Structure in European Oak Gallwasps (Hymenoptera: Cynipidae)***. PhD Thesis, University of Oxford.

Bennett KD, Tzedakis PC, Willis KJ (1991) Quaternary refugia of North European trees. ***Journal of Biogeography*** **18**, 103-115.

Bordács S, Popescu F, Slade D, Csaikl UM, Lesur I, Borovics A., Kezdy P, König AO, Gomory D, Brewer S, Burg K, Petit RJ (2002) Chloroplast DNA variation of white oaks in northern Balkans and in the Carpathian Basin. ***Forest Ecology and Management*** **156**, 197-209.

Brower AVZ, de Salle R (1994) Practical and theoretical considerations for choice of a DNA sequence region in insect molecular systematics, with a short review of published studies using nuclear gene regions. ***Annales of the Entomological Society of America*** **87**, 702-716.

Browicz K, Menitsky GL (1971) ***Fagaceae.*** Flora Iranica: Flora des Iranischen Hochlandes und der umrahmenden Gebirge, vol. 77 (ed. Rechinger KH). Akademische Druck und Verlagsanstalt, Graz.

Camus A (1936-1938) ***Les chênes. Monographie de genre Quercus***, vol. 1 (ed. Lechevalier P) Paris.

Camus A (1938-1939) ***Les chênes. Monographie de genre Quercus***, vol. 2 (ed. Lechevalier P) Paris.

Castri di F, Goodall DW, Specht RL (eds.) (1981) ***Mediterranean-type Shrublands***. Ecosystems of the World, vol 11. Elsevier, Amsterdam.

Castroviejo S, Laínz M, López González G, Montserrat P, Muñoz Garmendia F, Paiva J, Villar L (eds.) (1986) ***Lycopodiceae Papaveraceae.*** Flora Iberica: Plantas vasculares de la Península Ibérica e Islas Baleares, vol. 1. CSIC, Madrid.

Chodjai M (1980) Studies on the Cynipid Hymenoptera and gall inducing species in the fauna of the oak forests of Iran. ***Journal of the Entomological Society of Iran***, Supplement 3.

Chown SL, Gaston KJ (2000). Areas, cradles and museums: the latitudinal gradient in species richness. ***Trends in Ecology and Evolution*** **15**, 311-315.

Cole K (1985) Past rates of change, species richness, and a model of vegetational inertia in the Grand-Canyon, Arizona. ***American Naturalist*** **125**, 289-303.

Cook JM, Rokas A, Pagel M, Stone GN (2002) Evolutionary shifts between host oak species and host plant organs in *Andricus* gallwasps. ***Evolution*** **56**, 1821-1830.

Covas R, Blondel J (1998) Biogeography and history of the Mediterranean bird fauna. ***Ibis*** **140**, 395-407.

Csóka G, Stone GN, Atkinson R, Schönrogge K (1998) The population genetics of post-glacial invasions of northern Europe by cynipid gallwasps (Hymenoptera: Cynipidae). In: *The Biology of Gall-inducing Arthropods* (eds. Csóka G, Mattson W, Stone GN and Price PW), pp. 280-294. General Technical Report NC 199. USDA. North Central Research Station, St.Paul, Minnesota.

Dumolin-Lapègue S, Demesure B, Fineschi S, Le Corre V, Petit RJ (1997) Phylogeographic structure of white oaks throughout the European continent. *Genetics*, **146**, 1475-1487.

Dupre C (2000) How to determine a regional species pool: a study in two Swedish regions. *Oikos* **89**, 128-136.

Fauquette S, Suc JP, Guiot J, Diniz F, Feddi N, Zheng Z, Bessais E, Drivaliari A 1999 Climate and biomes in the West Mediterranean area during the Pliocene. *Palaeogeography - Palaeoclimatology - Palaeoecology* **152**, 15-36.

Ferris C, Oliver RP, Davy AJ, Hewitt GM (1993) Native oak chloroplasts reveal an ancient divide across Europe. *Molecular Ecology* **2**, 337-344.

Fineschi S, Taurchini D, Grossoni P, Petit RJ, Vendramin GG (2002) Chloroplast DNA variation of white oaks in Italy. *Forest Ecology and Management* **156**, 103-114.

Gaston KJ (2000) Global patterns in biodiversity. *Nature* **504**, 220-227

Goværts R, Frodin DG (1998) *World Checklist and Bibliography of Fagales*. Royal Botanic Gardens, Kew.

Hewitt GM (1996) Some genetic consequences of ice ages and their role in divergence and speciation. *Biological Journal of the Linnaean Society* **58**, 247-276.

Hewitt GM (1999) Post-glacial re-colonization of European biota. *Biological Journal of the Linnean Society* **68**, 87-112.

Jalas J, Suominen J (eds.) (1988) *Atlas Florae Europaeae. Distribution of vascular plants in Europe II. Angiosperms (part) 3: Salicaceae to Balanophoraceae; 4: Polygonaceae; 5: Chenopodiaceae to Basellaceae*. Cambridge University Press, Cambridge.

Kaya Z, Raynal DJ (2001) Biodiversity and conservation of Turkish forests. *Biological Conservation* **97**, 131-141.

King RA, Ferris C (1998) Chloroplast DNA phylogeography of *Alnus glutinosa* (L.) Gaertn. *Molecular Ecology* **7**, 1151-1161.

Konarov VL (ed.) (1936) *Flora of the USSR*. Izdatel'stva Akademii Nauk SSSR, Moscow-Leningrad.

Lopes AP (2002) *Genetic Variation and Population Differentiation in* Quercus suber *L (Fagaceae)*. PhD Thesis, Cambridge University.

Manos PS, Doyle JJ, Noxon KC (1999) Phylogeny, biogeography and processes of molecular differentiation in *Quercus* subgenus *Quercus* (Fagaceae). *Molecular Phylogenetics and Evolution* **12**, 333-349.

Médail F, Quézel P (1997) Hot-spots analysis for conservation of plant biodiversity in the Mediterranean basin. *Annals of the Missouri Botanical Garden* **84**, 112-127.

Melika G, Csóka G, Pujade-Villar J (2000) Check-list of oak gall wasps of Hungary, with some taxonomic notes (Hymenoptera:Cynipidae, Cynipinae, Cynipini). *Annales historico-naturales musei nationalis hungarici* **92**, 265-296.

Moghaddam M, Ehdaie B, Waines JG (2000) Genetic diversity in populations of wild diploid wheat *Triticum urartu* Tum. ex. Gandil. revealed by isozyme markers. *Genetic Resources and Crop Evolution* **47**, 323-334.

Muir G, Fleming CC, Schlötterer C (2000) Species status of hybridising oaks. ***Nature* 405**, 1016.

Nieves-Aldrey JL (2001) ***Hymenoptera: Cynipidae.*** Fauna Ibérica, vol. 16 (ed. Ramos RA). CSIC, Madrid.

Olalde M, Herran A, Espinel S, Goicoechea PG (2002) White oaks phylogeography in the Iberian Peninsula. ***Forest Ecology and Management* 156**, 89-102.

Ovington JD (ed.) (1983) ***Temperate Broad-leaved Evergreen Forests.*** Ecosystems of the World, vol 10. Elsevier, Amsterdam.

Pamilo P, Savolainen O (1999) Post-glacial colonization, drift, local selection and conservation value of populations: a northern perspective. ***Hereditas* 130**, 229-238.

Paulo OS, Dias C, Bruford MW, Jordan WC, Nichols RA (2001) The persistence of Pliocene populations through the Pleistocene climatic cycles: evidence from the phylogeography of an Iberian lizard. ***Proceedings of the Royal Society of London, B* 268**, 1625-1630.

Petit RJ, Brewer S, Bordács S, Burg K, Cheddadi R, Coart E, Cottrell J, Csaikl UM, van Dam B, Deans JD, Espinel S, Fineschi S, Finkeldey R, Glaz I, Goicoechea PG, Jensen J S, König AO, Lowe AJ, Madsen SF, Mátyás G, Munro RC, Popescu F, Slade D, Tabbener H, de Vries SGM, Ziegenhagen B, de Beaulieu JL, Kremer A (2002a) Identification of refugia and post-glacial colonization routes of European white oaks based on chloroplast DNA and fossil pollen evidence. ***Forest Ecology and Management* 156**, 49-74.

Petit RJ, Csaikl UM, Bordács S, Burg K, Coart E, Cottrell J, van Dam B, Deans JD, Dumolin-Lapègue S, Fineschi S, Finkeldey R, Gillies A, Glaz I, Goicoechea PG, Jensen JS, König AO, Lowe AJ, Madsen SF, Mátyás G, Munro RC, Olalde M, Pemonge M H, Popescu F, Slade D, Tabbener H, Taurchini D, de Vries SGM, Ziegenhagen B, Kremer A (2002b) Chloroplast DNA variation in European white oaks. Phylogeography and patterns of diversity based on data from over 2600 populations. ***Forest Ecology and Management* 156**, 5-26.

Posada D, Crandall KA (1998) Modeltest: testing the model of DNA substitution. ***Bioinformatics* 14**, 817-818.

Reid WV (1998) Biodiversity hotspots. ***Trends in Ecology and Evolution* 13**, 275-280.

Röhrig E, Ulrich B (eds.) (1991) ***Temperate Deciduous Forests.*** Ecosystems of the World, vol. 7. Elsevier, Amsterdam.

Rokas A (2001) ***Tree Thinking. Molecules and Gallwasps: Analysing Evolutionary Patterns in European Gallwasps Using a Molecular Phylogenetic Approach.*** PhD Thesis, University of Edinburgh.

Rokas A, Atkinson RJ, Brown G, West SA, Stone GN (2001) Understanding patterns of genetic diversity in the oak gallwasp *Biorhiza pallida*: demographic history or a *Wolbachia* selective sweep? ***Heredity* 2001**, 294-304.

Rosenzweig ML (1995) ***Species Diversity in Space and Time.*** Cambridge University Press, Cambridge.

Stone GN, Sunnucks P (1993) Genetic consequences of an invasion through a patchy environment; the cynipid gallwasp *Andricus quercuscalicis* (Hymenoptera:Cynipidiae). ***Molecular Ecology* 2**, 251-268.

Stone G, Atkinson R, Rokas A, Csóka G, Nieves-Aldrey JL (2001) Differential success in northwards range expansion between ecotypes of the marble gallwasp *Andricus kollari*: a tale of two lifecycles. ***Molecular Ecology* 10**, 761-778.

Stone GN, Atkinson RJ, Brown G & Rokas A (2002a) The population genetic consequences of range expansion: a review of pattern and process, and the value of oak gallwasps as a model system. *Biodiversity Science* **10**, 80-97.

Stone GN, Atkinson R, Rokas A, Brown G (2002b) The population genetic consequences of range expansion. In: *Ecological Dynamics of Genes*. Special Symposium Volume, The British Ecological Society. Blackwell, Oxford.

Stone GN, Schönrogge K, Atkinson RJ, Bellido D, Pujade-Villar J (2002c) The population biology of oak gallwasps (Hymenoptera: Cynipidae). *Annual Review of Entomology* **47**, 633-668.

Suc JP (1984) Origin and evolution of the Mediterranean vegetation and climate in Europe. *Nature* **307**, 429-432.

Taberlet P (1998) Biodiversity at the intraspecific level: the comparative phylogeographic approach. *Journal of Biotechnology* **64**, 91-100

Taberlet P, Fumagalli L, Wust-Saucy A-G, Cosson J-F (1998) Comparative phylogeography and postglacial colonization routes in Europe. *Molecular Ecology* **7**, 453-464.

Thompson RS, Fleming RF (1996) Middle Pliocene vegetation: reconstructions, paleoclimatic inferences, and boundary conditions for climate modeling. *Marine Micropaleontology* **27**, 27-49.

Townsend CC, Guest ER (eds.) (1980) *Cornaceae to Rubiaceae*. Flora of Iraq. vol. 4, part 1. Ministry of Agriculture and Agrarian Research, Baghdad.

Turrill WB (1929) *The Plant Life of the Balkan Peninsula: A Phytogeographical Study*. The Clarendon Press, Oxford.

Tzedakis PC (1993) Long-term tree populations in Northwest Greece through multiple Quaternary climatic cycles. *Nature* **364**, 437-440.

Van Andel TH (2002) The climate and landscape of the middle part of the Weichselian glaciation in Europe: the stage 3 project. *Quaternary Research* **57**, 2-8.

Van Andel TH, Tzedakis PC (1996) Paleolithic landscapes of Europe and environs, 150,000-25,000 years ago: an overview. *Quaternary Science Review* **15**, 481-500.

Willis KJ, Whittaker RJ (2000) Paleoecology: the refugial debate. *Science* **287**, 1406-1407.

Yaltirik F (ed.) (1982) *Fagaceae*. Flora of Turkey and the eastern Aegean Islands. Edinburgh University Press, Edinburgh.

Zohary M (ed.) (1966) *Equisetaeae to Moringaceae*. Flora Palaestina. vol. 1. The Israel Academy of Sciences and Humanities, Jerusalem.

Part III

Emerging phylogeographical patterns in a southern European refugium: the Iberian Peninsula

Chapter 5

Refugia within refugia: patterns of phylogeographic concordance in the Iberian Peninsula

Africa Gómez and David H. Lunt
*E-mail: a.gomez@hull.ac.uk

Abstract
The Iberian Peninsula was one of the most important Pleistocene glacial refugia in Europe. A number of recent studies have documented the phylogeography of Iberian taxa and their relationship to more widely distributed species that expanded from this southern European refugium. We use a comparative approach to review the literature that challenges the paradigm of Iberia as a single refuge during Pleistocene glacial maxima and instead supports the occurrence of several Iberian refugia for a range of flora and fauna. Some patterns of phylogeographic concordance were found between the refugial areas identified by different case studies and these broadly overlapped with previously recognized areas of high endemism in the Iberian Peninsula. Such patterns help to illustrate the internal complexity of the Iberian Peninsula as a glacial refugium, and show that for many species, populations with a high degree of genetic structure have existed throughout the Pleistocene. Importantly, the occurrence of these 'refugia-within-refugia' may confound the interpretation of phylogeographic patterns of European species, and can misleadingly support the occurrence of northern refugia. We discuss these and other consequences, especially when a limited number of samples from the southern European refugia are used.

Keywords: comparative phylogeography, phylogeography, areas of endemism, glacial refugia, Pleistocene, Ice Ages

Introduction

The Iberian Peninsula was one of the most important Pleistocene glacial refugia in the European subcontinent (Hewitt 1999, 2001). This claim is well supported by several lines of evidence. The persistence of temperate species throughout the Ice Ages is provided by paleontological, palynological and paleolimnological data (Huntley & Birks 1983; Bennett *et al.* 1991). The high level of endemism in both Iberian plants and animals (Gómez-Campo *et al.* 1984; Doadrio 1988; Moreno Saiz *et al.* 1998; Ribera 2000; García-Barros *et al.* 2002) indirectly suggests *in situ* long-term survival, differentiation and

S. Weiss and N. Ferrand (eds.), Phylogeography of Southern European Refugia, 155–188.

speciation. Additionally, an increasing number of phylogeographic studies of European flora and fauna depict the Iberian Peninsula not only as a cradle for genetic differentiation, but also a species repository for the northern latitudes of Europe after the Pleistocene Ice Ages. Thus, since the last reviews on European phylogeography (Comes & Kadereit 1998; Taberlet *et al.* 1998; Hewitt 1999), many species have been added to the wide array of organisms known to have colonized western and northern Europe from the Iberian Peninsula after the Ice Ages, including mammals such as the roe deer (*Capreolus capreolus*, Vernesi *et al.* 2002), woodmouse (*Apodemus sylvaticus*, Michaux *et al.* 1998, 2003) and field voles (*Microtus agrestis*, Jaarola & Searle 2002); birds (chaffinch, *Fringilla coelebs*, Griswold & Baker 2002), reptiles (pond turtle, *Emys orbicularis*, Lenk *et al.* 1999) amphibians (natterjack toad, *Bufo calamita*, Beebee & Rowe 2000) and plants (ivy, *Hedera* ssp., Grivet & Petit 2002) to name a few.

In spite of its geographically isolated position on the westernmost point of Europe, several characteristics favored survival in the Iberian Peninsula throughout the Pleistocene. First, the Iberian Peninsula possesses high physiographic complexity, with several large mountain ranges primarily oriented east-west. This mountain range orientation offers the highest microclimatic scope, and allows survival of populations by altitudinal shifts, tracking suitable microclimates up or down mountains as the general climate worsens or ameliorates (Hewitt 1996). Second, due to its geographical position, the Iberian Peninsula is under the influence of both the North Atlantic and the Mediterranean, and enjoys a wide range of climates, including desert, Mediterranean, Alpine, and Atlantic. Interestingly, these very same characteristics, together with its large area (580 000 km^2), make it unlikely that Iberia offered a single homogeneous and continuous refugial area throughout the Pleistocene. Instead, the differential distribution and fragmented nature of suitable habitats favor the occurrence of multiple glacial refugia isolated from one another by the harsh climate of the high central Iberian plateau.

Phylogeographic concordance refers to the non-random patterns of similarity in the geographic distribution of evolutionary lineages of codistributed species or species complexes (Avise 2000). Phylogeographic concordance has parallels to historical biogeography since it can reveal the role of vicariance in structuring biotas (Avise 2000; Riddle *et al.* 2000). Comparative phylogeography can shed light on the role of geography in speciation, the associations between climate cycles and species distributions, and can help to identify biodiversity hotspots and inform conservation policies (Avise 2000; Hewitt 2000; Riddle *et al.* 2000). Insights provided by comparative phylogeography can additionally aid sampling design when planning future studies. Comparative phylogeographic analyses have yielded numerous novel insights into the development of regional historical patterns of genetic lineages (Bernatchez & Wilson 1998; Schneider *et al.* 1998; Walker & Avise 1998; Riddle *et al.* 2000; Arbogast & Kenagy 2001). In Europe, the location of the major refugia, the main postglacial colonization routes and the suture zones for terrestrial species have been well established and show concordant patterns (Hewitt 1996, 1999; Taberlet 1998). Given the above, and the recent surge of phylogeographic studies centered in the Iberian Peninsula, it is opportune to undertake a comparative phylogeographic analysis on this area in order to understand the number and location of glacial refugia present, point to areas where more work is needed, and contribute to clarify the broader picture of European phylogeography.

Here, using a comparative phylogeographic approach, we: i) review phylogeographic and biogeographic evidence indicating the occurrence of multiple glacial refugia in Iberia, ii) investigate the patterns of concordance found, and iii) discuss the consequences of these findings for European phylogeographic research and historical biogeography. We have excluded marine taxa, and the Balearic Islands, focusing on mainland species. We review both single species and species complexes (with largely parapatric distributions) for a geographic component to their genetic diversity. We note that both the literature in general and this review are biased towards species with low dispersal abilities or limited gene flow, which are most likely to reflect historical discontinuities. We primarily review and discuss case studies with extensive sampling of the Iberian Peninsula. We are also aware that expansions into Iberia from other regions may give the appearance of multiple refugia and concentrate on endemics and systems for which there is adequate sampling outside of Iberia. Since different organismal dispersal and habitat requirements can lead to distinct patterns of phylogeographic structuring, we discuss freshwater/amphibian, terrestrial and lacustrine organisms in separate sections. Throughout the review we use the existence of this Type I phylogeographic structure (Avise 2000) as evidence of multiple refugia in the Iberian Peninsula.

Freshwater / amphibian habitats

1. Fish

Freshwater fish tend to show particularly clear phylogeographic structure as they do not normally disperse between river basins, and thus the distribution of their lineages tends to reflect the history of river drainages instead of contemporary dispersal (see reviews in Bernatchez & Wilson 1998; Avise 2000). The 11 main river basins in the Iberian Peninsula formed from the Upper-Oligocene to the Pliocene, together with the rise of the main mountain ranges. The current network of river systems was already formed by the Quaternary, and the general absence of canals between rivers has maintained the natural distribution of native fish barring intentional human introductions (Doadrio

1988). The Pyrenees constitute a formidable barrier for freshwater fishes as no river drainage crosses them, and, covered by an extensive ice cap, they were an even stronger barrier during glaciations. Therefore, due to its isolation, the Iberian fish fauna is not as species rich as that of central Europe (Bănărescu 1991; Elvira 1995). Not surprisingly over 45% of Iberia's native fish species are endemic. This high level of endemism is shared with the other southern European peninsulas, particularly the Balkans (Bănărescu 1991). Thus, Iberia behaved more as a cradle for freshwater fish endemics than as a refugium for the rest of Europe. Here we will illustrate the phylogeography of Iberian fishes with two relatively well-studied Iberian species complexes.

The *Luciobarbus* group

The subgenus *Luciobarbus* colonized the Iberian Peninsula from North Africa and gave rise to a complex of six endemic species (Machordom & Doadrio 2001; Doadrio *et al.* 2002). The phylogeography of these barbels has been investigated using mtDNA sequence variation (Zardoya & Doadrio 1998; Callejas & Ochando 2000; Machordom & Doadrio 2001; Doadrio *et al.* 2002). Analyses revealed a clear-cut geographic distribution of the different species, with one mainly Mediterranean clade, including *B. graellsi*, *B. guiraonis* and *B. microcephalus*, one southern clade of *B. sclateri* and one mainly Atlantic with *B. bocagei* and *B. comizo* (Figure 1). In the first two clades, species are distributed according to the main river basins, although some cases of sympatry due to secondary contact are present in the Tajo and Guadiana basins, most likely due to episodes of headwater river capture. The average sequence divergence among these seven species varies from less than 0.5% to 6.5%, and hybridization events have been reported in areas of sympatry (Callejas & Ochando 2000). Application of molecular clocks calibrated with the opening of the Gibraltar straits suggests that the radiation of these Iberian barbels began in the mid Pliocene and continued throughout the Pleistocene, with distinct lineages existing in separate river basin refugia (Machordom & Doadrio 2001).

The *Leuciscus* group

Several Iberian endemic species of the genus *Leuciscus*, mostly with parapatric distribution ranges, have been described. *Leuciscus carolitertii* inhabits northern Atlantic rivers, excluding the Cantabrian basin. The sister species *L. pyrenaicus* inhabits the southern Atlantic and Mediterranean Iberian rivers. Both species co-occur in the Tajo River, probably as a consequence of a river capture. An mtDNA analysis of populations of these species revealed four main lineages (Brito *et al.* 1997; Zardoya & Doadrio 1998), two of them coincident with the already described species. The other two highly divergent lineages are restricted to small river drainages in the southernmost tip of Portugal and they were proposed to be new species, *L. aradensis* and *L. torgalensis* (Figure 1) (Coelho *et al.* 1998). These species, with sequence divergences between 5 and 11% were estimated to have radiated between 2 and 3 mya, probably following the partition of river drainages from the Pliocene to the early Pleistocene (Brito *et al.* 1997).

The Brown Trout

The brown trout, *Salmo trutta*, is possibly the best studied European freshwater fish from a genetic and phylogeographic perspective (Antunes *et al.* 2001). Due to its anadromous life history, at least for some populations, the phylogeographic history of brown trout does not compare well to other Iberian fish. Admixture between different lineages is common, due to multiple colonizations and secondary contact zones (Ball-Llosera *et al.* 2002), therefore there is not a clear concordance between the distribution of genetic lineages and current river drainages. Of the five main genetic lineages found in the global distribution range of this species (Bernatchez 2001), three (the Atlantic, the Mediterranean, and the Adriatic) are naturally present in the Iberian Peninsula. These lineages display a large amount of genetic diversity (Machordom *et al.* 2000; Suárez *et al.* 2001). In the Atlantic rivers, Iberian brown trout display clear genetic subdivisions and populations show evidence of genetic isolation for at least 200 000 years (Weiss *et al.* 2000), pointing to the occurrence of a complex of glacial refugia in the area. In the Mediterranean clade, four main subgroups, probably the result of allopatry during the Pleistocene have been identified, but secondary admixture and the effects of drift have contributed to a complex current distribution of lineages (Ball-Llosera *et al.* 2002). In summary, brown trout display a high level of complexity in the Iberian Peninsula, pointing to the presence of several Iberian refugia. However, and in a similar fashion to other freshwater fish, brown trout lineages did not contribute to the most recent waves of postglacial colonization in northern Europe (Weiss *et al.* 2000). Even for a species that presumably has the ability to colonize across saline habitats, the Iberian Peninsula is nonetheless significant as a complex of refugia.

Other fish

Other phylogeography studies in Iberian fish indicate that refugial populations and local endemic species remained in several of the main Iberian river basins throughout the Pleistocene. The final configuration of the current drainage system seems to have been the historical process that generated barriers of gene flow isolating populations and initiating the process of differentiation in *Chondrostoma lemmingii*, *C. lusitanicum*, and *Anaecypris hispanica* (Carmona *et al.* 2000; Alves *et al.* 2001; Mesquita *et al.* 2001) (see Table 1). Another Iberian endemic fish, *Aphanius iberus*, which inhabits salt marshes, also shows deep phylogeographic patterns coinciding with the Atlantic and Mediterranean drainages (Perdices *et al.* 2001). In summary, the main river catchments, including the Guadalquivir, Guadiana, Tajo, Ebro and Duero, and a number of minor ones (significantly the southern Portuguese Mira and Arade rivers) seem to have harbored distinct fish lineages in suitable habitats throughout the Pleistocene's multiple Ice Ages.

2. Amphibians

Of the 28 native Iberian amphibian species, eight are endemic (Vargas *et al.* 1998), but the number of described species and subspecies is growing (García-París & Martínez-Solano 2001). Due to their mobility over land, amphibians are less confined to river basins as fishes, but they tend to show strong genetic variation among populations, and their patterns of genetic diversity do still tend to reflect historical rather than contemporary processes (see review in Avise 2000). Several of the endemic species have restricted ranges (around the Pyrenees *Rana pyrenaica* and *Euproctus asper*, in western Iberia *Rana iberica*, and *Triturus boscai*) suggesting long-term population persistence in these areas.

The *Discoglossus* toads

García-París and Jockusch (1999) investigated the phylogeography of the two Iberian *Discoglossus* endemics: *D. jeanneae* and *D. galganoi* (Figure 1), with a sampling design that aimed to pinpoint the contact zone between these morphologically very similar toads. Cytochrome b sequence divergence between these two species was 8.6%, and suggested that the two lineages started to diverge before 5 mya. The western species *D. galganoi* showed higher and geographically structured intraspecific diversity than *D. jeanneae*, with haplotypes south and north the Duero River estimated to have diverged over 1 mya (Figure 1) (García-París & Jockusch 1999; Martínez-Solano, pers. comm.). As for *D. jeanneae*, two very closely related haplotypes were found, which might suggest a recent episode of range expansion, and then a contraction, which left the patchy distribution we find today. The location of the refugium from which *D. jeanneae* expanded has been suggested to be in or nearby the Betic ranges (Barbadillo 1987).

The midwife toads

The midwife toads (genus *Alytes*) are thought to have evolved in the Iberian Peninsula in the late Miocene. Three mainland Iberian species, plus an endemic of the Balearic Islands are currently recognized. Two of the Iberian species are endemic, the other one, *A. obstetricans* is also present in western

Europe and a small region in North Africa. The phylogenetic and biogeographic relationships of these taxa were examined by Arntzen & García-París (1995) using morphological and allozyme variation. *Alytes cisternasii* is a quite divergent species that differentiated in the western half of the Iberian Peninsula. *Alytes dickhilleni* is present in the Betic mountains and it is thought to have evolved in this area (Arntzen & García-París 1995) coinciding with the opening of the Betic Straits and their isolation in the Betic-Riffean massif (5-6 mya). *Alytes obstetricans* presents a high level of geographic variation, which has led to the description of four subspecies with parapatric distributions based on allozyme and morphological differences (Arntzen & García-París 1995; García-París & Martínez-Solano 2001). More recently phylogeographic analyses using mtDNA sequence variation have established geographic differentiation in Portugal indicating the occurrence of multiple Pleistocene refugia (Fonseca *et al.* 2003). The extent to which these subspecies hybridize and the distribution limits need to be investigated further, but sequence divergence between them suggests that the radiation in *A. obstetricans* happened around 3.5 mya (Arntzen & García-París 1995; Fonseca *et al.* 2003). The location of putative glacial refugia is not yet known, although it seems reasonable to assume that each subspecies evolved in isolated Iberian refugia during glacial periods, and that range expansion happened during the Holocene, resulting in the contact zones observed today.

The golden-striped salamander

The golden-striped salamander, *Chioglossa lusitanica*, inhabits streams in humid forested areas in northwestern Iberia. A detailed allozyme and mtDNA variation study revealed a high degree of genetic subdivision (Figure 1), with populations south of the Mondego River being quite distinct from the northern populations (Alexandrino *et al.* 2000). The phylogeography of this salamander was further investigated over its entire geographical range using cytochrome *b* sequence variation (Alexandrino *et al.* 2002) and the pattern found (Figure 1) suggested the occurrence of a minimum of two refugia. The authors suggest that the level of divergence between the two main lineages (1.5%) reflects divergence from the late Pliocene to early Pleistocene (3 to 1.5 mya).

The fire salamander

Nine out of the sixteen described subspecies of the highly polytypic European fire salamander *Salamandra salamandra* occur in the Iberian Peninsula (Alcobendas *et al.* 1996; García-París *et al.* 2003). These morphologically recognized subspecies can be assigned to four mtDNA lineages, each of them apparently of different age (Dopazo *et al.* 1998; Steinfartz *et al.* 2000; García-París *et al.* 2003). First, a quite distinct and basal clade encompasses a single subspecies, *S. s. longirostris*, occurring in the western Betic ranges of southern

Spain. This clade, with 6.3% sequence divergence from the rest, evolved in isolation 2-4 mya, with a putative refugium existing throughout most of the Pleistocene around the Betic ranges (Steinfartz *et al.* 2000). A second, quite genetically complex clade, corresponds to *S. s. bernardezi*, which occurs in Picos de Europa, and is characterized by its viviparous reproduction (Dopazo *et al.* 1998). The situation of this clade is peculiar as its closest relative is a ovoviviparous subspecies found in the southernmost tip of Italy and the large genetic divergence between the Iberian and Italian subclades suggests that the ancestor of this clade was much more widespread in a past interglacial, and that both subclades have survived at or close to their current distribution ranges for at least several glacial cycles (Steinfartz *et al.* 2000). Therefore, viviparity is likely to have arisen in the ancestor of the Iberian subclade, possibly in a glacial cycle when the populations of the clade retreated to the Picos de Europa ranges, where viviparity was favored by the lack of suitable aquatic habitat for larvae. A third, geographically subdivided mtDNA lineage coincides with three recognized subspecies, one in the Algarve area (*crespoi*), another in the central mountain ranges (*almanzoris*) and the third north of the Guadalquivir River (*morenica*), thus further indicating more recent refugia for this salamander (García-París *et al.* 1998, 2003). Finally, the fourth mtDNA clade, comprising three subspecies, *fastuosa*, *gallaica* and *terrestris* is widespread in the Iberian Peninsula and also occurs in most of Europe. Despite its relatively low diversity, suggesting relatively recent colonization of the area, this clade exhibits substantial geographic structure with evidence of past fragmentation in the Iberian Peninsula, and the level of diversity is consistent with occupation of the Iberian Peninsula for at least one glacial cycle (García-París *et al.* 2003). The joint study of life history, allozymes and mtDNA variation revealed the complex population interactions between this clade, comprising both ovoviviparous and viviparous populations, and the northern viviparous populations of Picos de Europa (García-París *et al.* 2003). In summary, we can conclude that *S. salamandra* survived in the Iberian Peninsula in several glacial refugia, throughout the Pleistocene in the Betic ranges, through several glacial cycles in Picos de Europa and at least throughout the last glacial cycle in several other refugia.

Other amphibians

The marbled newt *Triturus marmoratus*, distributed in Iberia and western France, shows evidence of two Iberian refugia: *T. m. pygmaeus* is restricted to the southwestern part of the Iberian Peninsula and *T. m. marmoratus* to the northern half and most of France (García-París *et al.* 1993). Both subspecies are over 4% divergent in their mtDNA, which suggests independent evolution since the Pliocene (Wallis & Arntzen 1989). In addition, *Triturus boscai*, an endemic species inhabiting the western half of the Iberian Peninsula comprises two population groups showing chromosomal incompatibilities and hybrid sterility (Herrero 1991). The distribution of the subspecies of *Triturus helveticus* is also suggestive of past population fragmentation, although a phylogeographic analysis is needed.

In conclusion, most of the amphibian taxa investigated show evidence of multiple Iberian mtDNA lineages that are geographically structured indicating survival in different Pleistocene glacial refugia. Several of these species show compelling phylogeographic concordance with a putative refugium located in or near the southern Betic ranges (*Salamandra salamandra longirostris, Discoglossus jeanneae, Alytes dickhilleni*). The Atlantic side of the Peninsula also seems to have served as refugium or refugia for multiple species, and may often have limited the distribution of other taxa possibly due to its high humidity (*Chioglossa lusitanica, Alytes cisternasii, A. obstetricans boscai, Triturus boscai, Discoglossus galganoi*). Finally, several endemic amphibian species have restricted ranges suggestive of relict habitat in the present conditions following more extensive distributions in the past. Interestingly, such relictual ranges often coincide with areas known for their high level of endemism in plants.

3. Terrestrial habitats

3.1 Reptiles

Reptiles share the low mobility of amphibians but are often associated with drier and warmer climates. Glacial advances are very likely to have limited their distribution to the southernmost reaches of the European continent, and there, fragmentation of populations in suitable habitats should have led to geographic structuring. Of the 38 reptile species in the Iberian Peninsula eight are endemic (21%), and, as in amphibians, the number of described species is growing (Andreu *et al.* 1998). Several endemic species have restricted ranges indicating geographic structuring: *Podarcis bocagei* is distributed in northwestern Iberia, *Lacerta schreiberi* and *L. monticola* are restricted to northwestern Iberia and Central System mountains, *Lacerta bonnali* occurs in the Pyrenees, and *Algyroides marchi* occurs in the eastern end of the Betic ranges. A subspecies of the widespread *Lacerta lepida* (also present in France), *L. l. nevadensis* occurs in the Betic ranges of southern Spain. Three published studies will be presented here as case studies.

The Iberian wall lizards

The small Iberian wall lizard, *Podarcis hispanica*, has been shown to be a species complex containing several lineages with a minimum of six species, several still undescribed (Harris & Sa-Sousa 2001, 2002) (see Figure 1). The range of sequence divergence between the lineages (9.5 to 15.2%) indicates a pre-Pleistocene divergence. Most of the lineages are parapatric, although the geographic distribution and contact zones between them are only well known for western Iberia (Harris & Sa-Sousa 2001). In the cases investigated, interbreeding does not seem to happen, despite high morphological similarity. The genetic and geographic pattern found indicates survival of isolated populations in more than one Iberian refugium.

The Schreiber's lizard

The lizard *Lacerta schreiberi* is an Iberian endemic with a preference for humid temperate forests and mountain river valleys. The distribution range and the deep phylogeographic structure revealed by cytochrome *b* sequence variation is shown in Figure 1 (Paulo *et al.* 2001). Despite a remarkable morphological uniformity, the two major clades found, inland and coastal, display a sequence divergence of 4.7%, suggesting divergence before 2 mya (Paulo *et al.* 2001, 2002). Each of these clades was subdivided in a northern and a southern lineage, with divergence times of 1 and 0.6 mya, and none of the clades overlap geographically. This pattern suggests survival in separate glacial refugia, with a northern coastal, southern coastal, Central System and Montes de Toledo refugia. The phylogeography and diversification time frame of this lizard, particularly the coastal lineages, is remarkably concordant with the salamander *Chioglossa lusitanica* (Paulo *et al.* 2001; see also Alexandrino *et al.* this volume).

The viviparous lizard

The western oviparous form of the viviparous lacertid lizard *Zootoca vivipara*, displays geographic variation in sex-linked allozyme alleles and mtDNA (cytochrome *b*), with two maternal lineages, one distributed in the central and eastern Pyrenees and Aquitaine, and the other in Picos de Europa and southern slopes of the western Pyrenees (Figure 1) (Guillaume *et al.* 2000; Surget-Groba *et al.* 2001). Both lineages meet in a contact zone in the upper Ossau valley and Aquitaine (Guillaume *et al.* 2000; Surget-Groba *et al.* 2001). These authors explained their results as suggesting geographic isolation of the viviparous populations in two refugia, one near Picos de Europa and the other near the Pyrenees during glacial periods, and subsequent expansion with recent secondary contact during the Holocene.

The viperine snake

Some recent studies support the presence of several Iberian refugia in other Iberian reptiles. The Viperine snake, *Natrix maura*, which colonized western Europe from Iberia, has been shown to comprise several mtDNA European lineages with a 1.3% sequence divergence (Guicking *et al.* 2002). All the western European lineages are present in Iberia, which presents the highest genetic diversity, as well as the most ancestral haplotypes. The current preliminary data does not reveal the location of the putative Iberian refugia, but the data is suggestive that France and western Italy were colonized via two routes, one through the east and one through the west of the Pyrenees.

3.2 Mammals

There are 118 described mammal species in the Iberian Peninsula – and the highest level of endemicity for mammals in Europe (Baquero & Tellería 2001). However, the number of studies centered on Iberian mammals is quite small, most probably due to the difficulties involved in obtaining samples.

The European rabbit

The rabbit, *Oryctolagus cuniculus*, is a native Iberian species with a natural distribution that in the Middle Ages reached the Loire River in France (Queney *et al.* 2001; see also Ferrand & Branco, this volume). Mitochondrial cytochrome *b* RFLP and sequence variation of Iberian rabbits revealed the occurrence of two clades with a strong geographic partitioning. One is located in southwestern Iberia and the other in the northeast, with an area of secondary contact crossing the Peninsula from the southeast to the northwest (Branco *et al.* 2000, 2002). The haplotype distribution is concordant with survival of rabbit populations in two glacial refugia and subsequent postglacial range expansion as climate ameliorated. These refugia are hypothesized to be in northeastern Spain, probably in the Mediterranean coast, and in the southernmost part of Spain, close to the Gibraltar straits. The two lineages presented a sequence divergence estimated to date at approximately 2 mya, or the Pliocene-Pleistocene boundary (Biju-Duval *et al.* 1991).

The Iberian hares

Three species of hare inhabit the Iberian Peninsula, the European hare, *Lepus europaeus*, and two endemic species, *L. castroviejoi* and *L. granatensis*. The restricted endemic *L. castroviejoi* inhabits the western side of Picos de Europa, *L. granatensis* inhabits most of the Iberian Peninsula south of the Ebro River, and *L. europaeus* inhabits the area north of the Ebro River and the eastern strip of the Cantabrian area. An RFLP study on mtDNA variation was undertaken using Iberian samples from these species (Pérez-Suárez *et al.* 1994). The three species were quite distinct in their mtDNA, and did not share any of the 11 haplotypes found, although only one or two populations were sampled for each species. The parapatric ranges and the level of sequence divergence found between them (8-9%), suggests the isolation of these species in at least two Iberian refugia, one of them in or near Picos de Europa, and their independent evolution for 4.4 to 6 myr (Pérez-Suárez *et al.* 1994).

Other mammals

Several other species of mammals have been focus of European phylogeographic analyses (Taberlet *et al.* 1998; Seddon *et al.* 2001; Vernesi *et al.* 2002; Michaux *et al.* 2003). However, the number of samples remains insufficient to draw any clear-cut patterns within the Iberian Peninsula. A recent exception is constituted by the field vole (Jaarola & Tegelstrom 1995; Jaarola & Searle 2002). The southern mtDNA lineage of this rodent was postulated to have recolonized southern France, Switzerland, Slovenia and Hungary from an Iberian refugium (Jaarola & Searle 2002). Detailed sampling in the Iberian Peninsula revealed two lineages, one of them restricted to Serra da Estrela in Portugal (Jaarola & Searle, pers. com.). The authors hypothesized that the field vole survived at least through the last glaciation in two Iberian refugia.

3.3 Invertebrates

Although there have been a number of phylogeographic studies of European invertebrates – biogeographical data is particularly good for insects – relatively few studies have included detailed sampling in Iberia. Two cases are presented below, and a third, the maritime pine scale insect is included later with its host.

The leaf beetles of the *Timarcha goettingensis* complex

The beetles of the genus *Timarcha* are apterous insects favoring mountain regions distributed in central Europe and the northern half of the Iberian Peninsula. Nested clade analysis on mtDNA sequence variation on ten Iberian endemic species from the complex revealed the importance of population range expansions, but also of past population fragmentation in a number of Iberian refugia (Gómez-Zurita *et al.* 2000).

The meadow grasshopper

The meadow grasshopper, *Chorthippus parallelus*, has become a European phylogeography icon since the early 1990s (Cooper & Hewitt 1993; Cooper *et al.* 1995; Lunt *et al.* 1998). The species is subdivided, with different lineages in the three southern peninsulas. Expansion from a Balkan refugium has recolonized most of Europe and a hybrid zone occurs in the Pyrenees between the Balkan and Iberian Peninsula subspecies. This hybrid zone has been documented using sequence data, chromosomes, morphology, allozymes and behavior (see references in Hewitt 1993; Hewitt 1996, 1999, 2001). Sequence variation in an anonymous nuclear DNA sequence in European populations supported the separate origin for populations now found in France and Spain (Cooper & Hewitt 1993; Cooper *et al.* 1995). Additionally, the extensive geographic variation of the Spanish populations suggested the occurrence of at least two putative glacial refugia, one in the west, south of Picos de Europa, and the other possibly in the southeast (Hewitt 1996).

3.4 Plants

A number of European phylogeographic studies including the Iberian Peninsula present some evidence of Iberian refugia for plant taxa including ivy, white oaks, black alder, silver fir, *Aconitum lycoctonum*, and *Arabidopsis thaliana* (Konnert & Bergmann 1995; Ferris *et al.* 1998; King & Ferris 1998; Utelli *et al.* 1999; Muir *et al.* 2000; Sharbel *et al.* 2000; Grivet & Petit 2002). However, most of them include few Iberian localities, and therefore yield little information on the location and number of potential refugia within Iberia. The number of studies with extensive sampling in Iberia, has, however grown dramatically in recent years and here we review several of them.

The Scots Pine

The Scots pine, *Pinus sylvestris*, inhabits mountains in the Iberian Peninsula, usually in the altitudinal range of 1000 to 2000 m. It requires humid conditions, and it cannot withstand droughts. Several subspecies or races have been recognized in its natural distribution, including two in the Pyrenees (*P. s. catalaunica* and *P. s. pyrenaica*), another in the Iberian System, Maestrazgo, Central System and northern Portugal (*P. s. iberica*), and a relict subspecies in the south (Sierra de Baza, Sierra Nevada, *P. s. nevadensis*). Mitochondrial DNA and allozyme variation suggests that the Scots pine survived in the Iberian Peninsula during the Pleistocene glaciations, but most likely did not contribute to the postglacial colonization of northern Europe (Sinclair *et al.* 1999; Soranzo *et al.* 2000). Evidence of strong substructure in the Iberian Peninsula exists from mtDNA variation that coincides remarkably well with the recognized subspecies (Sinclair *et al.* 1999; Soranzo *et al.* 2000). This strongly suggests that the Scots pine survived in several refugia in Iberia: the Betic ranges, the Central System and the Pyrenees.

The maritime pine and its scale insect

The maritime pine, *Pinus pinaster*, forms scattered populations in the western Mediterranean basin. Allozyme analysis on twelve natural Iberian populations (Salvador *et al.* 2000) suggests the occurrence of an eastern refugium and a southern refugium possibly near or in the Betic ranges (see Martínez *et al.* this volume). Studies using cpDNA microsatellite markers (paternally transmitted in pines) failed to resolve the historical patterns of maritime pine in the Iberian Peninsula (Vendramin *et al.* 1998).

The bast scale insect, *Matsucoccus feytaudi*, is a specific pest of maritime pine, so its phylogeography may also inform us about the history of the maritime pine. A phylogeographic study of this insect in its natural range revealed three mtDNA lineages termed: Western, found throughout most of the species distribution range; Andalusian in the Betic ranges, and Moroccan (Burban *et al.* 1999). The haplotype diversity of the different populations suggests colonization of western Europe from a refugium in the west of Iberia, possibly in Portugal. This phylogeographic structure is remarkably similar to the phylogeography of its host.

The white oaks, *Quercus*

The deciduous oaks, *Quercus*, are widespread European forest trees, some of them of considerable economic importance. They are hermaphroditic and have heavy seeds (acorns), which are often edible, and disperse via birds. They produce abundant pollen that is dispersed via wind. Eight species are present in Europe. As oaks hybridize frequently, their maternally transmitted cpDNA haplotypes are often shared by several sympatric species (Dumolin-Lapègue *et al.* 1999). Although abundant information on the European phylogeography and population structure of the pedunculate oak, *Q. robur* and the sessile oak, *Q. petraea*, was available (Dumolin-Lapègue *et al.* 1997, 1999; Ferris *et al.* 1998) the sampling of the Iberian Peninsula remained quite poor, and several endemic species had not been sampled. In the context of an ambitious, Europe-wide project, ample information of cpDNA variation has recently become available for the six species of Iberian white oaks (Olalde *et al.* 2002). The data suggests Iberian refugia for four species, *Q. robur*, *Q. faginea*, *Q. canariensis* and *Q. pyrenaica*, but the data is not conclusive for the other two species (Olalde *et al.* 2002; Petit *et al.* 2002). In all, four mitochondrial lineages were detected, although two of them were represented by a single haplotype. The four lineages had a mostly parapatric distribution, one of them present in the northwestern half of the peninsula, the second one southeastern, and the other two near the Pyrenees (Olalde *et al.* 2002). At least two refugia were detected, one southwestern, the other one in the Betic mountains area, and a third one on the Mediterranean coast near the Ebro basin (Petit *et al.* 2002). Some other smaller refugia in the north of the peninsula were also proposed (Olalde *et al.* 2002).

Other plants

Analysis of cpDNA, allozyme and RAPD variation in the Iberian *Senecio gallicus* (Comes & Abbott 1998) revealed a complex phylogeographic pattern with coastal and inland groups. The coastal group with a high level of geographic variability, seems to be older, with putative glacial refugia in the southernmost tip of Iberia and SW Portugal, and has given rise to the inland group.

The hoary plantain, a common perennial plant of grasslands was investigated using RFLP on cpDNA (Van Dijk & Bakx-Schotman 1997). This plant is present in northern Spain, the Pyrenees and scattered localities in mountain areas of southern Spain. Of the nine haplotypes found, three are present in Iberia. A contact zone between two of these lineages (one diploid and the other tetraploid) occurs in the Pyrenees. The data suggests two refugia in Iberia.

The variation in chloroplast DNA in heather, *Calluna vulgaris*, is congruent with pollen records that suggest its survival in refugia in northern Spain (Picos de Europa) and southern France. In addition, a second Iberian refugium in the Pyrenees was suggested by the genetic data (Rendell & Ennos 2002).

Support for the occurrence of two Iberian refugia is provided by a genetic analysis of Holm oak, *Quercus ilex* (Lumaret *et al.* 2002). The distribution of the two cpDNA oak lineages in Iberia roughly coincides with the two recognized morphs, one distributed along the eastern Mediterranean coast and the second in the rest of Iberia suggesting the occurrence of two distinct southern refugia for this species.

4. Lacustrine fauna

Rotifers

A recent phylogeographic appraisal in the Iberian Peninsula focused on the rotifer *Brachionus plicatilis*, a species complex distributed globally in salt lakes and coastal lagoons. Six of the species in the complex are present in Spain (Ortells *et al.* 2000; Gómez *et al.* 2002), the most common of them is *B. plicatilis s.s.* (Ciros-Pérez *et al.* 2001). Rotifers reproduce by parthenogenesis, but when the environment deteriorates they produce sexual resting eggs, which can be sampled from lake sediments. Using such samples of *B. plicatilis*, mtDNA sequence variation revealed 21 haplotypes. The haplotypes clustered in two groups with 2.8% divergence between them and a strong northern and southern geographic component overlapping in just two of the 18 lakes (Gómez *et al.* 2000) (Figure 1). The high genetic diversity and geographic structuring of the northern clade indicates that it had survived in the Iberian Peninsula for several glacial cycles. However, the southern lineage shows signs of recent expansion and although the presence of a second Iberian refuge is likely, colonization from northern Africa cannot be ruled out. The location of the hypothesized northern refugium would be in the Ebro river basin or on the northeastern Iberian coast.

A recent study confirmed the pattern of southern and northern refugia in another widespread species from the *Brachionus plicatilis* complex, *B.* 'Manjavacas' (Gómez *et al.*, unpublished data). In this species, sequence variation of the mitochondrial gene COI revealed seven haplotypes in two geographically segregated lineages (Figure 1).

Discussion

Refugia within refugia

We have shown that many species and species complexes show strong genetic subdivisions in the Iberian Peninsula indicative of past population isolation. These taxa inhabit a range of habitats from rivers, humid forests and Mediterranean evergreen forests, to salt lakes and mountains. The current distribution of these lineages is largely parapatric and the genetic divergence between them is consistent with extended periods of isolation. These observations can be most easily interpreted by survival throughout the Pleistocene Ice Ages in separate glacial refugia (Table 1). To our knowledge, the earliest discussion on the occurrence of distinct isolated populations (refugia) within the southern Mediterranean peninsulas at the height of glacial advances was by Hewitt in his 1996 review (p. 265), as suggested by Cooper & Hewitt's (1993) results on *Chorthippus parallelus* and palynological data. However, in general, both palynological data and paleoclimatic reconstructions supported the paradigm of single refugia in the southern extremes of the Mediterranean peninsulas (see Olalde *et al.* 2002 for an excellent discussion), and evidence for the actual widespread occurrence of multiple refugia has only begun to accumulate and be acknowledged more recently, mostly based on phylogeographic information. The realization that many species display a strong population substructure within the Iberian glacial refugium, or were actually composed of completely isolated populations in distinct Iberian refugia has important consequences for the interpretation of European phylogeography and refugial diversity. The consequences of this 'refugia-within-refugia' scenario for the interpretation of European phylogeographic patterns need to be explored further. In our literature review we failed to find a single case in which a *detailed* sampling program recovered evidence for a single refugium in the Iberian Peninsula. This supports the idea that the presence of separate refugia is a general phenomenon, and reflects a common vicariant history of Iberian flora and fauna. We have deliberately avoided classifying the taxonomic level (e.g. populations, subspecies, species) at which to document phylogeographic structure.

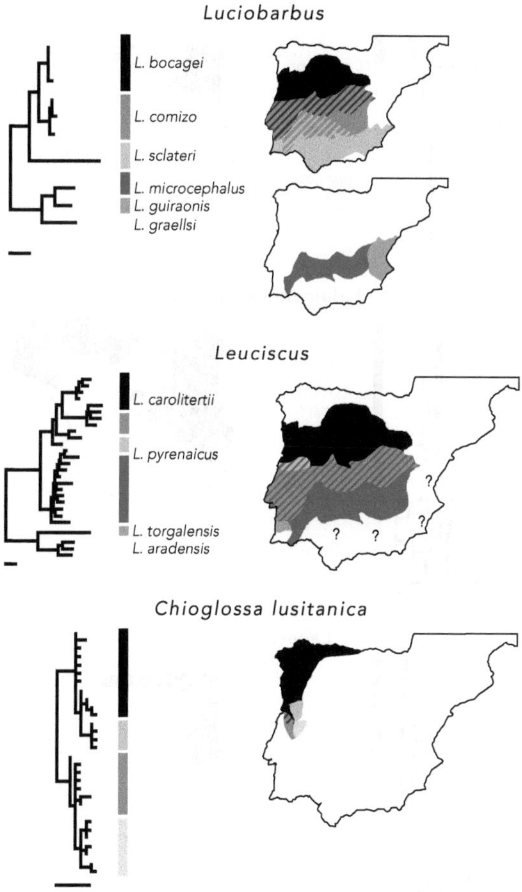

***Figure 1**. Phylogeographic structure of some of the case studies discussed in the text. See main text for references. The scale bar under each tree represents 1% sequence divergence.*

Continued on next page

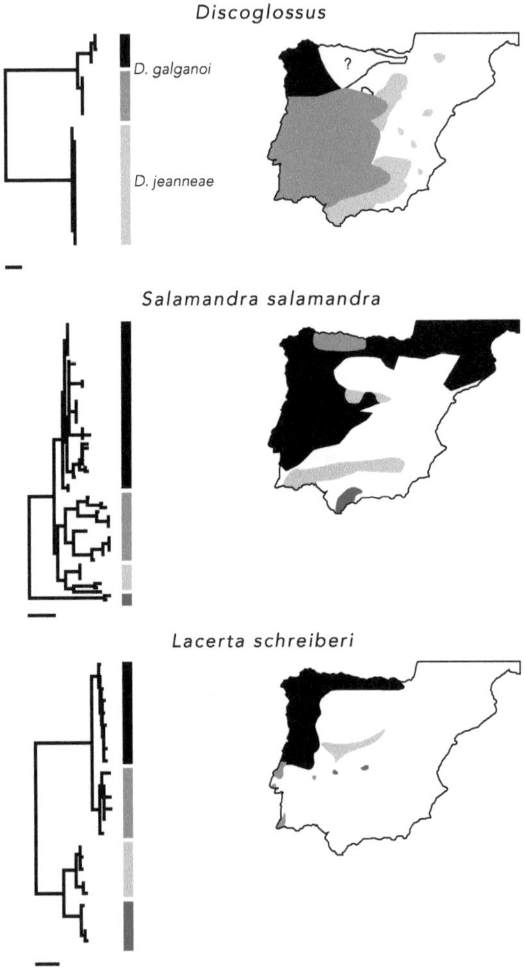

Figure 1. Continued.

Continued on next page

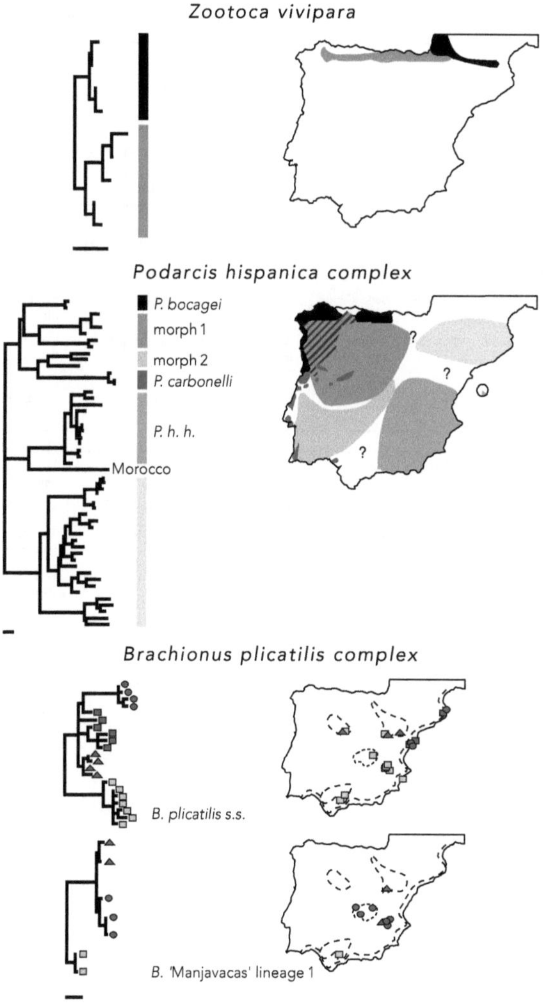

***Figure 1**. Continued.*

***Table 1**. Species and species groups showing some evidence for two or more Iberian refugia. The putative number and location of refugia, and the methods used are indicated.*

Taxon	Putative number and location of Iberian refugia	Remarks	Methods	Reference
Plants				
Calluna vulgaris	2 Northern, Pyrenees		RFLP cpDNA	Rendell & Ennos 2002
Quercus ilex	2 Eastern, southern		RFLP cpDNA	Lumaret *et al.* 2002
Hedera complex	3?*H. hibernica*, *H. helix*, *H. maderensis*	5 Iberian samples	RFLP cpDNA, ITS RFLP, microsats.	Grivet & Petit 2002
Senecio gallicus	SW Portugal, S Spain		RFLP cpDNA, Allozymes, RAPD	Comes & Abbot 1998, 2000
Pinus pinaster	2, East, Betic		Allozymes	Salvador *et al.* 2000
Pinus sylvestris	3, Pyrenees, Betic, Central		RFLP mtDNA	Soranzo *et al.* 2000; Sinclair *et al.* 1999
Plantago media	2?, C. Spain, Pyrenees	Reduced sampling in scattered southern populations	RFLP cpDNA	Van Dijk & Bakx-Schotman 1997
Quercus (white oaks)	2-4	6 species	RFLP cpDNA	Olalde *et al.* 2002; Petit *et al.* 2002
Mammals				
Microtus agrestis	2 Serra da Estrela and other		mtDNA	Jaarola *et al.* 2002, unpublished
Apodemus sylvaticus	2?	Scattered sampling	mtDNA	Michaux *et al.* 2003
Sorex coronatus, *S. granarius*	Western, northern	Few samples from 2 populations.	mtDNA	Taberlet *et al.* 1994; Fumagalli *et al.* 1996
Lepus sp.	2-3, S Ebro R., Picos de Europa	1 or 2 populations sampled per species.	RFLP mtDNA	Pérez-Suárez *et al.* 1994
Oryctolagus cuniculus	2, SW and NE		RFLP mtDNA	Branco *et al.* 2000, 2002
Reptiles and amphibians				
Lacerta schreiberi	4		mtDNA	Paulo *et al.* 2001, 2002
Podarcis hispanica	>2		mtDNA	Harris & Sa-Sousa 2000, 2002
Zootoca vivipara	2, Picos de Europa, Pyrenees		Allozymes, mtDNA	Guillaume *et al.* 2000; Surget-Groba *et al.* 2001
Triturus marmoratus/pygmaeus	2, Betic, North	Only 5 indivs. from 2 pops.	RFLP mtDNA	Wallis & Arntzen 1989
Salamandra salamandra	2, N and S		mtDNA	Steinfartz *et al.* 2000; Garcia-Paris *et al.* 2003
Discoglossus	2		mtDNA	García-París & Jockusch 1999
Alytes	>2		Allozymes, mtDNA	Arntzen & García-París 1995; Fonseca *et al.* 2003
Chioglossa lusitanica	2, NW, N Portugal		Allozymes, mtDNA	Alexandrino *et al.* 2000, 2002

Table 1. *Continued.*

Taxon	Putative number and location of Iberian refugia	Remarks	Methods	Reference
Fish				
Barbus (Luciobarbus)	6, most major river systems		mtDNA	Zardoya & Doadrio 1998; Machordom & Doadrio 2001; Callejas & Ochando 2000;
Leuciscus	4, North, South, Arade, Mira		mtDNA	Brito *et al.* 1997; Zardoya & Doadrio 1998
Anaecypris hispanica	4, Guadiana		mtDNA	Alves *et al.* 2001
Chondrostoma lemingii	4, Guadiana, Guadalquivir, Tajo, Duero		Allozymes, mtDNA	Carmona *et al.* 2000
Chondrostoma lusitanicum	>3, Mira-Arade, Sado, Tajo-Samarra		mtDNA RFLP	Mesquita *et al.* 2001
Aphanius iberus	2, Mediterranean, Atlantic		Allozymes, mtDNA	Perdices *et al.* 2001
Salmo trutta	>= 2, Duero, Tajo-Guadiana		mtDNA	Machordom *et al.* 2000; Suárez *et al.* 2001
Invertebrates				
Timarcha goettingensis	Several N		mtDNA	Gómez-Zurita *et al.* 2000
Chorthippus parallelus erythropus	2?	Only 4 pops. sampled	nDNA,	Cooper & Hewitt 1993; Cooper *et al.* 1995
Brachionus plicatilis	1-2, SW and NE		mtDNA	Gómez *et al.* 2000
Brachionus 'Manjavacas'	1-2	Limited sampling	mtDNA	Gómez *et al.* 2002
Matsucoccus feytaudi	2, W and Betic		MtDNA?	Burban *et al.* 1999

It is well established that genetic diversity has a hierarchical nature, and phylogeographic lineages are both part of higher level phylogenies as well as finer-scaled population structure (Avise 2000). In addition, taxonomic level is not standardized across taxa. Therefore we do not feel it especially useful (even if it were possible) to try to standardize a molecular clock and to determine which separations have their origins within the Pleistocene. Instead we view the lack of mixing of phylogeographic lineages as evidence for a physically separate organismal history throughout the Pleistocene even if the last common ancestor of the molecules studied predated this boundary. It is this lack of gene flow between phylogeographic lineages represented by geographically structured clades that points to distinct Pleistocene 'refugia-within-refugia.'

One implication of this refugia-within-refugia scenario is that the high genetic diversity of the south, that is, the contrast 'Northern-Purity, Southern-Richness' (reviewed in Hewitt 2000) will most likely be due to two levels of variation. First, the higher demographic stability of populations in the southern refugia and the loss of diversity associated with the colonization process in the north can lead to a higher intrapopulation polymorphism in the south (Hewitt 1996). Second, this diversity can be highly structured geographically due to allopatric differentiation between populations and therefore contribute to the higher diversity of southern areas (Guillaume *et al.* 2000). Since the rapid northward expansion with climatic amelioration will approximate serial bottlenecks, with the leading edge contributing disproportionately to the genetic composition of the northern populations (Ibrahim *et al.* 1996), not all of the southern diversity will be represented. This is especially significant if the diversity is structured into divergent allopatric lineages, as we suggest in this review, since loss of a lineage during the expansion will lose more biological information than a comparable level of unstructured diversity. Moreover, since different refugia will have different access to trans-Pyrenean dispersal routes, one lineage may effectively exclude a second from later northward expansion and reinforce the disparity between north and south. It may not be surprising therefore that there has not yet been a clear demonstration of separate Iberian lineages recolonizing Europe by distinct routes (east and west of the Pyrenees), although for the viperine snake *Natrix maura* there is some evidence that separate Iberian lineages might have dispersed around each edge of the Pyrenees (Guicking *et al.* 2002). Much more rigorous sampling designs are required to assess the potential generality of this double-colonization pattern for other taxa.

Consequences of pre-Pleistocene differentiation

The consequences of the refugia-within-refugia scenario are made still more complex by the cyclical nature of the Pleistocene Ice Ages. It is quite possible that not all the refugia were equally suitable during each glaciation. In addition, and as suggested by the data reviewed, all the refugia were not suitable for all species. Although patterns of pre-Pleistocene differentiation must be taken into account to interpret the likely original causes of population differentiation, the fact that multiple populations that diverged prior to the Pleistocene are found today in the area suggests persistent availability of suitable habitat in or near the putative refugial areas through the Pleistocene.

Consequences of inadequate sampling of southern refugia

Special consideration of the southern Mediterranean refugia has been recommended when attempting the reconstruction of European phylogeography patterns (Taberlet 1998). However, this recommendation is not always easy to follow, due to problems associated with sampling particular taxa. In addition, some authors have explicitly assumed that single mtDNA sequences from particular geographical regions, including refugia, are good representatives of the genetic variability of such regions (e.g. Bilton *et al.* 1998). As the data reviewed here shows, this assumption is frequently violated. In addition, as the different lineages present in southern refugia need not be sister taxa (i.e. a mono-phyletic group) due to complex patterns of population range contractions and expansions throughout the different Ice Ages (Steinfartz *et al.* 2000), monophyly of a random sample from the putative refugium and of samples from postglacial recolonized areas cannot necessarily be assumed.

One particularly serious consequence of the failure to recognize the true phylogeographic structure of the southern peninsulas is the incorrect inference of northern refugia. In this context 'northern' refers to areas north of the Pyrenees or Alps. Studies may, for example, infer a refuge in southern France because this locality contains haplotypes very distinct from those sampled from Iberia, Italy or the Balkans. Without extensive sampling of Iberia, however, it would not be possible to distinguish a northern refugium scenario from range expansion into southern France from a second non-sampled Iberian refugium. Since Iberian multiple refugia have been shown to be so common, care must be taken to sample extensively in southern Europe before concluding that haplotype distributions indicate Ice Age survival in northern regions. Unfortunately, this is an aspect that studies on northern refugia fail to address (Bilton *et al.* 1998; Stewart & Lister 2001). Finally, as Mediterranean refugia often display high levels of genetic diversity, and, if the occurrence of several refugia within refugia in the south is common, biased conclusions could be drawn if sampling is poor in these areas, or misses whole areas containing putative refugia. These problems are more likely to be widespread in terrestrial organisms, where the Iberian Peninsula is more likely to have harbored the populations that served for the colonization of northern Europe, than for freshwater organisms, which are often endemic to the Peninsula.The European hedgehog can illustrate the problems that incomplete sampling of southern refugia can pose for the interpretation of European phylogeographic patterns. The hedgehog mtDNA lineage sampled in the Iberian Peninsula, one of the two presumed glacial refugia in western Europe, was involved in the colonization of France, the UK and Ireland (Santucci *et al.* 1998; Seddon *et al.* 2001). A recent study including additional samples and using a larger mtDNA fragment identified two strongly supported monophyletic lineages in this clade (Seddon *et al.* 2001). One of these lineages was present in Iberia and southern France, the other one in France, UK and Ireland, and this geographic distribution suggests survival in different glacial refugia. One of the refugia occupied by this clade seems to have been in the Iberian Peninsula, as most of the diversity, including basal haplotypes, is present there, but the location of the other one is unclear. Although the basal haplotypes of the latter group are present in northern France, the current absence of samples from southwestern France and northeastern Spain precludes a conclusion about the location of the refugium for this clade though it could well be in Iberia. Although the occurrence of northern refugia should not be discarded a priori, and seems supported by fossil data in some species (Stewart & Lister 2001), exhaustive sampling of the southern distribution range where refugia are more likely, is imperative.

Consequences of refugial extinction

The reduction or extinction of populations in putative refugia is another factor that can complicate interpretation in a refugia-within-refugia scenario. Distribution ranges and population sizes of many species have often contracted in historical times due to anthropogenic effects, as forests were cleared for farming or due to direct human exploitation, and also due to climate warming and the disappearance of suitable habitat in the Holocene, as documented for the brown bear (Taberlet & Bouvet 1994) and Atlantic salmon, *Salmo salar* (Consuegra *et al.* 2002). Therefore, refugial populations, especially along the southern fringes of their distribution, might have declined considerably to isolated relicts (Paulo *et al.* 2001) or even disappeared altogether during the Holocene. In any case, the remaining population relicts do not necessarily harbor the levels of genetic diversity expected for a refugium, as 1) they may have suffered genetic bottlenecks and loss of genetic diversity, and 2) an already extinct refugia within the Iberian Peninsula could have given rise to the currently more northern European population. If this scenario is suspected or documented, the only way ahead might be the analysis of ancient DNA in subfossil material or historical samples that can be compared with the DNA of the living descendants. Indeed, despite the limitations of ancient DNA analysis, the few phylogeographic studies that have used this technique have yielded surprising results (Leonard *et al.* 2000; Consuegra *et al.* 2002).

Patterns of phylogeographic concordance: how many refugia and where?

Comparative phylogeography can help to pinpoint refugia common to several species, and serve as an aid to disclose cryptic events of historical vicariance. The suggestion of common refugial areas for distinct phylogeographic lineages (phylogeographic concordance aspect III according to Avise 2000) can be reinforced if such areas also show evidence of high endemism (phylogeographic concordance aspect IV, Avise 2000). Although we found considerable heterogeneity in the number and location of putative glacial refugia of the case studies investigated, this was most probably due to different habitat requirements. Yet clear patterns of phylogenetic congruence were apparent. For terrestrial taxa at least seven putative glacial refugia were recognized (Figure 2). In freshwater fish, the refugia coincide with the main river basins and a number of minor ones (Figure 3). Although most of the studies lack geo-

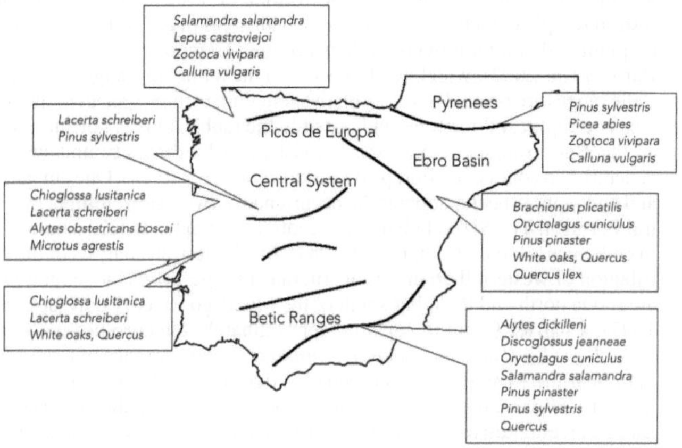

Figure 2*. Map of the Iberian Peninsula showing the approximate location of putative inferred terrestrial refugia and the fauna that supports them.*

Figure 3*. Map of the Iberian Peninsula showing the location of river basins and the freshwater fish species that support their status as glacial refugia.*

graphic precision pinpointing the localization of refugia, the suggested terrestrial refugia coincide quite strikingly with areas of endemism recognized by plant and animal biogeographers (Gómez-Campo *et al.* 1984; García-Barros *et al.* 2002), which tend to occur in the mountain ranges. Thus, the Betic ranges are pointed out as a putative refugium in three species of amphibians and reptiles, three trees and possibly the rabbit. This is not surprising, as the Betic ranges display the highest plant biodiversity and level of endemism in continental Europe (Gómez-Campo *et al.* 1984). One amphibian and one reptile species support the occurrence of two glacial refugia in Central Portugal. The Serra da Estrela in central Portugal is indeed an area rich in endemics (Gómez-Campo *et al.* 1984). In addition, an analysis of the distribution of western Iberian earthworm fauna suggests two biogeographical areas one north and the other south of the Mondego River (Rodríguez *et al.* 1997), a barrier found to separate approximately the current lineages of *Chioglossa lusitanica* and *Lacerta schreiberi*. The occurrence of a refugium in or near the Pyrenees is supported by a tree species, *Pinus sylvestris*, the heather *Calluna vulgaris*, the lizard *Zootoca vivipara* and the presence of a European refugium for the silver fir, *Abies alba* (Konnert & Bergmann 1995). The Pyrenees is indeed an area rich in animal and plant endemics, which suggests that in spite of being covered by an ice cap during glacial maxima, at least areas in or near the Pyrenean range served as glacial refugia (Gómez-Campo *et al.* 1984; García-Barros *et al.* 2002). Other areas of endemism also likely to have acted as glacial refugia are Picos de Europa and the Central System. An eastern refugium on the Mediterranean coast of Spain, possibly close to the Ebro valley, has been suggested for rotifers (Gómez *et al.* 2000), rabbits (Branco *et al.* 2000) *Quercus ilex* (Lumaret *et al.* 2002) and *Pinus pinaster* (Salvador *et al.* 2000). The Ebro River basin has indeed been recognized as a center for endemism for steppe fauna, and the continuity of its flora and fauna since the Tertiary is also well supported (Ribera & Blasco-Zumeta 1998). Further studies can help to delimit and possibly increase the number of separate glacial refugia for Iberian flora and fauna, and to sharpen our perception of which habitats were present.

Conclusions and recommendations

To paraphrase Hewitt (1996), the Iberian Peninsula has been shown to be an excellent 'theatre for phylogeographic analysis'. In spite of its small size compared to other regions used for comparative phylogeography at a regional level, and the limited number of case studies available, the Iberian Peninsula clearly comprised a number of separate glacial refugia during the Pleistocene Ice Ages. Evidence of concordant patterns, suggesting shared regional refugia, was found, although more studies are needed to allow better analysis of these patterns. These putative refugia are remarkably congruent with centers of endemism for the current flora and fauna in Iberia, a conclusion that fits well with the continuum between phylogeography and biogeography (Avise 2000). Neglecting the occurrence of multiple Iberian lineages can result in poor or biased sampling and lead to misleading conclusions when attempting European phylogeographic studies. Finally, the acknowledgement of the higher biodiversity awarded by the persistence of these separate refugia throughout the Pleistocene Ice Ages can be highly informative in the design of conservation areas in the Iberian Peninsula and can contribute to the understanding of the historical patterns that have given shape to the rich biodiversity of this corner of the planet.

Acknowledgements

We are grateful to Godfrey Hewitt, J.W. Arntzen, Bernd Hænfling and Natacha Mesquita for helpful comments and discussion on previous versions of this manuscript. We are also indebted to D. Guicking, J. Harris, M. Jaarola, and Y. Surget-Groba for sharing unpublished material with us. We thank Steven Weiss for his editing help and numerous comments, and all the organizers for a most inspiring and fruitful *Phylogeography in Southern European Refugia* meeting in Oporto.

References

Alcobendas M, Dopazo H, Alberch P (1996) Geographic variation in allozymes of populations of *Salamandra salamandra* (Amphibia: Urodela) exhibiting distinct reproductive modes. ***Journal of Evolutionary Biology*** 9, 83-102.

Alexandrino J, Froufe E, Arntzen JW, Ferrand N (2000) Genetic subdivision, glacial refugia and postglacial recolonization in the golden-striped salamander, *Chioglossa lusitanica* (Amphibia: Urodela). ***Molecular Ecology*** 9, 771-781.

Alexandrino J, Arntzen JW, Ferrand N (2002) Nested clade analysis and the genetic evidence for population expansion in the phylogeography of the golden-striped salamander, *Chioglossa lusitanica* (Amphibia: Urodela). ***Heredity*** **88**, 66-74.

Alves MJ, Coelho H, Collares-Pereira MJ, Coelho MM (2001) Mitochondrial DNA variation in the highly endangered cyprinid fish *Anaecypris hispanica*: importance for conservation. ***Heredity*** 87, 463-473.

Andreu A, Bea A, Braña F, Galán P, López Jurado LF, Pérez Mellado V, Pleguezuelos JM, Salvador A (1998) ***Reptiles.*** Fauna Iberica, vol. 10. CSIC, Madrid.

Antunes A, Faria R, Weiss S, Alexandrino P (2001) Complex evolutionary history in the brown trout: insights on the recognition of conservation units. ***Conservation Genetics*** **2**, 337-347.

Arbogast BS, Kenagy GJ (2001) Comparative phylogeography as an integrative approach to historical biogeography. ***Journal of Biogeography*** **28**, 819-825.

Arntzen JW, García-París M (1995) Morphological and allozyme studies of midwife toads (genus *Alytes*), including the description of two new taxa from Spain. ***Contributions to Zoology*** **65**, 5-34.

Avise JC (2000) ***Phylogeography: The History and Formation of Species*** Harvard University Press, Cambridge.

Ball-Llosera NS, García-Marín JL, Pla C (2002) Managing fish populations under mosaic relationships. The case of brown trout (*Salmo trutta*) in peripheral Mediterranean populations. ***Conservation Genetics*** **3**, 385-400.

Bănărescu P (1991) ***Zoogeography of Fresh Waters: Distribution and Dispersal of Freshwater Animals in North America and Eurasia.*** vol. 2. AULA-Verlag, Wiesbaden

Baquero RA, Tellería JL (2001) Species richness, rarity and endemicity of European mammals: a biogeographical approach. ***Biodiversity and Conservation*** **10**, 29-44.

Barbadillo LJ (1987) ***La Guia de Incafo de los Anfibios y Reptiles de la Península Ibérica, Islas Baleares y Canarias.*** Incafo, Spain.

Beebee TJC, Rowe G (2000) Microsatellite analysis of natterjack toad *Bufo calamita* Laurenti populations: consequences of dispersal from a Pleistocene refugium. ***Biological Journal of the Linnean Society*** **69**, 367-381.

Bennett KD, Tzedakis PC, Willis KJ (1991) Quaternary refugia of north European trees. ***Journal of Biogeography*** **18**, 103-115.

Bernatchez L (2001) The evolutionary history of brown trout (*Salmo trutta* L.) inferred from phylogeographic, nested clade, and mismatch analyses of mitochondrial DNA variation. ***Evolution*** **55**, 351-379.

Bernatchez L, Wilson CC (1998) Comparative phylogeography of nearctic and palearctic fishes. ***Molecular Ecology*** **7**, 431-452.

Biju-Duval C, Ennafaa H, Dennebouy N, Monnerot M, Mignotte F, Soriguer R, El Gaied A, El Hili A, Monoulou JC (1991) Mitochondrial DNA evolution in Lagomorphs: origin of systematic heteroplasmy and organization of diversity in European rabbits. ***Journal of Molecular Evolution*** **33**, 92-102.

Bilton DT, Mirol PM, Mascheretti S, Fredga K, Zima J, Searle JB (1998) Mediterranean Europe as an area of endemism for small mammals rather than a source for northwards postglacial colonization. ***Proceedings of the Royal Society of London, B*** **265**, 1219-1226.

Branco M, Ferrand N, Monnerot M (2000) Phylogeography of the European rabbit (*Oryctolagus cuniculus*) in the Iberian Peninsula inferred from RFLP analysis of the cytochrome b gene. ***Heredity*** **85**, 307-317.

Branco M, Monnerot M, Ferrand N, Templeton AR (2002) Postglacial dispersal of the European rabbit (*Oryctolagus cuniculus*) on the Iberian Peninsula reconstructed from nested clade and mismatch analyses of mitochondrial DNA genetic variation. ***Evolution*** **56**, 792-803.

Brito RM, Briolay J, Galtier N, Bouvet J, Coelho MM (1997) Phylogenetic relationships within the genus *Leuciscus* (Pisces, Cyprinidae) in Portuguese fresh waters, based on mitochondrial DNA cytochrome b sequences. ***Molecular Phylogenetics and Evolution*** **8**, 435-442.

Burban C, Petit RJ, Carcreff E, Jactel H (1999) Rangewide variation of the maritime pine bast scale *Matsucoccus feytaudi* Duc. (Homoptera : Matsucoccidae) in relation to the genetic structure of its host. ***Molecular Ecology*** **8**, 1593-1602.

Callejas C, Ochando MD (2000) Recent radiation of Iberian Barbel fish (Teleostei, Cyprinidae) inferred from cytochrome b genes. ***The Journal of Heredity*** **91**, 283-288.

Carmona JA, Domínguez J, Doadrio I (2000) Congruence between allozyme and cytochrome b gene sequence data in assessing genetic differentiation within the Iberian endemic *Chondrostoma lemingii* (Pisces: Cyprinidae). ***Heredity*** **84**, 721-732.

Ciros-Pérez J, Gómez A, Serra M (2001) On the taxonomy of three sympatric sibling species of the *Brachionus plicatilis* (Rotifera) complex from Spain, with the description of *B. ibericus* n. sp. ***Journal of Plankton Research*** **23**, 1311-1328.

Coelho MM, Bogutskaya NG, Rodrigues JA, Collares-Pereira MJ (1998) *Leuciscus torgalensis* and *L. aradensis*, two new cyprinids from Portuguese fresh waters. ***Journal of Fish Biology*** **52**, 937-950.

Comes HP, Abbott RJ (1998) The relative importance of historical events and gene flow on the population structure of a Mediterranean ragwort, *Senecio gallicus* (Asteraceae). ***Evolution*** **52**, 355-367.

Comes HP, Abbott RJ (2000) Random amplified polymorphic DNA (RAPD) and quantitative trait analyses across a major phylogeographic break in the Mediterranean ragwort *Senecio gallicus* VILL. (Asteraceae). ***Molecular Ecology*** **9**, 61-76.

Comes HP, Kadereit JW (1998) The effect of quaternary climatic changes on plant distribution and evolution. ***Trends in Plant Science*** **3**, 432-438.

Consuegra S, De Leaniz CG, Serdio A, González-Morales M, Straus LG, Knox D, Verspoor E (2002) Mitochondrial DNA variation in Pleistocene and modern Atlantic salmon from the Iberian glacial refugium. ***Molecular Ecology*** **11**, 2037-2048.

Cooper SJB, Hewitt GM (1993) Nuclear DNA sequence divergence between parapatric subspecies of the grasshopper *Chorthippus parallelus*. ***Insect Molecular Biology*** **2**, 185-194.

Cooper SJB, Ibrahim KM, Hewitt GM (1995) Postglacial expansion and genome subdivision in the European grasshopper *Chorthippus parallelus*. ***Molecular Ecology*** **4**, 49-60.

Doadrio I (1988) Delimitation of areas in the Iberian Peninsula on the basis of freshwater fishes. ***Bonner Zoologische Beiträge*** **39**, 113-128.

Doadrio I, Carmona JA, Machordom A (2002) Haplotype diversity and phylogenetic relation-ships among the Iberian barbels (*Barbus*, Cyprinidae) reveal two evolutionary lineages. ***The Journal of Heredity*** **93**, 140-147.

Dopazo H, Boto L, Alberch P (1998) Mitochondrial DNA variability in viviparous and ovoviviparous populations of the urodele *Salamandra salamandra*. ***Journal of Evolutionary Biology*** **11**, 365-378.

Dumolin-Lapègue S, Demesure B, Fineschi S, LeCorre V, Petit RJ (1997) Phylogeographic structure of white oaks throughout the European continent. ***Genetics*** **146**, 1475-1487.

Dumolin-Lapègue S, Kremer A, Petit RJ (1999) Are chloroplast and mitochondrial DNA variation species independent in oaks? ***Evolution*** **53**, 1406-1413.

Elvira B (1995) Conservation status of endemic freshwater fish in Spain. ***Biological Conservation*** **72**, 129-136.

Ferris C, King RA, Vainola R, Hewitt GM (1998) Chloroplast DNA recognizes three refugial sources of European oaks and suggests independent eastern and western immigrations to Finland. ***Heredity*** **80**, 584-593.

Fonseca A, Arntzen JW, Crespo E, Ferrand N (2003) Regional differentiation in the common midwife toad (*Alytes obstetricans*) in Portugal: a picture from mitochondrial DNA. ***Zeitschrift für Feldherpetologie*** **10**, 83-89.

Fumagalli L, Hausser J, Taberlet P, Gielly L, Stewart DT (1996) Phylogenetic structures of the Holarctic *Sorex araneus* group and its relationship with *S. samniticus*, as inferred from mtDNS sequences. ***Hereditas*** 125, 191-199.

García-Barros E, Gurrea P, Luciañez MJ, Cano JM, Munguira ML, Moreno JC, Sainz H, Sanz MJ, Simón JC (2002) Parsimony analysis of endemicity and its application to animal and plant geographical distributions in the Ibero-Balearic region (western Mediterranean). ***Journal of Biogeography*** **29**, 109-124.

García-París M, Jockusch EL (1999) A mitochondrial DNA perspective on the evolution of Iberian *Discoglossus* (Amphibia: Anura). ***Journal of Zoology, London*** **248**, 209-218.

García-París M, Martínez-Solano I (2001) Nuevo status taxonómico para las poblaciones Ibero-Mediterraneas de *Alytes obstetricans* (Anura: Discoglossidae). ***Revista Española de Herpetologia*** **15**, 99-113.

García-París M, Herrero P, Martín C, Dorda J, Esteban M, Arano B (1993) Morphological characterization, cytogenetic analysis, and geographical distribution of the pygmy marbled newt *Triturus marmoratus pygmaeus* (Wolterstoff, 1905) (Caudata, Salamandridae). ***Bijdragen tot de Dierkunde*** **63**, 3-14.

García-París M, Alcobendas M, Alberch P (1998) Influence of the Guadalquivir River on mitochondrial DNA evolution of *Salamandra salamandra* (Caudata: Salamandridae) from Southern Spain. ***Copeia*** **1998**, 173-176.

García-París M, Alcobendas M, Buckley D, Wake DB (2003) Dispersal of viviparity across contact zones in Iberian populations of fire salamanders (Salamandra) inferred from discordance of genetic and morphological traits. ***Evolution*** **57**, 129-143.

Gómez A, Carvalho GR, Lunt DH (2000) Phylogeography and regional endemism of a passively dispersing zooplankter: mitochondrial DNA variation in rotifer resting egg banks. ***Proceedings of the Royal Society of London, B*** **267**, 2189-2197.

Gómez A, Serra M, Carvalho GR, Lunt DH (2002) Speciation in ancient cryptic species complexes: evidence from the molecular phylogeny of *Brachionus plicatilis* (Rotifera). ***Evolution*** **56**, 1431-1444.

Gómez-Campo C, Bermudez-de-Castro L, Cagiga MJ, Sanchez-Yelamo MD (1984) Endemism in the Iberian Peninsula. ***Webbia*** **38**, 709-714.

Gómez-Zurita J, Petitpierre E, Juan C (2000) Nested cladistic analysis, phylogeography and speciation in the *Timarcha goettingensis* complex (Coleoptera, Chrysomelidae). ***Molecular Ecology*** **9**, 557-570.

Griswold CK, Baker AJ (2002) Time to most recent common ancestor and divergence times of populations of common chaffinches (*Fringilla coelebs*) in Europe and North Africa: insights into Pleistocene refugia and current levels of migration. ***Evolution*** **56**, 143-153.

Grivet D, Petit RJ (2002) Phylogeography of the common ivy (*Hedera* sp.) in Europe: genetic differentiation through space and time. ***Molecular Ecology*** **11**, 1351-1362.

Guicking D, Joger U, Wink M (2002) Molecular phylogeography of the Viperine snake (*Natrix maura*) and the Dice snake (*Natrix tessellata*): first results. ***Biota*** **3**, 47-57.

Guillaume GP, Heulin B, Arrayago MJ, Bea A, Brana F (2000) Refuge areas and suture zones in the Pyrenean and Cantabrian regions: geographic variation of the female MPI sex-linked alleles among oviparous populations of the lizard *Lacerta (Zootoca) vivipara*. ***Ecography*** **23**, 3-10.

Harris DJ, Sa-Sousa P (2001) Species distinction and relationships of the western Iberian *Podarcis* lizards (Reptilia, Lacertidae) based on morphology and mitochondrial DNA sequences. ***Herpetological Journal*** **11**, 129-136.

Harris DJ, Sa-Sousa P (2002) Molecular phylogenetics of the Iberian wall lizards (*Podarcis*): is *Podarcis hispanica* a species complex? ***Molecular Phylogenetics and Evolution* 23**, 75-81.

Herrero P (1991) Polytypic chromosomal variation in *Triturus boscai* (Urodela, Salamandridae). ***Genetics Selection Evolution* 23**, 263-272.

Hewitt GM (1993) After the ice: *Parallelus* meets *Erythropus* in the Pyrenees. In: ***Hybrid zones and the Evolutionary Process*** (ed. Harrison RG), pp. 140-164. Oxford University Press, Oxford.

Hewitt GM (1996) Some genetic consequences of ice ages, and their role in divergence and speciation. ***Biological Journal of the Linnean Society* 58**, 247-276.

Hewitt GM (1999) Post-glacial re-colonization of European biota. ***Biological Journal of the Linnean Society* 68**, 87-112.

Hewitt GM (2000) The genetic legacy of the Quaternary ice ages. ***Nature* 405**, 907-913.

Hewitt GM (2001) Speciation, hybrid zones and phylogeography - or seeing genes in space and time. ***Molecular Ecology* 10**, 537-549.

Huntley B, Birks JDS (1983) ***An Atlas of Past and Present Pollen Maps for Europe: 0-13,000 years ago***. Cambridge University Press, Cambridge.

Ibrahim KM, Nichols RA, Hewitt GM (1996) Spatial patterns of genetic variation generated by different forms of dispersal during range expansion. ***Heredity* 77**, 282-291.

Jaarola M, Searle JB (2002) Phylogeography of field voles (*Microtus agrestis*) in Eurasia inferred from mitochondrial DNA sequences. ***Molecular Ecology* 11**, 2613-2621.

Jaarola M, Tegelstrom H (1995) Colonization history of North European Field Voles (*Microtus agrestis*) revealed by mitochondrial DNA. ***Molecular Ecology* 4**, 299-310.

King RA, Ferris C (1998) Chloroplast DNA phylogeography of *Alnus glutinosa* (L.) Gaertn. ***Molecular Ecology* 7**, 1151-1161.

Konnert M, Bergmann F (1995) The geographical distribution of genetic variation in silver fir (*Abies alba*, Pinaceae) in relation to its migration history. ***Plant Systematics and Evolution* 196**, 19-30.

Lenk P, Fritz U, Joger U, Wink M (1999) Mitochondrial phylogeography of the European pond turtle, *Emys orbicularis* (Linnaeus 1758). ***Molecular Ecology* 8**, 1911-1922.

Leonard JA, Wayne RK, Cooper A (2000) Population genetics of Ice age brown bears. ***Proceedings of the National Academy of Science, USA*** 97, 1651-1654.

Lumaret R, Mir C, Michaud H, Raynal V (2002) Phylogeographical variation of chloroplast DNA in holm oak (*Quercus ilex* L.). ***Molecular Ecology* 11**, 2327-2336.

Lunt DH, Ibrahim KM, Hewitt GM (1998) MtDNA phylogeography and postglacial patterns of subdivision in the meadow grasshopper *Chorthippus parallelus*. ***Heredity* 80**, 633-641.

Machordom A, Doadrio I (2001) Evidence of a cenozoic Betic-Kabilian connection based on freshwater fish phylogeography (*Luciobarbus*, Cyprinidae). ***Molecular Phylogenetics and Evolution* 18**, 252-263.

Machordom A, Suárez J, Almodóvar A, Bautista JM (2000) Mitochondrial haplotype variation and phylogeography of Iberian brown trout populations. ***Molecular Ecology* 9**, 1325-1338.

Mesquita N, Carvalho GR, Shaw P, Crespo E, Coelho MM (2001) River basin-related genetic structuring in an endangered fish species, *Chondrostoma lusitanicum*, based on mtDNA sequencing and RFLP analysis. ***Heredity* 86**, 253-264.

Michaux JR, Libois R, Ramalhinho MG, Maurois C (1998) On the mtDNA restriction patterns variation of the Iberian wood mouse (*Apodemus sylvaticus*). Comparison with other west Mediterranean populations. ***Hereditas*** **129**, 187-194.

Michaux JR, Magnanou E, Paradis E, Nieberding C, Libois R (2003) Mitochondrial phylogeography of the Woodmouse (*Apodemus sylvaticus*) in the Western Palearctic region. ***Molecular Ecology*** **12**, 685-697.

Moreno Saiz JC, Castro Parga I, Sainz Ollero H (1998) Numerical analyses of distribution of Iberian and Balearic endemic monocotyledons. ***Journal of Biogeography*** **25**, 179-194.

Muir G, Fleming CC, Schlötterer C (2000) Species status of hybridizing oaks. ***Nature*** **405**, 1016.

Olalde M, Herran A, Espinel S, Goicoechea PG (2002) White oaks phylogeography in the Iberian Peninsula. ***Forest Ecology and Management*** **156**, 89-102.

Ortells R, Snell TW, Gómez A, Serra M (2000) Patterns of genetic differentiation in resting egg banks of a rotifer species complex in Spain. ***Archiv für Hydrobiologie*** **149**, 529-551.

Paulo OS, Dias C, Bruford MW, Jordan WC, Nichols RA (2001) The persistence of Pliocene populations through the Pleistocene climatic cycles: evidence from the phylogeography of an Iberian lizard. ***Proceedings of the Royal Society of London, B*** **268**, 1625-1630.

Paulo OS, Jordan WC, Bruford MW, Nichols RA (2002) Using nested clade analysis to assess the history of colonisation and the persistence of populations of an Iberian lizard. ***Molecular Ecology*** **11**, 809-819.

Perdices A, Carmona JA, Fernández-Delgado C, Doadrio I (2001) Nuclear and mitochondrial data reveal high genetic divergence among Atlantic and Mediterranean populations of the Iberian killifish *Aphanius iberus* (Teleostei: Cyprinodontidae). ***Heredity*** **87**, 314-324.

Pérez-Suárez G, Palacios F, Boursot P (1994) Speciation and paraphyly in Western Mediterranean hares (*Lepus castroviejoi, L. europaeus, L. granatensis* and *L. capensis*) revealed by mitochondrial DNA phylogeny. ***Biochemical Genetics*** **32**, 423-436.

Petit RJ, Brewer S, Bordács S, Burg K, Cheddadi R, Coart E, Cottrell J, Czaikl UM, van Dam B, Deans JD, Espinel S, Fineschi S, Finkeldey R, Glaz I, Goicoechea PG, Jensen JS, König AO, Lowe AJ, Madsen SF, Mátyás G, Munro RC, Popescu F, Slade D, Tabbener H, deVries SGM, Ziegenhagen B, de Beaulieu J-L, Kremer A (2002) Identification of refugia and post-glacial colonisation routes of European white oaks based on chloroplast DNA and fossil pollen evidence. ***Forest Ecology and Management*** **156**, 49-74.

Queney G, Ferrand N, Weiss S, Mougel F, Monnerot M (2001) Stationary distributions of microsatellite loci between divergent population groups of the European rabbit (*Oryctolagus cuniculus*). ***Molecular Biology and Evolution*** **18**, 2169-2178.

Rendell S, Ennos RA (2002) Chloroplast DNA diversity in *Calluna vulgaris* (heather) populations in Europe. ***Molecular Ecology*** **11**, 69-78.

Ribera I (2000) Biogeography and conservation of Iberian water beetles. ***Biological Conservation*** **92**, 131-150.

Ribera I, Blasco-Zumeta J (1998) Biogeographical links between steppe insects in the Monegros region (Aragon, Spain), the Eastern Mediterranean and Central Asia. ***Journal of Biogeography*** **25**, 969-986.

Riddle BR, Hafner DJ, Alexander LF, Jaeger JR (2000) Cryptic vicariance in the historical assembly of a Baja California Peninsular Desert biota. ***Proceedings of the National Academy of Sciences, USA*** **97**, 14438-14443.

Rodríguez T, Trigo D, Díaz Cosín D (1997) Biogeographical zonation of the western Iberian peninsula on the basis of the distribution of earthworm species. ***Journal of Biogeography* 24**, 893-901.

Salvador L, Alia R, Agundez R, Gil L (2000) Genetic variation and migration pathways of maritime pine (*Pinus pinaster* Ait.) in the Iberian peninsula. ***Theoretical Applied Genetics* 100**, 89-95.

Santucci F, Emerson BC, Hewitt GM (1998) Mitochondrial DNA phylogeography of European hedgehogs. ***Molecular Ecology*** 7, 1163-1172.

Schneider CJ, Cunningham M, Moritz C (1998) Comparative phylogeography and the history of endemic vertebrates in the Wet tropics rainforests of Australia. ***Molecular Ecology*** 7, 487-498.

Seddon JM, Santucci F, Reeve NJ, Hewitt GM (2001) DNA footprints of European hedgehogs, *Erinaceus europaeus* and *E. concolor*. Pleistocene refugia, postglacial expansion and colonization routes. ***Molecular Ecology* 10**, 2187-2198.

Sharbel TF, Haubold B, Mitchell-Olds T (2000) Genetic isolation by distance in *Arabidopsis thaliana*: biogeography and postglacial colonization of Europe. ***Molecular Ecology* 9**, 2109-2118.

Sinclair WT, Morman JD, Ennos RA (1999) The postglacial history of Scots pine (*Pynus sylvestris* L.) in western Europe: evidence from mitochondrial DNA variation. ***Molecular Ecology* 8**, 83-88.

Soranzo N, Alia R, Provan J, Powell W (2000) Patterns of variation at a mitochondrial sequence-tagged-site locus provides new insights into the postglacial history of European *Pinus sylvestris* populations. ***Molecular Ecology* 9**, 1205-1211.

Steinfartz S, Veith M, Tautz D (2000) Mitochondrial sequence analysis of *Salamandra* taxa suggests old splits of major lineages and postglacial recolonizations of Central Europe from distinct source populations of *Salamandra salamandra*. ***Molecular Ecology* 9**, 397-410.

Stewart JR, Lister AM (2001) Cryptic northern refugia and the origins of the modern biota. ***Trends in Ecology and Evolution* 16**, 608-613.

Suárez J, Bautista JM, Almodóvar A, Machordom A (2001) Evolution of the mitochondrial control region in Palaearctic brown trout (*Salmo trutta*) populations: the biogeographical role of the Iberian Peninsula. ***Heredity* 87**, 198-206.

Surget-Groba Y, Heulin B, Guillaume GP, Thorpe RS, Kupriyanova L, Vogrin N, Maslak R, Mazzotti S, Venczel M, Ghira I, Odierna G, Leontyeva O, Monney JC, Smith N (2001) Intraspecific phylogeography of *Lacerta vivipara* and the evolution of viviparity. ***Molecular Phylogenetics and Evolution* 18**, 449-459.

Taberlet P (1998) Biodiversity at the intraspecific level: the comparative phylogeographic approach. ***Journal of Biotechnology* 64**, 91-100.

Taberlet P, Bouvet J (1994) Mitochondrial-DNA polymorphism, phylogeography, and conservation genetics of the brown bear *Ursus arctos* in Europe. ***Proceedings of the Royal Society of London, B* 255**, 195-200.

Taberlet P, Fumagalli L, Hausser J (1994) Chromosomal versus mitochondrial DNA evolution: tracking the evolutionary history of the southwestern European populations of the *Sorex araneus* group (Mammalia, Insectivora). ***Evolution* 48**, 623-636.

Taberlet P, Fumagalli L, Wust-Saucy AG, Cosson JF (1998) Comparative phylogeography and postglacial colonization routes in Europe. ***Molecular Ecology*** 7, 453-464.

Utelli AB, Roy BA, Baltisberger M (1999) History can be more important than 'pollination syndrome' in determining the genetic structure of plant populations: the case of *Aconitum lycoctonum* (Ranunculaceae). ***Heredity* 82**, 574-584.

Van Dijk P, Bakx-Schotman T (1997) Chloroplast DNA phylogeography and cytotype geography in autopolyploid ***Plantago media*. *Molecular Ecology* 6**, 345-352.

Vargas JM, Real R, Guerrero JC (1998) Biogeographical regions of the Iberian Peninsula based on freshwater fish and amphibian distributions. ***Ecography* 21**, 371-382.

Vendramin GG, Anzidei M, Madaghiele A, Bucci G (1998) Distribution of genetic diversity in *Pinus pinaster* Ait. as revealed by chloroplast microsatellites. ***Theoretical Applied Genetics* 97**, 456-463.

Vernesi C, Pecchioli E, Caramelli D, Tiedermann R, Randi E, Bertorelle G (2002) The genetic structure of natural and reintroduced roe deer (*Capreolus capreolus*) populations in the Alps and central Italy, with reference to the mitochondrial DNA phylogeography of Europe. ***Molecular Ecology* 11**, 1285-1297.

Walker D, Avise JC (1998) Principles of phylogeography as illustrated by freshwater and terrestrial turtles in the southeastern United States. ***Annual Review of Ecology and Systematics* 29**, 23-58.

Wallis GP, Arntzen JW (1989) Mitochondrial-DNA variation in the crested newt superspecies: limited cytoplasmic gene flow among species. ***Evolution* 43**, 88-104.

Weiss S, Antunes A, Schlötterer C, Alexandrino P (2000) Mitochondrial haplotype diversity among Portuguese brown trout *Salmo trutta* L. populations: relevance to the post-Pleistocene recolonization of northern Europe. ***Molecular Ecology* 9**, 691-698.

Zardoya R, Doadrio I (1998) Phylogenetic relationships of Iberian cyprinids: systematic and biogeographical implications. ***Proceedings of the Royal Society of London, B* 265**, 1365-1372.

Chapter 6

Historical biogeography and conservation of the golden-striped salamander (*Chioglossa lusitanica*) in northwestern Iberia: integrating ecological, phenotypic and phylogeographic data

João Alexandrino*, José Teixeira, Jan W. Arntzen and Nuno Ferrand
*E-mail: jalex@rc.unesp.br

Abstract

The golden-striped salamander (*Chioglossa lusitanica*) is an endemic species inhabiting streamside habitats in mountainous areas in the northwestern Iberian Peninsula. This salamandrid is listed in the IUCN Red Data Book as a threatened species. The combination of bioclimatic modeling of the species distribution and multivariate analysis of genetic and phenotypic data strengthens previous hypotheses concerning the historical biogeography of *C. lusitanica*: the Pleistocene subdivision of the species' range and a process of postglacial recolonization. Discrepancies between bioclimatic modeling predictions and the present-day distribution suggest that the species may still be expanding its range northwards. We propose the identification of two distinct units for the conservation of the species and suggest that this information should be taken into account in defining key areas for conservation in the Iberian Peninsula.

Keywords: allozymes, *Chioglossa lusitanica*, ecological modeling, evolutionary significant units, golden-striped salamander, glacial refugia, mitochondrial DNA, phenotypic variation, postglacial range expansion

Introduction

Fluctuations in spatial distribution and abundance are commonplace for many plant and animal species. The biotic effects of Pleistocene glaciations exemplify how climate changes influence species distributions – by alternately inducing southward range contractions with northward expansions (Hewitt 1996, 1999, 2000). Southern Europe contained refugia where many species survived during glacial periods and thus represents a center of origin for many postglacial recolonizations. The geographic patterns resulting from these processes differ with the varying dispersal abilities and ecological requirements of each species (Hewitt 1996; Taberlet *et al.* 1998).

S. Weiss and N. Ferrand (eds.), Phylogeography of Southern European Refugia, 189–205.

Patterns of species distribution and diversity have been represented through plotting presence/absence data on grid systems at various scales. Spatial modeling techniques may predict the distribution of species (Walker 1990; Pereira & Itami 1991; Brito *et al.* 1996) and individuals (Austin *et al.* 1996), can be used to estimate suitable habitat (Mladenoff & Sickley 1998) or aid in the design of conservation plans (Velázquez & Bocco 1994). One important limitation of these models is that they fail to capture historical patterns of population persistence.

The geographical distribution of genetic diversity in species may be used to reconstruct historical biogeographies (Avise 1994, 1998; Bermingham & Moritz 1998). Phylogeography seeks to reveal historical biogeography of species through i) qualitative spatial association of alleles with geography, and ii) quantitative estimates of historical population size (Avise 2000; Emerson *et al.* 2001; Hare 2001; Templeton 2002). Ideally, phylogeographic inferences should be accompanied by evidence from independent sources such as the fossil record or paleoecology. Recently, a novel approach was explored using paleoclimatological models of species distributions in conjunction with phylogeography (Hugall *et al.* 2002).

One important caveat of phylogeographic studies is that they have mostly relied on single-locus data, usually from mitochondrial DNA. Because nuclear gene based research is only now emerging (Hare 2001), multi-locus allele frequency data such as those obtained from allozyme polymorphism studies are invaluable in phylogeographic inference. Multivariate analysis of allele frequency data (e.g. data from allozyme variation) may summarize variation at several genes with a few independent synthetic variables that can be represented in geographical maps (Menozzi *et al.* 1978). Such analysis in humans has been used to construct maps that helped reveal demic expansions and determine centers of origin (Cavalli-Sforza *et al.* 1993, 1994; Ray *et al.* 2005). While this approach has been applied extensively in human studies (Menozzi *et al.* 1978; Piazza *et al.* 1981; Cavalli-Sforza *et al.* 1993, 1994; Bosch *et al.* 1997; Ray *et al.* 2005) it has rarely been used on other organisms (Guinand & Easteal 1996; Le Corre *et al.* 1998).

The golden-striped salamander, *Chioglossa lusitanica*, is an Iberian endemic listed in the IUCN Red Data Book. Its range is restricted to the northwestern corner of the Iberian Peninsula, where it lives around small brooks in fairly mountainous terrain. We recently analyzed allozyme and mitochondrial DNA variation and uncovered two genetically distinct groups of populations that are geographically separated by the river Mondego in central Portugal (Alexandrino *et al.* 1997, 2000, 2002). The two groups represent lineages that separated in the early Pleistocene, probably as a result of climate change in combination with local environmental conditions. The secondary contact zone between the groups is shaped by neutral gene exchange but with limited spatial introgression (Sequeira *et al.* 2005). We further inferred that the northern part of the present range was colonized from a refuge located between the Mondego and Douro rivers, and that major rivers such as the

Douro and the Minho acted as barriers to dispersal, lowering genetic diversity through sequential bottlenecking of northward expanding populations.

We combine ecological models of the distribution of *C. lusitanica* (Teixeira *et al.* 2001) with both mtDNA data and synthetic genetic maps constructed from multivariate analysis of allozyme genetic data. Our aims are i) to discuss hypothesized historical biogeographic scenarios, and ii) to propose areas for the long-term conservation of the species based on present habitat suitability and historical population persistence.

Debate in the field of conservation genetics has revolved around the relative importance of molecular versus quantitative genetic and phenotypic characterization of diversity and the resulting assignment of systematic management units. It was recently suggested that greater clarity would be achieved by partitioning genetic diversity into two components: that arising from adaptive evolution and that resulting from long-term historical isolation (Moritz 2002). The former can be estimated through analysis of phenotypic variation, while the latter is readily assayed through molecular phylogeography. Both approaches have their place, but measure different components of intraspecific diversity. It has been suggested that long-term historical isolation and persistence of populations should be given more emphasis in conservation because the genetic variation arising through such processes represents an irreplaceable component of intraspecific biodiversity. As phenotypic diversity is also highly relevant to conservation efforts, we also compare overall patterns of genetic and morphological variation in *C. lusitanica*.

Materials and methods

The distribution model

The spatial model used to predict the distribution of *C. lusitanica* was recently constructed based on presence/absence data in Portugal and a set of environmental parameters, using logistic regression (Teixeira *et al.* 2001). The probability of the species' occurrence across the Iberian Peninsula was estimated with a 93% success rate, based on total annual precipitation (PRET), slope (SLOP), altitude (ALTI), and mean July temperatures (TJUL), $g(x) = -0.087 + 0.131 \times PRET + 0.063 \times SLOP - 0.063 \times ALTI - 0.052 \times TJUL$) (Teixeira *et al.* 2001).

Allozyme data

Genetic data consisted of allele frequencies at six polymorphic allozyme loci (PGM1, PEPB, PEPC, PEPD, ADH and PGD) scored from 17 populations distributed across the entire species' range (Alexandrino *et al.* 2000). The most common allele at five nearly di-allelic loci (PEPB*1, PEPC*1, PEPD*1,

PGD*1 and ADH*1) and four out of five alleles at the highly polymorphic PGM1 locus (PGM1*1F, PGM1*1S, PGM1*2 and PGM1*3F) were used for PCA. This selection of alleles emphasizes the major components of variation, and the exclusion of rare alleles (frequency <0.05) decreases the effect of sampling error. Allele frequencies were spatially interpolated by a linear distance weighting model (kriging default) to generate 400 allele frequency values regularly distributed within a grid delimited by parallels 39-44° N and meridians 4-9° W, using the Surfer 6.0 geostatistical software (Golden Software 1996). Interpolated allele frequencies for each allele were then used as input variables in a principal component analysis with the software package Statistica/w 4.5 (StatSoft 1993). The factor scores for the first two principal components (PC) were used to construct geographical maps with the kriging interpolation procedure in Surfer 6.0 (Golden Software 1996). Following Menozzi *et al.* (1978), maps were overlaid with a weighting function reflecting the percent of total variance explained, for the area within the *C. lusitanica* range. The software used for the manipulation of maps was Idrisi for Windows v. 2.0 (Eastman 1997).

Mitochondrial DNA

Genetic data consisted of mtDNA haplotypes scored from the same 17 populations as noted above (Alexandrino *et al.* 2002). A geographic map summarizing mtDNA variation was constructed by overlaying the results from nested clade analysis (NCA) (Templeton 1998) and population expansion tests (see Alexandrino *et al.* 2002).

Multivariate analysis of phenotypic data

Morphometric and dorsal color pattern variation in *C. lusitanica* were previously described (Alexandrino 2000; Alexandrino *et al.*, unpublished results). Data for eight morphometric measurements in 18 populations were analyzed for males and females separately using principal component analysis (PCA). Color pattern data for 420 individuals classified into six distinct types was analyzed by Correspondence Analysis (CA). Trend surface maps were generated for both PCA and CA axes by kriging under default settings in Surfer 6.0 (Golden Software 1996) geostatistical software.

Results and discussion

Genetic diversity

The first axis from the PCA based on allozyme frequencies explained 70% of the total variation, showing a south-to-north cline in variation from Serra do Muradal in central Portugal to Salas in northern Spain (Table 1; Figure 1A). The steep geographic gradient along the Mondego valley likely reflects recent secondary contact between two formerly isolated groups and implies the existence of both southern and northern glacial refugia (Alexandrino *et al.* 2000, 2002). The second axis explained 25% of the total variation (Figure 1B). It shows a diffusion gradient, suggesting range expansion from a center of origin following genetic isolation (Menozzi *et al.* 1978).

Table 1*. Factor loadings for the first and second Principal Component (PC) axis for nine allele frequencies observed at six allozyme loci in 17 populations of* Chioglossa lusitanica *(Alexandrino* et al. *2000)*

Allele	PC 1	PC 2
PGM1*1F	0.88	0.15
PGM1*1S	0.79	0.48
PGM1*2	0.01	-0.96
PGM1*3F	-0.97	0.09
PEPB*1	-0.05	0.96
PEPC*1	0.96	-0.14
PEPD*1	0.97	-0.02
PGD*1	0.94	-0.01
ADH*1	0.98	-0.05
Variance explained	70 %	25 %

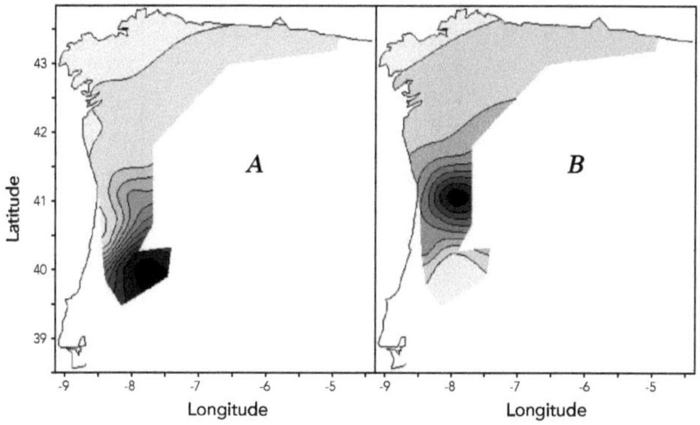

Figure 1*. Principal Component Analysis of nine independent allele frequencies at the loci PGM1, PEPB, PEPC, PEPD, PGD and ADH in 17 populations of* Chioglossa lusitanica *across its documented range (Arntzen 1999):* ***A)*** *synthetic map for the first principal component, representing 70% of the total variation;* ***B)*** *synthetic map for the second principal component, representing 25% of the total variation.*

PC1 and PC2 in combination clearly show the two differentiated population groups to the south and north of the Mondego River (Figure 2). We infer that the northern range expanded from a glacial refuge in the Serra de Montemuro, northwards to occupy most of Galicia and Asturias in Spain and southwards to the Mondego river valley. The genetic uniformity exhibited in populations in the northernmost range of the species, suggests that these territories were either recently colonized or reflect founder effects (Alexandrino *et al.* 2000, 2002). Two divergent mtDNA lineages, distributed south and north of the Mondego valley, were observed in populations of *C. lusitanica*, implying the same past fragmentation processes promoting divergence at nuclear allozyme loci (Figure 3) (Alexandrino *et al.* 2000, 2002). Levels of local population mtDNA diversity revealed that populations near the river valley had a more stable demography in the past when compared with other populations to the south and to the north which may have undergone demographic and range expansions (Alexandrino *et al.* 2002).

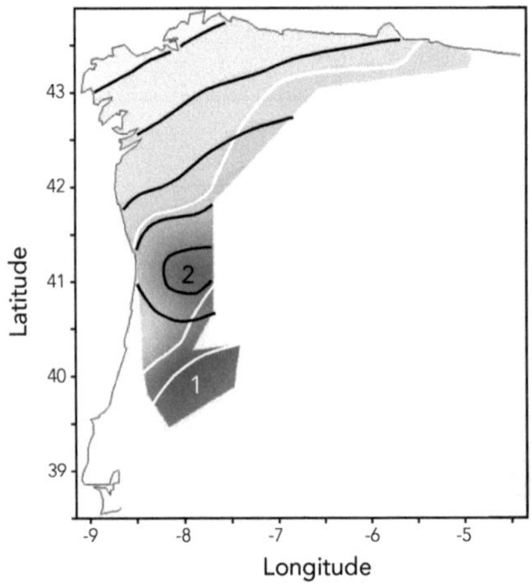

Figure 2*. Superimposition of maps A and B of Figure 1, weighting the variance explained by each of the two PCs. The first principal component is depicted by white contour lines and the second principal component by dark contour lines with the numbers 1 and 2 representing the density center of each component's gradient. The two components correspond to two distinct evolutionary lineages in* Chioglossa lusitanica.

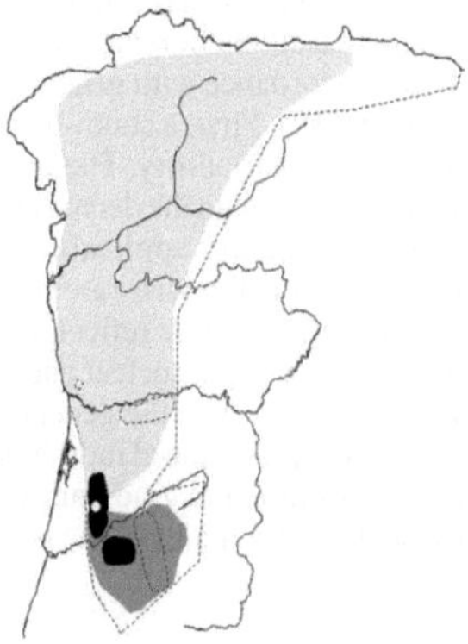

***Figure 3**. Geographic distribution of the two mtDNA clades observed in* Chioglossa lusitanica. *The two clades are represented with distinct gray shading. The two entire clades and dotted-line areas may have undergone recent population expansions while black shaded areas represent historically more stable populations, according to mtDNA diversity analysis (see text for details).*

Phenotypic vs. genetic diversity

Morphometric variation in *C. lusitanica* was found to be consistent with documented genetic differentiation. Populations south of the Mondego River are characterized by shorter digits than populations to the north, as revealed by PCA (Figure 4A) (Alexandrino 2000; Alexandrino *et al.*, unpublished). However, a finer scaled assessment reveals a stepped south-to-north cline of increasing limb, toe- and finger length. We suggest that both historical isolation (vicariance) and selection account for the observed variation. Short appendages, with a low volume to surface ratio, may represent an adaptation to xeric environments (Nevo 1972; Lee 1993). *C. lusitanica* is a terrestrial streamside salamander extremely dependent on moist habitats and indeed the level of annual precipitation is the main predictor of its range in Portugal (Teixeira *et al.* 2001). Given that southern populations appear to occupy a more xeric environment than northern populations (Arntzen & Alexandrino 2004) and assuming that rainfall gradients in the past paralleled those found today, selection could have produced the documented (stepped) clines. Neither the pattern nor variability in color was associated with group membership or with geographic distances between populations (Alexandrino 2000; Alexandrino *et al.*, unpublished). However, color pattern variability was higher within the contact zone than elsewhere, suggesting that the mixing of differentiated gene pools increased phenotypic variation. Two additional phenotypic characters show concordance with genetic variation within the northern population group (Figure 4B). First, a south-to-north decrease was observed in genetic and color pattern variability. The processes of sequential bottlenecking and drift invoked to explain the decrease in genetic variation (Alexandrino *et al.* 2000) appear equally applicable to morphological variation. Secondly, the dominance of an otherwise rare color type in populations immediately south of the Douro River may reflect a separate historical refuge, as suggested by the presence of unique nuclear and mtDNA alleles (Alexandrino *et al.* 2000, 2002). Overall, however, the genetic subdivision of *C. lusitanica* is not matched by an equally pronounced morphological differentiation. Selection operating along environmental gradients appears to be more important in shaping phenotypic diversity than genetic isolation.

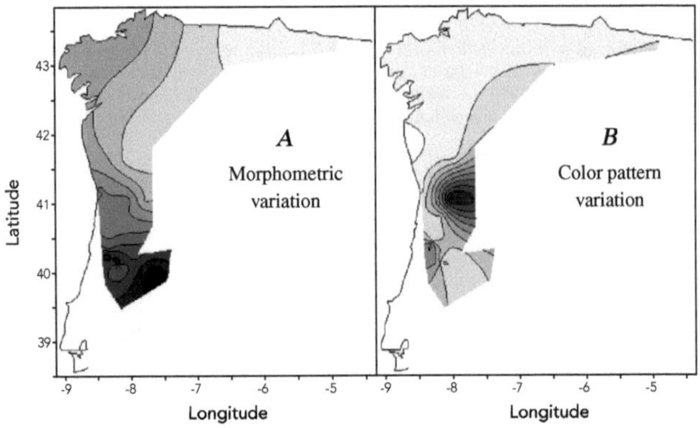

Figure 4*. Phenotypic diversity in* Chioglossa lusitanica*: **A)** trend surface map for morphometric variation generated by the kriging of mean factor scores of the first Principal Component axis (47% of total variation) of 18 populations of female C. lusitanica; **B)** trend surface map for color pattern variation generated by the kriging of mean factor scores of the first Correspondence Analysis axis (61% of total variation) of 20 populations of* C. lusitanica.

Insights from the ecological models and comparison with *Lacerta schreiberi*

Ecological models for the past and present day distribution of *C. lusitanica* help to understand the historical biogeography of the species. First, the eastern

Mondego valley receives low annual precipitation and is thus poor habitat exhibiting a correspondingly low probability for *C. lusitanica* occurrence according to the ecological model (Figure 5A). This region effectively separates southern and northern populations, at least in the central and eastern part of the species range. Under adverse climatic conditions of the Pleistocene, isolation may have been complete, supporting the hypothesis of vicariance across the Mondego river valley (Teixeira & Arntzen 2002).

Second, a discrepancy exists between the current distribution of *C. lusitanica* and the model for northern Spain from eastern Cantabria to the Pyrenees (Figure 5B). Parameters invoked to explain species absence in this area have been soil type and the presence of a competing species (Vences 1997; Teixeira *et al.* 2001). However, the one Iberian amphibian that has similar habitat characteristics – *Euproctus asper* – and could possibly outcompete *C. lusitanica*, is confined to the Pyrenees and thus the two species do not co-occur (Teixeira *et al.* 2001). An alternative explanation to a limiting ecological factor would be that *C. lusitanica* is still in the process of expanding its range. However, with the rafting of larvae (Thiesmeier 1994) and substantial migration distances of several hundreds of meters overnight (Arntzen 1981, 1994), low dispersal ability is an unlikely explanation for the unoccupied area.

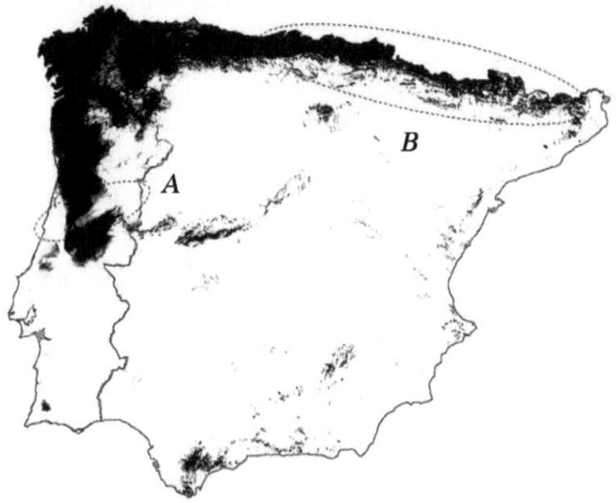

Figure 5. *Predictive distribution map of* Chioglossa lusitanica *in the Iberian Peninsula following the data presented in Teixeira* et al. *(2001).* ***A)*** *Mondego valley;* ***B)*** *northeastern Spain, from eastern Cantabria to the Pyrenees, where* C. lusitanica *does not occur.*

Third, the documented distribution of *C. lusitanica* appears continuous across the range, with the exception of an isolated population in the Serra de Sintra (see below). However, several potentially suitable areas are shown in the model to the south of the current distribution (e.g. Monchique and the central Spanish system). These isolated mountain ranges may have been out of range for colonization, the local populations may have gone extinct, or species presence remains unrecorded. The Sintra population may have resulted from an introduction in the past century (Arntzen 1999). Its rediscovery and subsequent investigation may prove highly informative. Genetic variation across *C. lusitanica*'s range is substantial, providing the possibility to distinguish between introduced and native occurrences. If introduced, it would show the survival of a population over six decades, confirming the general habitat suitability of the area, as predicted from the model. If native, it would indicate that *C. lusitanica* had a historical distribution that was more extensive than presently recognized.

Lacerta schreiberi has a range similar to that of *C. lusitanica* but occurs in several isolated mountain ranges of Portugal and central Spain (Brito *et al.* 1998). Ecological modeling applied to *L. schreiberi* (Brito *et al.* 1996) and the genetic structuring of populations (Paulo *et al.* 2002; Godinho *et al.* 2003, 2006a,b) suggests that the isolated populations resulted from range fragmentation due to climate warming since the last glacial maximum. This species, possessing broader ecological tolerance than *C. lusitanica* and possibly a higher dispersal ability, may have reached northeastern Iberia through postglacial recolonization (Paulo *et al.* 2002). *L. schreiberi* meets the congeneric *L. bilineata* in northern Spain close to the French border in a parapatric contact zone, the boundaries of which are maintained by interspecific competition (Barbadillo *et al.* 1999).

An historical biogeographical scenario

The joint interpretation of two independent data sets – biogeographical and genetic – strongly supports a vicariant scenario for the history of populations south and north of the Mondego River. Climatic change in the Pleistocene resulted in subdivision of the species range, with refugia located south and north of the river (Figure 6A). Sequential glacial and interglacial periods would have provoked recurrent range contraction and expansion (Figures 6A & 6B). After the last glacial maximum (18 kyr BP) range expansion shaped the present-day geographical distribution of *C. lusitanica*. In the south, a secondary contact zone resulted from the expansion of the two putative refugia and was observed near the Mondego valley (Figure 6C; Alexandrino *et al.* 2000). The Mondego valley appears to have played an important role as a dispersal barrier for *C. lusitanica*, either serving as a complete barrier to gene flow during some period in the past, or as a zone of low population density that limited introgression between the two groups

(Figure 6C). To the north, the stepwise decrease in genetic variability as measured by heterozygosity, allelic richness and haplotype diversity (Alexandrino *et al.* 2000) is consistent with a major postglacial colonization originating from a southern Douro refuge.

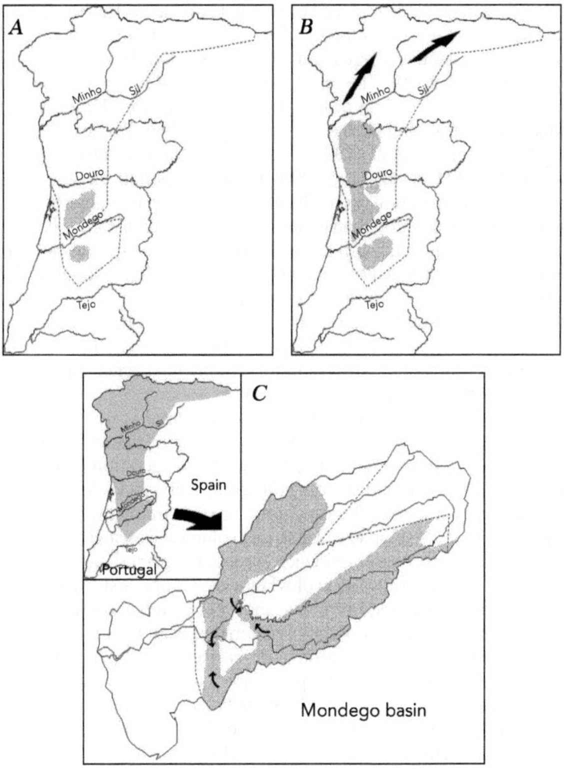

Figure 6*. Historical biogeographical scenario for* Chioglossa lusitanica*: **A)** separation of two Pleistocene refugia (dotted areas). The present day species range is after Arntzen (1999); **B)** postglacial expansion from refugial areas (dotted areas and arrows); **C)** secondary contact limited by a zone of less favorable habitat near the valley of the Mondego River (undotted area within range).*

Implications for conservation

Synthetic genetic maps complemented by ecological models of past and present-day species distribution identified a southern and a northern center of genetic variation. These areas could be associated with Pleistocene glacial refugia. The population groups are distinguished by largely concordant variation across several nuclear and cytoplasmic genes, as well as morphological variation. This suggests a long-standing evolutionary divergence between groups. Introgression is limited and spatially restricted (Alexandrino 2000; Alexandrino *et al.* 2000). The management status of the two groups should be determined. The concept of Evolutionary Significant Units (ESU) was introduced to help answer such a question (Ryder 1986; Waples 1991; Moritz 1994). Following Peatkau (1999), we favor a holistic definition of ESU (Bernatchez 1995; Crandall *et al.* 2000) over more restrictive criteria that may be problematic at the intraspecific level. Accordingly, we support the recognition of two conservation management units in *C. lusitanica*. The observation that almost all genetic variation observed across the species range is also found between the Muradal Mountains and the Douro River suggests that this area should be central in conservation planning.

It is now well established that the Iberian Peninsula served as a major refuge during Pleistocene Ice Ages (Hewitt 1999). However, the consequences of this fact are still a matter of discussion. Some researchers emphasize the role this and other European peninsulas had as a source of postglacial migrations and associated recolonizations at a continental scale for a variety of organisms (Hewitt 1999), while others suggest instead that those phenomena promoted long-term fragmentation followed by speciation and endemism (Bilton *et al.* 1998). Notwithstanding this controversy, it is clear that the present-day patterns of Iberian flora and fauna are not simply explained by these alternative models (see Gómez & Lunt, this volume). Research on the golden-striped salamander provides evidence for an unexpected natural history of populations and begs the question of whether other organisms with different, less explicit ecological requirements, show the same or similar patterns of fragmentation and dispersal. Parallel patterns of regional diversity in Iberia could have profound implications for conservation planning. Various researchers working on Iberian herpetofauna have indeed presented remarkably concordant results, some of which are listed as follows:

1) In *L. schreiberi*, two highly divergent mtDNA lineages were described, revealing an ancient split (Paulo *et al.* 2001). However, nuclear gene data obtained from both electrophoretic analysis of allozyme variation and the sequencing of nuclear genes (Godinho *et al.* 2003, 2006a,b) showed highly discrepant patterns when compared with mtDNA, suggesting a more complex history of populations. The latter authors also predict processes of hybridization and admixture along contact zones between the two divergent mtDNA lineages, and a postglacial expansion to the north.

2) Similarly, two highly divergent mtDNA lineages are detected in the natterjack toad, *Bufo calamita*, (Rowe *et al.* 2006) and the marbled newt, *Triturus marmoratus* (García-París *et al.* 2001) separating the north and the south of Iberia, thus suggesting a common phylogeographic history.

3) Using genetic data, as well as external morphology and morphometrics, Sánchez-Herráiz *et al.* (2000) described a new Pelodytidae species (*Pelodytes ibericus*) from the southern Iberian Peninsula (previously considered the parsley frog, *Pelodytes punctatus*). A contact zone and admixture is also predicted between the two taxa. This set of results in different amphibian and reptile species from the Iberian Peninsula clearly suggests that Pleistocene climatic oscillations associated with the high persistence times of species occupying a southern European refuge acted in combination to cause i) a pattern of fragmentation and deep subdivision in most Iberian amphibians and reptiles (Gómez & Lunt, this volume) that now generally show two evolutionary lineages, ii) a post-Würm expansion from glacial refugia, leading to the formation of secondary hybrid zones, and iii) a wide variety of scenarios of hybridization and admixture, some of which may be described as incipient or completed endemic speciation events.

Acknowledgements

This work was partially supported by the 'Instituto da Conservação da Natureza' (ICN), the 'LIFE' program of the European Community and the 'Fundação para a Ciência e Tecnologia' (PRAXIS XXI/BD/5917/95 and BD/3376/2000 PhD grants to JA and JT respectively, and research project POCTI/BSE/40987/2001). Licensing was provided by ICN for Portugal and by the Galician and Asturian 'Junta de Medio Ambiente' for Spain. We thank Steve Weiss for comments and helpful suggestions that improved an earlier version of the manuscript.

References

Alexandrino J (2000) ***Diversidade genética e morfológica na Salamandra lusitânica,* Chioglossa lusitanica *(Amphibia: Urodela): biogeografia histórica, implicações taxonómicas e conservação.*** PhD Thesis, University of Oporto.

Alexandrino J, Ferrand N, Arntzen JW (1997) Genetic variation in some populations of the golden-striped salamander, *Chioglossa lusitanica* (Amphibia: Urodela), in Portugal. ***Biochemical Genetics*** **35**, 371-381.

Alexandrino J, Froufe E, Arntzen JW, Ferrand N (2000) Genetic subdivision, glacial refugia and postglacial recolonization in the golden-striped salamander, *Chioglossa lusitanica* (Amphibia: Urodela). ***Molecular Ecology*** **9**, 771-781.

Alexandrino J, Arntzen JW, Ferrand N (2002) Nested clade analysis and the genetic evidence for population expansion in the phylogeography of the golden-striped salamander, *Chioglossa lusitanica* (Amphibia: Urodela). ***Heredity*** **88**, 66-74.

Arntzen JW (1981) Ecological observations on *Chioglossa lusitanica* (Caudata, Salamandridae). ***Amphibia-Reptilia*** **1**, 187-203.

Arntzen JW (1994) Speedy salamanders: sedentariness and migration of *Chioglossa lusitanica*. ***Revista Española Herpetología*** **8**, 81-86.

Arntzen JW (1999) *Chioglossa lusitanica* Bocage, 1864 – Goldstreifensalamander. In: ***Handbuch der Amphibien und Reptilien Europas*** (eds. Grossenbacher K, Thiesmeier B), pp. 301-321. AULA-Verlag, Wiesbaden.

Arntzen JW, Alexandrino J (2004) Ecological modelling of genetically differentiated forms of the Iberian endemic Golden-striped salamander, *Chioglossa lusitanica*. ***Herpetological Journal*** **14**, 137-141.

Austin GE, Thomas CG, Houston DC, Thompson DBA (1996) Predicting the spatial distribution of the buzzard *Buteo buteo* nesting areas using Geographical Information Systems and Remote Sensing. ***Journal of Applied Ecology*** **33**, 1541-1550.

Avise JC (1994) ***Molecular Markers, Natural History and Evolution.*** Chapman & Hall, New York.

Avise JC (1998) The history and purview of phylogeography: a personal reflection. ***Molecular Ecology*** **7**, 371-379.

Avise JC (2000). ***Phylogeography: The History and Formation of Species.*** Harvard University Press, Cambridge.

Barbadillo LJ, Lacomba JI, Pérez-Mellado V, Sancho V, López-Jurado LP (1999) ***Anfibios y reptiles de la Península Ibérica, Baleares y Canarias: Guía ilustrada para conocer todas las especies.*** Editorial GeoPlaneta S.A., Barcelona.

Bermingham E, Moritz C (1998) Comparative phylogeography: concepts and applications. ***Molecular Ecology*** **7**, 367-369.

Bernatchez L (1995) A role for molecular systematics in defining evolutionary significant units in fishes. ***American Fisheries Society Symposium*** **17**, 114-132.

Bilton DT, Mirol PM, Mascheretti M, Fredga K, Zima J, Searle JB (1998) Mediterranean Europe as an area of endemism for small mammals rather than a source for northwards postglacial colonization. ***Proceedings of the Royal Society of London, B*** **265**, 1219-1226.

Bosch E, Calafell F, Perez-Lezaun A, Comas D, Mateu E, Bertranpetit J (1997) Population history of North Africa: evidence from classical genetic markers. ***Human Biology*** **69**, 295-311.

Brito JC, Brito e Abreu F, Paulo OS, Rosa HD, Crespo EG (1996) Distribution of Schreiber's green lizard (*Lacerta schreiberi*) in Portugal: a predictive model. ***Herpetological Journal*** **6**, 43-47.

Brito JC, Paulo OS, Crespo EG (1998) Distribution and habitats of Schreiber's green lizard (*Lacerta schreiberi*) in Portugal. ***Herpetological Journal*** **8**, 187-194.

Cavalli-Sforza L, Menozzi P, Piazza A (1993) Demic expansions and human evolution. ***Science*** **259**, 639-646.

Cavalli-Sforza L, Menozzi P, Piazza A (1994) ***The History and Geography of Human Genes.*** Princeton University Press, Princeton.

Crandall KA, Bininda-Emonds ORP, Mace GM, Wayne RK (2000) Considering evolutionary processes in conservation biology. ***Trends in Ecology and Evolution*** **15**, 290-295.

Eastman JR (1997) ***Idrisi for Windows: User's Guide. Version 2.0.*** Clark University, Worcester, USA.

Emerson BC, Paradis E, Thébaud C (2001) Revealing the demographic histories of species using DNA sequences. ***TREE* 16**, 707-716.

García-París M, Arano B, Herrero P (2001) Molecular characterization of the contact zone between *Triturus pygmaeus* and *T. marmoratus* (Caudata: Salamandridae) in Central Spain and their taxonomic assessment. ***Revista Española de Herpetología* 15**, 115-126.

Godinho R, Paulo OS, Ferrand N, Luis C, Rosa HD, Crespo EG (2003) Major patterns of population differentiation in the Iberian Schreiber's Green lizard (*Lacerta schreiberi*) inferred from Protein polymorphism. ***Herpetological Journal* 13**, 35-42.

Godinho R, Domingues V, Crespo EG, Ferrand V (2006a) Extensive intraspecific polymorphism detected by SSCP at the nuclear C-*mos* gene in the endemic Iberian lizard, *Lacerta schreiberi*. ***Molecular Ecology* 15**, 731-738.

Godinho R, Mendonça B, Crespo EG, Ferrand N (2006b) Genealogy of the nuclear β-fibrinogen locus in a highly structured lizard species: comparison with mtDNA and evidence for intragenic recombination in the hybrid zone. ***Heredity***, doi:10.1038/sj.hdy.6800823.

Golden Software (1996) ***Surfer for Windows (Win32): Surface Mapping System, Version 6.04.*** Golden Software, Inc., Golden.

Guinand B, Easteal S (1996) Multivariate patterns of genetic differentiation support complex colonization schemes in *Bufo marinus* populations. ***Evolution* 50**, 944-951.

Hare MP (2001) Prospects for nuclear gene phylogeography. ***Trends in Ecology and Evolution* 16**, 700-706.

Hewitt GM (1996) Some genetic consequences of ice ages, and their role in divergence and speciation. ***Biological Journal of the Linnean Society* 58**, 247-276.

Hewitt GM (1999) Post-glacial re-colonization of European biota. ***Biological Journal of the Linnean Society* 68**, 87-112.

Hewitt GM (2000) The genetic legacy of the Quaternary ice ages. ***Nature* 405**, 907-913.

Hugall A, Moritz C, Moussalli A, Stanisic J (2002) Reconciling paleodistribution models and comparative phylogeography in the Wet tropics. ***Proceedings of the National Academy of Sciences, USA* 99**, 6112-6117.

Le Corre V, Roussel G, Zanetto A, Kremer A (1998) Geographical structure of gene diversity in *Quercus petraea* (Matt.) Liebl. III. Patterns of variation identified by geostatistical analyses. ***Heredity* 80**, 464-473.

Lee JC (1993) Geographic variation in size and shape of neotropical frogs: a precipitation gradient analysis. ***Occasional Papers of the Museum of Natural History, the University of Kansas* 163**, 1-20.

Menozzi P, Piazza A, Cavalli-Sforza L (1978) Synthetic maps of human gene frequencies in Europeans. ***Science* 201**, 786-792.

Mladenoff DJ, Sickley TA (1998) Assessing potential grey wolf restoration in the northeastern United States: a spatial prediction of favourable habitat and potential population levels. ***Journal of Wildlife Management* 62**, 1-10.

Moritz C (1994) Defining 'Evolutionary Significant Units' for conservation. ***Trends in Ecology and Evolution* 9**, 373-375.

Moritz C (2002) Strategies to protect biological diversity. ***Systematic Biology* 21**, 238-254.

Nevo E (1972) Climatic adaptation in size of the green-toad, *Bufo viridis*. ***Israel Journal of Medical Science*** **54**, 1010.

Paetkau D (1999) Using genetics to identify intraspecific conservation units: a critique of current methods. ***Conservation Biology*** **13**, 1507-1509.

Paulo OS, Dias C, Bruford MW, Jordan WC, Nichols RA (2001) The persistence of Pliocene populations through the Pleistocene climatic cycles: evidence from the phylogeography of an Iberian lizard. ***Proceedings of the Royal Society of London, B*** 268, 1625-1630.

Paulo OS, Jordan WC, Bruford MW, Nichols RA (2002) Using nested clade analysis to assess the history of colonization and persistence of populations of an Iberian Lizard. ***Molecular Ecology*** **11**, 809-819.

Pereira J, Itami R (1991) GIS-based habitat modelling using logistic multiple regression: a study of the Mt. Graham Red Squirrel. ***Photogrammetric Engineering & Remote Sensing*** 57, 1475-1486.

Piazza A, Menozzi P, Cavalli-Sforza L (1981) The making and testing of geographic gene-frequency maps. ***Biometrics*** **37**, 635-659.

Ray N, Currat M, Berthier P, Excoffier L (2005) Recovering the geographic origin of early modern humans by realistic and spatially explicit simulations. ***Genome Research*** **15**, 1161-1167.

Rowe G, Harris DJ, Beebee TJC (2006) Lusitania revisited: a phylogeographic analysis of the natterjack toad *Bufo calamita* across its entire biogeographical range. ***Molecular Phylogenetics and Evolution*** **39**, 335-346.

Ryder OA (1986) Species conservation and systematics: the dilemma of subspecies. ***Trends in Ecology and Evolution*** **1**, 9-10.

Sánchez-Herráiz MJ, Barbadillo J, Machordom A, Sanchíz B (2000) A new species of Pelodítid frog from the Iberian Peninsula. ***Herpetologica*** **56**, 105-118.

Sequeira F, Alexandrino J, Rocha R, Arntzen JW, Ferrand N (2005) Genetic exchange across a hybrid zone within the Iberian endemic golden-striped salamander, *Chioglossa lusitanica*. ***Molecular Ecology*** **14**, 245-254.

StatSoft (1993) ***STATISTICA for Windows, Release 4.5.*** StatSoft Inc., Tulsa.

Taberlet P, Fumagalli L, Wust-Saucy A, Cosson J (1998) Comparative phylogeography and postglacial colonization routes in Europe. ***Molecular Ecology*** **7**, 453-464.

Teixeira J, Arntzen JW (2002) Potential impact of climate warming on the distribution of the Golden-striped salamander, *Chioglossa lusitanica*, on the Iberian Peninsula. ***Biodiversity and Conservation*** **11**, 2167-2176.

Teixeira J, Ferrand N, Arntzen JW (2001) Biogeography of the golden-striped salamander, *Chioglossa lusitanica*: a field survey and spatial modelling approach. ***Ecography*** **24**, 618-624.

Templeton AR (1998) Nested clade analyses of phylogeographic data: testing hypotheses about gene flow and population history. ***Molecular Ecology*** 7, 381-397.

Templeton AR (2002) Out of Africa again and again. ***Nature*** **416**, 45-51.

Thiesmeier B (1994) Trophische Beziehungen und Habitatpräferenzen sympatrisch lebender *Salamandra salamandra* und *Chioglossa lusitanica*-Larven. ***Abhandlungen und Berichte für Naturkunde*** **17**, 119-126.

Velázquez A, Bocco G (1994) Modelling conservation alternatives with ILWIS: a case study of the volcano rabbit. ***ITC Journal*** **1994**, 197-204.

Vences M (1997) *Chioglossa lusitanica* Barboza du Bocage, 1864. Salamandra rabilarga; Salamandra-dourada. In: ***Distribución y biogeografía de los anfibios y reptiles en España y Portugal*** (ed. Pleguezuelos JM), pp. 103-105. Universidad de Granada & A.H.E., Granada.

Walker PA (1990) Modelling wildlife distributions using geographic information system: kangaroos in relation to climate. ***Journal of Biogeography*** **17**, 279-289.

Waples RS (1991) Pacific salmon, *Oncorhynchus* spp., and the definition of 'species' under the endangered species act. ***Marine Fisheries Review*** **53**, 11-22.

Chapter 7

The evolutionary history of the European rabbit (*Oryctolagus cuniculus*): major patterns of population differentiation and geographic expansion inferred from protein polymorphism

Nuno Ferrand and Madalena Branco
E-mail: nferrand@mail.icav.up.pt

Abstract
The patterns of population differentiation and geographical expansion of the European rabbit (*Oryctolagus cuniculus*) remain largely unknown. Using gene frequency data for 20 polymorphic protein loci (102 alleles), we investigated the evolutionary history of the rabbit through the analysis of 13 representative populations and the use of both the neighbor-joining (NJ) and the unweighted pair-group method with arithmetic mean (UPGMA) trees. We also conducted a separate analysis comparing one domestic and one wild population with previously published results. Our data indicate that an ancient split separated southwestern Iberian populations from all others, including domestic breeds, and that this division may have corresponded to the emergence of the subspecies *O.c. algirus* and *O.c. cuniculus*. Separation times between the two major groups of populations were estimated with Nei's genetic distance and were found to be highly discrepant with the mtDNA divergence estimate. The southwestern Iberian populations (*algirus* group) are more polymorphic than northern populations (*cuniculus* group), the latter displaying more than simply a subset of southern alleles. These results are thus compatible with the isolation of a marginal population or with a smaller long-term population size in the north. The high degree of genetic differentiation between the two subspecies allows the reconstruction of rabbit geographical expansion. France, Britain and other European countries, as well as Australia, were colonized by animals belonging to the *cuniculus* group, from which domestic breeds are exclusively derived. In contrast, Azorean island populations represent an expansion of the *algirus* group and show evidence of a strong bottleneck effect.

Keywords: protein polymorphism, phylogenetic trees, major groups of populations, geographical expansion, European rabbit, *Oryctolagus cuniculus*

Introduction

The European rabbit *Oryctolagus cuniculus* is a fascinating species for a variety of reasons. First, it constitutes one of the most remarkable geographical expansions of a mammal. Originally restricted to the Iberian Peninsula

S. Weiss and N. Ferrand (eds.), Phylogeography of Southern European Refugia, 207–235.

and the Mediterranean area of France (Callou 1995), it was probably introduced into North Africa during historical times (Dobson 1998), arrived in Britain around the 11th century, and later expanded into most of central and northeastern Europe (19th and 20th centuries). It was introduced into Australia, New Zealand, Chile and Argentina, as well as more than 800 islands throughout the world. This very successful process of transport and colonization began with the first contacts made by Mediterranean navigators (*ca.* 1000 years BC, e.g. the Balearic Islands, Spain, or Zembra, Tunisia) and still continues today (Flux & Fullagar 1992). Second, this successful colonizer now occupies regions with a remarkable diversity of ecological contexts, from subtropical to sub-Antarctic climates and, dependent on the various communities to which it is now adapted is viewed as a pest (Williams *et al.* 1995), a biological invader (Barrett & Richardson 1986) or a key species on which a variety of threatened predators depend to survive (Rogers *et al.* 1994; Villafuerte 1994). Third, throughout history man has expressed curiosity for the rabbit: Latin and Greek historians like Varro, Strabo or Pliny the Old, described the maintenance of animals in closed parks (*leporaria*), the introduction of animals on islands, ferret hunting or the consumption of rabbit embryos by humans. This perpetual interest in the rabbit culminated with western Europe's only successful attempt at animal domestication – a feat of ethnozoological significance. Today, the rabbit is an important laboratory animal for biomedical research (Weisbroth *et al.* 1974), serving as a highly valued model for research on the mammalian immunological system (Dubiski 1987; van der Loo 1993; Su & Nei 1999) and for human arteriosclerosis (Beaty *et al.* 1992). Further, rabbits are severely affected by epizootics like myxomatosis and the rabbit viral hemorrhagic disease (RVHD), offering opportunities for studying host/parasite coevolution (Lewontin 1970; Anderson & May 1982; Langman 1989) and its effects on the genetic diversity of wild populations (Queney *et al.* 2000).

The fossil record of the rabbit is very scarce: a single tooth found in the region of Granada (Andalusia, Spain) marks the appearance of the genus *Oryctolagus* six mya (Lopez-Martinez 1989). Other more complete fossil remains have been attributed to two species with intermediate characteristics between present day rabbits and hares: *O. laynensis* and *O. lacosti*. The first species seems to have occupied the whole of Spain about 2-3 mya, while the second reached southern France and existed about 1.8 to 2 mya (Lopez-Martinez 1989). *O. cuniculus* appears 900 000 years ago in southern Spain, but much later in southern France (approx. 300 000 years ago). More recent evidence is provided by archaeozoological research: in Portugal, the rabbit is the most abundant mammal in Mesolithic sites from the Tejo and Sado valleys (Lentacker 1986; Arnaud 1987), and remarkable bone ornaments in the shape of rabbits date to the second half of the 4th millenium BC (Leisner 1983). Early introductions of rabbits to some Mediterranean islands are also documented in archaeozoological findings, like Menorca, in the Balearics (1400-1300 BC,

Reumer & Sanders 1984) and Zembra (end of the Neolithic to the 3rd century, Vigne 1988). The situation is less clear in the Mediterranean region of France where the rabbit seems to have maintained a fundamental role in the human diet until the end of the Palaeolithic (Pages 1980). Subsequently, rabbits may have been extirpated from this region only to be reintroduced during the Roman period or, alternatively, may have persisted in small refugia from where recolonization followed. A recent analysis of all archaeozoological evidence obtained in France south of river Loire seems to favor this last hypothesis (Donard 1982; Callou 1995).

Differentiation of rabbit ecto-parasites may provide insights on coevolutionary processes between hosts and parasites and has deserved some attention in the past (Beaucournu 1980). Three specific rabbit fleas were found to be especially important in the analysis of rabbit evolution: *Xenopsylla cunicularis*, *Odontopsyllus quirosi* and *Caenopsylla laptevi*. The first genus, *Xenopsylla*, contains a high number of species that mainly occur in tropical Africa where they parasitize Gerbilidae rodents (Beaucournu & Launay 1990). In contrast, *X. cunicularis* is specific to the rabbit, almost exclusively occurring in the Iberian Peninsula and southern France. Launay & Beaucournu (1982) suggest that *X. cunicularis* results from the capture and derivation of an ancestral flea that parasitized those rodents, whose fossorial habits are similar to those of the rabbit. This hypothesis is supported by the fact that Gerbilidae fossils, contemporary with *Oryctolagus* were found in the Iberian Peninsula (Launay & Beaucournu 1982). Additionally, the occurrence of *X. cunicularis* in wild rabbit populations from North Africa was interpreted as a secondary event resulting from a recent expansion of the rabbit (Launay & Beaucournu 1982). The other species, *O. quirosi* and *C. laptevi*, can be each divided into two distinct subspecies that correspond to different groups of rabbit populations. Accordingly, *O. q. quirosi* and *C. l. ibera* were found in the central and southern Iberian Peninsula, whereas *O. q. episcopalis* and *C. l. relicta* are limited to the French Mediterranean basin (Beaucournu & Launay 1990). Taken as a whole, ecto-parasitological evidence allowed a more precise definition of hypotheses arising from paleontological and archaeozoological data because i) the origin of *O. cuniculus* is unequivocally placed in the European side of the Mediterranean basin, and ii) two main rabbit population groups are identified, one in the Iberian Peninsula and the other in southern France, their separation being probably caused by Quaternary glaciations.

Historically, the first genetic evidence contributing to our knowledge of rabbit history came from the discovery of immunoglobulin allotypes (Mage *et al.* 1973). Extensive serological studies developed in domestic breeds revealed genetic polymorphism at the IgKC1, IgVH1 and IgGCH2 loci (van der Loo 1987; van der Loo *et al.* 1987; Cazenave *et al.* 1987). When wild rabbit populations from central Europe (especially France, England, Holland and Belgium) and Australia were investigated, the most remarkable result was the high degree of genetic identity with domestic populations: no new alleles were described and the allelic distribution profiles were very similar (Curtain *et al.* 1973; van der Loo 1987; van der Loo *et al.* 1987). However, these results also revealed the existence of a significant and systematic linkage disequilibrium between the IgKC1 and IgGCH2 loci in all populations, indicating the probable occurrence of selective mechanisms promoting particular genotype combinations (van der Loo *et al.* 1987). The extension of this investigation to wild rabbit populations from the Iberian Peninsula showed a completely different scenario: the loci were found to harbor much higher variation with seven to eight new alleles in IgKC1 (Cazenave *et al.* 1987; van der Loo *et al.* 1991) and at least 10 new alleles in IgVH1 (Cazenave *et al.* 1987), but no polymorphism was described for IgGCH2. Overall the results obtained so far clearly support an Iberian origin of the rabbit, but are not informative in the discrimination of animals outside of Iberia. Moreover, difficulties resulting from the application of serological techniques in the definition of immunological loci (Nei 1975), the population and molecular evidence for the occurrence of strong selective mechanisms involved in the structural divergence of alleles (van der Loo 1987, 1993; van der Loo & Verdoodt 1992) and also the fact that they often constitute transspecific polymorphisms (Cazenave *et al.* 1987; van der Loo *et al.* 1999; Su & Nei 1999; Esteves *et al.* 2005) raises doubts about the feasibility of obtaining a reliable phylogenetic tree of rabbit populations.

Monnerot and coworkers followed a different approach by investigating patterns of sequence variation in rabbit mitochondrial DNA (mtDNA) (see Monnerot *et al.* 1994 for a review). The results show the occurrence of two very divergent mtDNA lines (4%) (Biju-Duval *et al.* 1991), indicating that an ancestral molecule may have existed more than 2 mya, long before the first known fossils of *O. cuniculus*. One of these mtDNA lineages (type A) is circumscribed to southwest Iberia, while the other (type B) occurs in northern Spain, France, England, the rest of Europe, Australia and all domestic breeds (Biju-Duval *et al.* 1991; Monnerot *et al.* 1994). Recently, a comprehensive survey of Iberian wild rabbit populations showed that the two mtDNA lineages are essentially allopatric, with a very limited overlap along a northwest-southeast gradient that divides the peninsula (Branco *et al.* 2000). This investigation also confirmed that Iberian rabbits are characterized by high levels of inter- and intra-population variability, while French rabbits do not show intra-population polymorphism (Monnerot *et al.* 1994). While these data are compatible with an older age of southwestern Iberian populations and a more recent occupation of southern France, important aspects associated with the use of the single mtDNA gene tree may limit the usefulness of this genetic system for understanding rabbit evolution (Nei 1987; Pamilo & Nei 1988).

The principal aim of the present study was to survey the natural range of the species for the presence of main patterns of population differentiation and, secondarily, to compare populations from introduced and natural ranges in the context of the documented history of rabbit expansion (Thompson & King 1994). Previous studies with rabbit electrophoretic polymorphisms were of little use because a very limited number of both populations and markers were investigated (Richardson *et al.* 1980; Hartl 1987; Arana *et al.* 1989; Peterka & Hartl 1992). When population-level phenomena are investigated, hypotheses must be tested across many loci (Pamilo & Nei 1988; Zhivolovsky & Feldman 1995). This strategy has been particularly effective for reconstructing the natural history of *Drosophila melanogaster* (Singh & Rhomberg 1987), and the house mouse *Mus musculus* (Boursot *et al.* 1993), as well as gaining an understanding of modern human origins (Nei & Roychoudhury 1982, 1993; Cavalli-Sforza *et al.* 1988, 1994; Nei 1995; Chikhi *et al.* 1998). We have therefore used gene frequency data for 102 alleles corresponding to 20 polymorphic loci to perform a phylogenetic analysis of a set of representative rabbit populations and propose, for the first time, a global scale scenario for the evolutionary history of *O. cuniculus*.

Materials and methods

Sampling

The sampling was designed to include representative populations in the natural range of the species (Iberian Peninsula and southern France), but avoiding hybrid populations occurring in central Iberia (Branco *et al.* 2000), because they are known to violate the fundamental principles of phylogenetic reconstruction (Felsenstein 1982; Nei 1987). Accordingly, we sampled seven populations from the southwestern Iberian Peninsula (Huelva, Doñana National Park, Las Lomas, Vila Viçosa, Santarém, Badajoz, and Idanha), one population from northeastern Spain (Tudela), and one population from the French Mediterranean region (Camargue National Park). In addition, we included samples that are known to represent recent expansions of the species: northern France (Versailles) and two Azorean islands (S. Jorge and Flores) (Figure 1). Finally, a sample of Portuguese domestic rabbits was also collected. This option is justified because these animals are not heavily selected and are thought to be the closest representatives of the initial stages of rabbit domestication (Ferrand, unpublished results).

A total of 546 blood samples was collected during 1990-1995 in EDTA-tubes from the marginal ear vein or, alternatively, by direct cardiac puncture. Red cells were separated from sera by quick centrifugation and mixed (1:2) with a glycerolated Tris-citric buffer pH 8.0. Both red cells and sera were stored at -20^0 until the electrophoretic analysis was conducted.

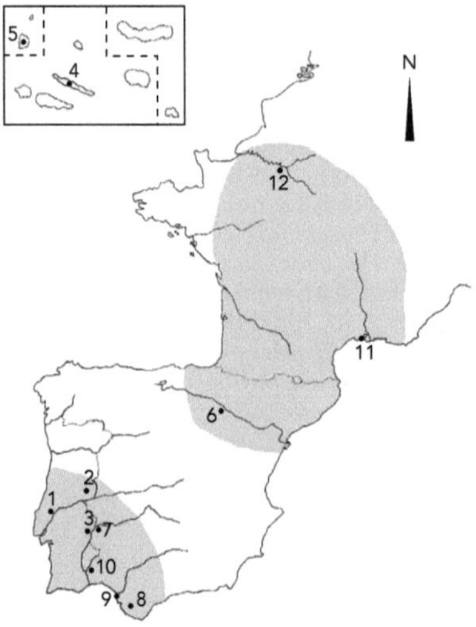

Figure 1. *Sampling locations of wild rabbit populations. 1 Santarém, 2 Idanha, 3 Vila Viçosa, 4 S. Jorge, 5 Flores, 6 Tudela, 7 Badajoz, 8 Las Lomas, 9 Doñana National Park, 10 Huelva, 11 Camargue, 12 Versailles.*

Protein analysis

The use of a limited number of electrophoretic systems to assay protein genetic variation is known to prevent the description of a considerable amount of hidden variability (Ramshaw *et al.* 1979; Brown *et al.* 1981), which severely limits phylogenetic inference (Lewontin 1991). However, the development of iso-electric focusing both in carrier ampholytes and immobilized pH gradients coupled with immuno-blotting detection techniques has substantially improved the ability to effectively discriminate electromorphs in their constituent alleles (Altland *et al.* 1987; Righetti *et al.* 1989; Righetti 1990). In this study, we chose a balanced strategy that included i) a detailed search for the adequate buffer systems and pH conditions for markers separated by conventional electrophoresis, ii) the analysis of denatured or chemically modified proteins when necessary, and iii) the use of high resolution iso-electric focusing systems whenever possible and sometimes in combination with conventional electrophoresis for the detection of subtypes.

A total of 20 polymorphic loci was examined. Conventional electrophoresis was used for typing i) peptidases (PEPA, PEPB, PEPC, EC 3.4.11/13; and PEPD, EC 3.4.13.9) following Branco *et al.* (1999), ii) nucleoside phosphorylase (NP, EC 2.4.2.1), adenosine deaminase (ADA, EC 3.5.4.4) and phosphogluconate dehydrogenase (PGD, EC 1.1.1.44) according to the technique described by Amorim *et al.* (1982), iii) carbonic anhydrase I and the types of carbonic anhydrase II (CAI/CAII, EC 4.2.1.1) following the systems originally proposed by Branco & Ferrand (2003), iv) mannose-6-phosphate isomerase (MPI, EC 5.3.1.8) as described in Vieira & Ferrand (1995), v) galactose-1-phosphate uridyltransferase (GALT, EC 2.7.7.12) following the methods of Siebert *et al.* (1980) and Ferrand (1995), and vi) transferrin (TF) (Ferrand *et al.* 1988). In addition, genetic variation in superoxide dismutase (SOD, EC 1.15.1.1) was routinely assayed in the same starch gel system used for peptidases (Ferrand, unpublished results). Extremely acid starch gel electrophoresis was employed in the analysis of the genetic polymorphisms of α- and β-globin genes (HBA and HBB, Ferrand 1989, 1990) after previous denaturation of hemoglobin. Iso-electric focusing in carrier ampholytes was used for typing i) albumin (ALB) (Ferrand & Rocha 1992), ii) acid phosphatase 3 (ACP3, EC 3.1.3.2) according to Branco & Ferrand (1998), iii) hemopexin (HPX) (Branco & Ferrand 2002), iv) vitamin-D binding protein (GC) (Ferrand 1995), v) glucose-phosphate isomerase (GPI, EC 5.3.1.9) following the technique established by Azevedo & Ferrand (unpublished results), and for the determination of subtypes in the loci ADA, GALT and TF (Ferrand 1995). Finally, hybrid iso-electric focusing was used for the separation of CAII subtypes (Branco & Ferrand 2003).

Samples were diluted sera or haemolysates for ACP3, ADA, ALB, GPI, MPI and SOD, as appropriate. When assaying GALT, NP, PGD and PEP's, samples were first reduced with dithiothreitol 120mM (5:1) for 1h at 37° C. The same procedure was used for the conventional separation of CAI/CAII, whereas for CAII subtypes samples were alternatively alkylated with iodoacetic acid 40mM in a 1:1 proportion. Globins (HBA and HBB) were obtained from diluted (1:5) haemolysates treated with acid acetone (Ferrand 1989). For GC, sera were treated with neuraminidase (*Clostridium perfringens*, SIGMA type V, 1.8 U/ml) in a 1:3 proportion for 18h at 37° C, while for HPX sera were diluted in distilled water in a proportion of 1:1. Finally, discrimination of TF alleles by IEF were made after partial purification with rivanol for two sets of samples: the first being iron-free due to an EDTA treatment, and the second being iron-saturated with ferric ammonium sulphate following (Zapolski & Princiotto 1980).

The recipes for enzyme staining of PGD, MPI, SOD, GALT, NP, ADA and GPI were adapted from Harris & Hopkinson (1976). Visualization of PEP's was done according to a procedure firstly described by Sugiura *et al.* (1977). CAI and CAII were previously identified by specific CO_2 hydration coupled with bromothymol blue detection; for routine assays, however, we used a standard Coomassie R-250 staining for CAI, and a fluorescein diacetate based detection method adapted from Harris & Hopkinson (1976) for CAII. Detection of ACP3 was made following Harris & Hopkinson (1976), but on nitrocellulose membranes according to an enzyme blotting procedure (Branco & Ferrand 1998). HBA, HBB, CAII subtypes and TF types were detected with a standard Coomassie R-250 staining, while for ALB and TF subtypes a standard Coomassie G-250 staining was used. Finally, both GC and HPX were detected after an immuno-blotting procedure described in Ferrand (1995).

Data analysis

Our set of 13 rabbit populations was analyzed for genetic distances using D_A, a modified Cavalli-Sforza distance suggested by Nei *et al.* (1983). These authors have shown that although not linearly related with time, D_A is more efficient in the separation of closely related populations, an assumption that is certainly verified for most of the samples used in this work. We therefore used the DISPAN package of Ota (1993). Phylogenetic reconstruction was done both with NJ and UPGMA methods. Although UPGMA trees are probably the most commonly used in the population genetics literature (Nei 1987), NJ trees are known to be more efficient than most other methods of phylogenetic reconstruction (Saitou & Imainishi 1989; Rzhetsky & Nei 1992). Reliability of nodes in the trees were assessed by bootstrapping loci (1000 replicates) using the DISPAN software, as above.

Various researchers have been involved in the investigation of rabbit evolution (Richardson *et al.* 1980; Arana *et al.* 1989; Peterka & Hartl 1992). Although differing in the number of loci and alleles used, these studies suggested that rabbit populations show low levels of differentiation and, notably, the last two concluded that the genetic distance between wild and domestic samples was not necessarily higher than that existing between two wild populations or between two domestic breeds. We therefore compared our results with those described earlier adopting the following strategy: i) selecting one wild rabbit population (Santarém) and the Portuguese domestics as representatives of our sampling, ii) constructing allele x population matrices for each of the works mentioned above but including the two Portuguese samples and accepting missing values for loci not investigated in our work, and iii) obtaining phylogenetic trees with and without the Portuguese samples following the exact procedures (genetic distances and reconstruction methods) adopted by the authors. Briefly, we used loci ADA, CAII, ES1 and PGD for comparisons made with the data of Richardson *et al.* (1980), ADA, CAII, DIA, ES1, ES3, HB, PGD and TF for comparisons made with the data of Arana *et al.* (1989), and ES1, GPI, MPI, PGD and PGM2 for comparisons made with the data of Peterka & Hartl (1992). Accordingly, we also typed the markers ES1, ES3, DIA and PGM2 for the two Portuguese samples (Santarém and domestic) in addition to the 20 electrophoretic polymorphisms analyzed in the present study (results not shown). Because of different separation techniques used by the various research groups, subtypes of ADA, CAII and TF were collapsed in conventional electromorphs for comparison purposes. Likewise, HB was treated as a single locus.

A hierarchical gene diversity analysis following the method described by Nei (1973, 1987) was done considering relevant levels of population subdivision: populations and subspecies (*O.c. algirus* and *O.c. cuniculus*). The Portuguese domestic stock was excluded from the analysis. In these conditions, total gene diversity (H_T) is partitioned into its components so that:

$$H_T = H_P + D_{PS} + D_{ST},$$

where H_P is average gene diversity within populations, D_{PS} is gene diversity between populations within subspecies and D_{ST} corresponds to diversity between subspecies. The relative importance of the three components are expressed in terms of G values (G_P, G_{PS}, and G_{ST}), that are obtained from the ratio of each component to H_T.

Measures of genetic variability (mean expected heterozygosity, H, mean number of alleles, n_a, and proportion of polymorphic loci, P, were obtained with the package BIOSYS, version 1.7 (Swofford & Selander 1989).

Results

Genetic distances and phylogenetic trees

The matrix of D_A genetic distances calculated after 20 polymorphic loci (102 alleles) for all 13 rabbit populations studied in this work is shown in Table 1, whereas the allelic frequencies are given in the appendix of this chapter (Table A1). The phylogenetic trees were constructed by applying both NJ (Saitou & Nei 1987) (Figure 2) and UPGMA (Sneath & Sokal 1973) (Figure 3) methods.

There are two major clusters of rabbit populations: A) all rabbit populations from the southwest of the Iberian Peninsula and the two samples from the Azorean islands of São Jorge and Flores, and B) the population from the northeast of Spain, the two French populations and the Portuguese domestic breed. The root of the NJ tree was found after evaluating the midpoint of the longest branch between two populations (Figure 2) (Farris 1972). This was necessary because the use of a closely related species as an outgroup (e.g. the Iberian hare, *Lepus granatensis*) is presently impossible due to the lack of adequate data. The complete separation of population groups A and B exhibits high bootstrap support in both the UPGMA (99%) and NJ (100%) trees. The major difference in the topology of the trees is associated with the position of the Azorean island populations: while in the UPGMA tree they correspond to the first split within group A (95% bootstrap value) and are separated from the southwestern populations, in the NJ tree they are clearly associated with a central Portuguese population (Vila Viçosa) and well inside group A, showing long branch lengths. The explanation for this difference involving populations from the islands of São Jorge and Flores is based in the assumptions for each grouping method. It is well known that rabbits were introduced in the Azorean Islands by Portuguese navigators in the beginning of the 15th century, most likely corresponding to a genetic bottleneck followed by high genetic drift. This is equivalent to an acceleration of the evolutionary rate and implicates large genetic distances that may be properly analyzed by the NJ method (Saitou & Nei 1987; Nei & Roychoudhury 1993) but incorrectly so with a UPGMA tree.

***Table 1**. Estimates of* D_A *distances for 13 representative rabbit populations based on gene frequency data for 20 polymorphic protein data. Values are multiplied by 100.*

Population	1	2	3	4	5	6	7	8	9	10	11	12	13
1 Santarém	-												
2 Idanha	8.00	-											
3 Vila Viçosa	3.27	7.47	-										
4 S. Jorge	7.37	11.48	6.31	-									
5 Flores	8.49	12.81	8.18	3.73	-								
6 Tudela	13.71	15.58	20.68	20.79	-								
7 Badajoz	6.43	7.95	5.53	9.12	9.95	16.34	-						
8 Las Lomas	6.01	8.58	6.34	10.18	12.42	14.51	7.03	-					
9 Doñana	5.88	7.18	5.56	9.81	11.24	13.63	9.04	5.93	-				
10 Huelva	4.62	7.71	4.44	8.05	9.45	12.91	6.48	5.03	4.77	-			
11 Camargue	16.40	19.31	18.15	20.84	22.82	9.66	18.63	16.91	16.55	15.82	-		
12 Versailles	17.63	20.47	19.75	22.23	21.85	11.23	19.53	18.42	17.79	16.48	2.87	-	
13 Domestic	19.44	21.78	22.36	25.67	24.20	10.27	20.48	20.77	20.59	19.32	5.35	3.31	-

Comparison with other datasets

Figure 4 shows the topology of the trees obtained with the data published by Richardson *et al.* (1980), Peterka & Hartl (1992) and Arana *et al.* (1989), and modified by the inclusion of two representative rabbit populations (Santarém and Portuguese domestic breed). In all three cases, the wild rabbit population from Santarém representing the A group as defined above is very different from all others and always corresponds to the first split in the tree. On the other hand, and most notably, Portuguese domestic rabbits are always included well within the main group of each tree – probably corresponding to the B group – and hardly distinguishable from a variety of other domestic breeds (e.g. New Zealand, Chinchilla, Californian and Spanish Giant), but also from wild populations of Australia, England, Austria, northern and southern France, and northern and central Spain. As the presence of missing values in the comparative allele x population matrices may influence the topology of trees (Nei & Roychoudhury 1993), the whole procedure was repeated again using only common markers. No significant differences were obtained (results not shown).

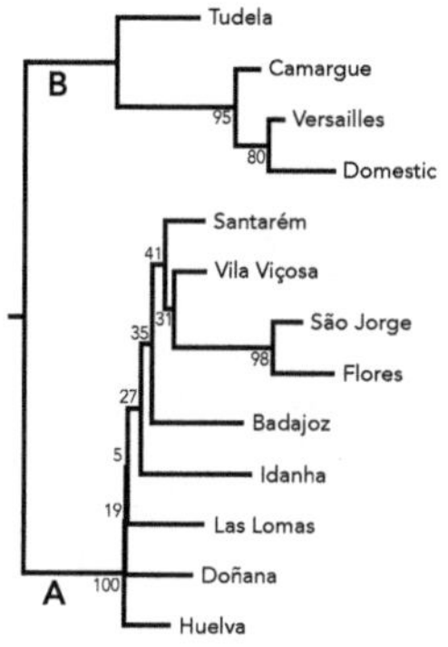

Figure 2. *NJ phylogenetic tree for 13 representative rabbit populations obtained from data in* D_A *values in Table 1. The bootstrap probabilities were obtained with Ota's (1993) computer program DISPAN. The two major groups of rabbit populations correspond to the subspecies* ***A)*** O.c. algirus *and* ***B)*** O.c. cuniculus.

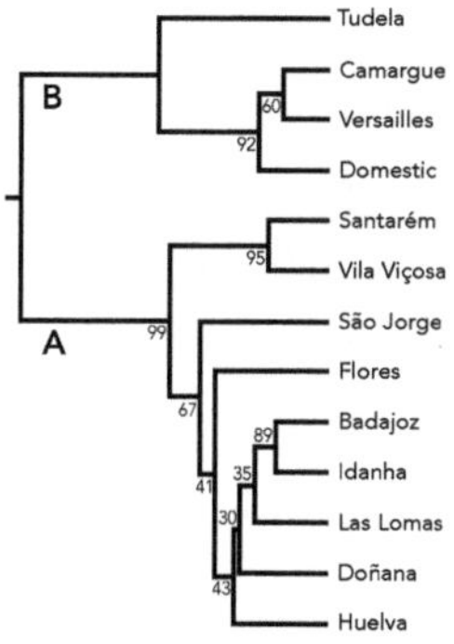

Figure 3. *UPGMA phylogenetic tree for 13 representative rabbit populations obtained from data in* D_A *values in Table 1. The bootstrap probabilities were obtained with Ota's (1993) computer program DISPAN. The two major groups of rabbit populations correspond to the subspecies* ***A)*** O.c. algirus *and* ***B)*** O.c. cuniculus.

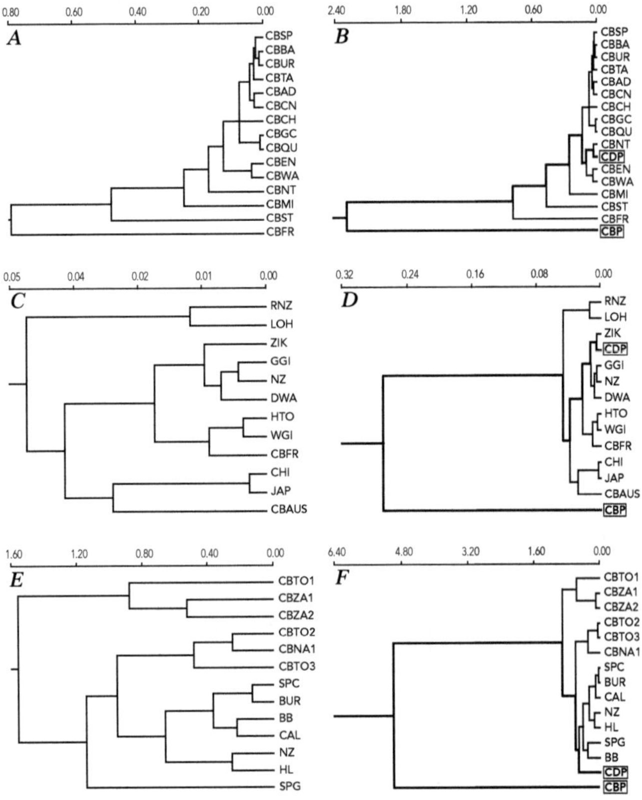

Figure 4. *Comparison of the results described by Richardson* et al. *(1980), Peterka & Hartl (1992) and Arana* et al. *(1989) with the data presented in this paper. All dendrograms are UPGMA trees using Rogers' distance (A and B, Richardson* et al. *1980), Nei's distance (C and D, Peterka & Hartl 1992) and Cavalli-Sforza's distance (E and F, Arana* et al. *1989). Trees B, D and F include two Portuguese rabbit populations (Santarém - CBP - and domestic - CDP) that are highlighted. Wild rabbit populations are always preceded by 'CB', the other samples being from domestic breeds. For the abbreviations see the original references.*

Gene diversity analysis

The results of the gene diversity analysis are summarized in Table 2. It is clear that the major part of the total variability is found within populations (78.5%). The other fractions are asymmetrically distributed: 6.4% is attributable to differences between populations and 15.1% to differences between the subspecies *algirus* and *cuniculus*. When an intermediate level of population structure (regions: Portugal, Spain, France and Azores) is included in the analysis, only a very small fraction of the genetic variability is obtained, thus indicating that regions are not meaningful sources of genetic structuring (results not shown).

Table 2. *Absolute* (H_T) *and relative* (G*'s*) *gene diversities for 12 rabbit populations based on 20 protein polymorphic loci.*

	H_T	Relative distribution (%)		
		G_P	G_{SP}	G_{ST}
Twenty electrophoretic loci	0.345	78.5	6.4	15.1

G_P – within population; G_{SP} – between populations within subspecies; G_{ST} – between subspecies.

Table 3. *Genetic variability measures obtained in 13 rabbit populations based on 20 protein polymorphic loci.*

Population	N	n_a	P	H
O.c. algirus **(continental populations)**				
Santarém	102.7 (13.3)	2.7 (0.2)	90.0	0.346 (0.042)
Idanha	25.1 (2.1)	2.4 (0.2)	80.0	0.264 (0.047)
Vila Viçosa	36.8 (1.4)	2.5 (0.2)	85.0	0.329 (0.043)
Badajoz	15.3 (0.6)	2.4 (0.2)	85.0	0.320 (0.049)
Las Lomas	35.5 (3.4)	2.5 (0.2)	85.0	0.344 (0.048)
Doñana	36.5 (8.9)	2.3 (0.2)	75.0	0.294 (0.054)
Huelva	57.2 (6.4)	2.7 (0.2)	90.0	0.305 (0.052)
O.c. algirus **(insular populations)**				
São Jorge	50.9 (1.8)	1.6 (0.2)	50.0	0.227 (0.056)
Flores	51.2 (2.9)	1.6 (0.1)	55.0	0.226 (0.051)
O.c. cuniculus				
Tudela	15.6 (2.5)	1.8 (0.2)	50.0	0.200 (0.054)
Camargue	28.5 (1.2)	1.7 (0.1)	60.0	0.184 (0.051)
Versailles	53.2 (2.6)	1.7 (0.2)	50.0	0.218 (0.057)
Domestic	98.1 (11.7)	1.7 (0.2)	45.0	0.170 (0.053)

N – mean sample size per locus; n_a – mean number of alleles per locus; P – percentage of polymorphic loci; H - expected heterozygosity. Standard errors are indicated in parentheses.

Polymorphism, heterozygosities and allele diversities

The proportion of polymorphic loci (P), average heterozygosities (H) and mean number of alleles per locus (n_a) are presented in Table 3. These values must be interpreted taking into consideration that each of the 20 investigated loci are polymorphic in at least one population. There is a significant difference in the H values of both subspecies, varying between 0.294 and 0.346 for *O.c. algirus* and between 0.170 and 0.218 for *O.c. cuniculus* (continental populations). Azorean populations show intermediate values (0.226/0.227). For both P and n_a only two groups of populations are apparent: the first comprises the *algirus* group (75.0% < P < 90.0%, and $2.4 < n_a < 2.7$) while the second is formed by the *cuniculus* group and the Azorean populations (45.0% < P < 60.0%, and $1,6 < n_a < 1.8$).

Discussion

Evolutionary differentiation of *O. cuniculus*

We identify the major patterns of population differentiation during rabbit evolution using a set of representative samples and 20 polymorphic loci. Both NJ and UPGMA trees showed a very clear first split that is supported by a high bootstrap value (99% to 100%) and corresponding to two groups: A – rabbits of southwestern Iberia and Azorean Islands – and B – rabbits of northeast Spain, France and domestic breeds. When comparing data from previous authors (Richardson *et al.* 1980; Arana *et al.* 1989; Peterka & Hartl 1992) it is remarkable to observe i) the consistency of the first split that has occurred between Portuguese wild rabbits and all others, and ii) the fact that the Portuguese domestic stock is always deeply nested within group B populations, thus showing consistent close genetic relationships with rabbits from central and northern Iberia, the rest of Europe, Australia and all domestic breeds (Figure 4). This observation is much strengthened by the fact that both the number and type of electrophoretic systems varied considerably between those studies, confirming the deep divergence between the two groups that were already apparent in the phylogenetic trees (Figures 2 & 3). The comparison with mtDNA polymorphism show a clear correspondence between the two sets of data. Briefly, southwestern Iberian populations along with Azorean Islands show haplogroup A, while populations from northern Spain, France, the rest of Europe, Australia and domestic breeds present haplogroup B (Biju-Duval *et al.* 1991; Monnerot *et al.* 1994). Taken together, these results strongly suggest that group A populations correspond to the *algirus* subspecies, whereas group B populations correspond to the *cuniculus* subspecies.

O. cuniculus is currently treated as consisting of two subspecies (Lopez-Martinez 1989), although controversy exists (Sharples *et al.* 1996). The recognition of subspecies was initially based mainly on size and characteristics of fur and led Cabrera (1914) to state that *algirus* is distributed in the Iberian Peninsula and North Africa while *cuniculus* occupies the rest of Europe and Australia. Difficulties in using morphological characteristics to define subspecies pushed Sharples *et al.* (1996) to contest this subdivision, suggesting instead a continuous gradient across the Iberian Peninsula. In a more extreme position, Gibb (1990) suggested that autochthonous rabbit populations exist only in Iberia, with all others being feral and derived from domestic animals released in different places. Our results combined with mtDNA variation as well as with immunoglobulin polymorphism (van der Loo *et al.* 1991, 1999) provide solid evidence in favor of two major population groups that have been evolving independently for a long period of time, thus conforming with a modern concept of subspecies (Avise & Hamrick 1997). However, it is also clear that the distribution boundaries settled by Cabrera (1914) for the subspecies *algirus* and *cuniculus* do not coincide with those of the present genetic analysis. This is because the morphological classification of rabbit subspecies has severe limitations and the genetic changes affecting those traits are not necessarily in direct relation to phylogeny. As a whole, the genetic evidence now presented allows a more accurate definition of the distribution areas of *algirus* and *cuniculus*.

The age of separation between the two rabbit subspecies is a more difficult issue to address. In the present study we used D_A, a distance measure that is not linear with time preventing the calculation of a divergence time between *algirus* and *cuniculus* (Nei *et al.* 1983). We therefore used the allelic frequency data of Table A1 together with a set of 10 monomorphic loci to have a random sample of the genome and calculate Nei's genetic distance (D_N) between the two subspecies (results not shown). This value (0.11) is well placed among other genetic distances described for subspecies and compiled by Nei (1987), and maybe used in the formula $t = 5 \times 10^6 \times D_N$ to calculate the divergence time (Nei 1978). However, this formulation assumes that only one-fourth of the amino acid substitutions in proteins are detectable by electrophoresis, which may not be appropriate in our study due to the extensive use of various buffers and pH values, high resolution iso-electric focusing systems and chemical modification of proteins (see material and methods). If we tentatively admit that detectability is 50%, then the formula for estimating t becomes $2.5 \times 10^6 \times D_N$. When both calculations are made, we obtain a window for the divergence time *algirus-cuniculus* that varies between 275 000 and 550 000 years. This is in sharp contrast with the 2 myr estimated by Biju-Duval *et al.* (1991) on the basis of mtDNA RFLPs of the whole molecule and recently confirmed with cytb RFLP data (Branco *et al.* 2000). Two different hypotheses may explain this discrepancy. The first is that divergence time maybe grossly underestimated due to the possible role of natural selection in keeping allelic frequencies constant across protein loci, thus preventing differentiation of isolated populations. Karl & Avise (1992) studied oysters from the Pacific and the Atlantic with a set of allozyme loci and found very little differentiation of the two groups. However, the very same populations show a deep divergence when studied at the DNA level, both with mtDNA and nuclear markers. If this is the case of our study, then genetic distances based on protein polymorphism are not useful for the estimation of the time since the two subspecies diverged. Alternatively, it maybe noted that mtDNA represents a gene tree rather than a population tree (Pamilo & Nei 1988), and thus the divergence time of 2 myr relates to the separation of two molecules but does not necessarily corresponds to the split of *algirus* and *cuniculus*. In this case, the coexistence of two highly divergent mtDNA haplotypes in the rabbit species maybe be due to ancestral polymorphism and/or capture of one of the types. In natural populations several similar situations have been described for a variety of species including the house mouse, voles and tree frogs (see Avise 1994 for a revision). In this respect, we can speculate that around 2 mya at least two other rabbit species occurred in the Iberian Peninsula: *O. lacosti* and *O. laynensis* (Lopez-Martinez 1989). In any case, only the sequencing of different nuclear genes may resolve this issue in the future.

Partitions of genetic diversity

The distribution pattern of genetic diversity as measured by P, H and n_a indicate that *algirus* populations are characterized by a higher genetic variability than that observed for *cuniculus* populations. This results from the fact that southwestern Iberian rabbits harbor a higher number of polymorphic loci (for example GALT, GC, NP, PEPA, PEPD and TF) and also because there is a higher number of alleles per locus especially associated with the occurrence of many private alleles (for example GALT, GC and TF). These results are compatible with an Iberian origin for the species and also suggest that the separation and long geographic isolation between the two subspecies may have led to an asymmetrical partition of genetic diversity within *O. cuniculus*. It is thus possible that the populations now identified as *cuniculus* result from the isolation of a marginal population that would represent only a small fraction of the total genetic diversity of the rabbit. A different possibility may be associated with long-term population size after fragmentation. It has been proposed that Quaternary glaciations in Iberia recurrently left two main refugia: one large area in the southwest and a second, much smaller, in the eastern Mediterranean coast (Cooper *et al.* 1995; Comes & Abbott 1998; Branco *et al.* 2000). If the availability of refugia and habitats would correspond to population numbers, then it is to be expected that *algirus* show higher levels of genetic variability than *cuniculus*.

When a gene diversity analysis is conducted it is observed that the major part of the total variability was found within populations. However, a considerable fraction of that variability (G_{ST} = 15.1%) may be attributed to differences between *algirus* and *cuniculus*, confirming the extension of the genetic differentiation between the two subspecies. If compared with results obtained in studies of other species (see Nei 1987 for a review), our values are generally higher and explain well the consistency of the first split of rabbit populations. Other components of genetic diversity are less important, but it may be stressed that G_{PS} (differences between populations within subspecies) is always higher than the component attributed to differences between regions (Portugal, Spain, France and Azores; see results section). When only southwestern Iberian populations are analyzed (results not shown) this difference is even more important, a fact that may have a biological meaning due to the social structure that characterizes rabbit populations (Ferrand & Branco, unpublished results).

Geographic expansion

Today, the rabbit shows a continuous distribution in Europe, including the Iberian Peninsula. In this region, the geographic expansion of both *algirus* and *cuniculus* groups of populations led to the establishment of a relatively narrow hybrid zone based on mtDNA analysis (Branco *et al.* 2000). These authors show that this zone follows a northwest-southeast direction that divides the Iberian Peninsula and where populations exhibit both maternal lines in similar frequencies. On the contrary, preliminary data on nuclear markers in the same populations suggest that introgression is much more important (M. Branco & N. Ferrand, unpublished results), which maybe expected because i) effective population size of nuclear markers is higher than mtDNA, and ii) males disperse more than females (Kunkele & Von-Holst 1996).

The evidence presented in this work also suggests that wild rabbits from France are the result of a recent colonization originating in the northeast of Spain and associated with a relatively well-marked bottleneck. In fact, genetic variability of French populations is a clear subset of that observed in northeastern Spain and may be explained by the difficulties in overcoming the geographical barrier of the Pyrenees. Alternatively, this process may have been due to human transportation, but the two hypotheses are not testable due to a complete lack of other sources of information. A third possibility is associated with the fact that rabbits have been present in the Mediterranean region of France for at least 300 000 years (Pages 1980). Notwithstanding a progressive decrease in rabbit abundance based on archaeozoological data, Donard (1982) suggests that relict populations may have persisted close to the Mediterranean and were at the origin of recolonization. In the light of the genetic evidence now described, this hypothesis seems very unlikely. From southern France rabbits expanded to the north, arriving in Britain during the 11th century and progressing to the northeastern limits of their range during the 19th and 20th centuries (Flux 1994).

Different authors have suggested that all populations outside Iberia would be feral and derived from animals domesticated in the Iberian Peninsula during

Roman occupation (Clutton-Brock 1987, 1992; Flux & Fullagar 1992; Flux 1994). Our data are not in agreement with this scenario and suggest instead that the hypothesis first proposed by Zeuner (1963) may be correct. According to this author, the geographical expansion of the rabbit is due to two independent processes, namely i) translocation of wild rabbits and ii) domestication. Translocation of rabbits in central Europe and Britain were common during the Middle Ages for hunting purposes and certainly allowed the escape of animals kept in *leporaria*. The allelic distributions of ALB and CAI clearly support this hypothesis: French populations of Camargue and Versailles, as well as a population from southern England (Ferrand & Rocha 1992), exhibit the allele ALB*2 in high frequencies, while the same does not happen in domestic breeds that show the fixation of ALB*1. On the other hand, CAI is polymorphic in wild *cuniculus* populations only (including Britain; Ferrand, unpublished results), thus implicating that the dispersion of the CAI*2 allele marks the expansion of wild rabbits in central Europe. Still associated with this is the colonization of Australia in thc 19th ccntury originating from a limited number of wild rabbits imported from Britain (Flux 1994). The available genetic evidence both at the nuclear (Richardson *et al.* 1980; Ferrand, unpublished results) and mtDNA levels are compatible with the historical documentation. The second process, domestication, did not take place earlier than the 15th century (Zeuner 1963), but it may have overlapped to some degree with translocation of wild animals, during the colonization of central and northern France, as suggested from the analysis of ancient DNA (Hardy *et al.* 1995). Releasing domesticated animals to the wild may have been common and certainly helped the final stages of rabbit colonization in Europe. Additionally, the recent invasion of important areas of South America (Chile and Argentina) has also been attributed to that practice, a hypothesis that is strongly supported by a genetic analysis of Chilean rabbits (Vieira 1993).

A remarkable aspect of rabbit geographical expansion is its presence in more than 800 islands throughout the world, where they were transported by man since historical times (Flux & Fullagar 1992). In this investigation, we have analyzed two Azorean populations (islands of São Jorge and Flores) and concluded that the colonization of the archipelago was achieved with rabbits belonging to the *algirus* group. Moreover, they show close genetic relationships with Portuguese continental populations (see Figure 2), thus being in agreement with historical documentation referring to its introduction in the period of Portuguese discoveries (15th century). It is probable that the colonization of other Atlantic islands (e.g. Madeira and Porto Santo, and the Canary Islands) happened during the same historical period, but probably involving other sources. In the Mediterranean, where human navigation started much earlier, the history of rabbit colonization was quite different. The introduction of the rabbit in Menorca (Balearic Islands) is dated between 1400 and 1300 BC (Reumer & Sanders 1984), while Zembra (Tunisia) was invaded between the end of the Neolithic and the 3rd century of our time (Vigne 1988).

This last case has been studied in detail by Ben Amor (1998) using protein polymorphism and mtDNA sequencing, and the comparison with our data and that described in Hardy *et al.* (1994) and Branco *et al.* (2000) strongly suggests an origin on the eastern Mediterranean coast of Spain. The arrival of the rabbit in North Africa dates probably to this period and very limited genetic data (Ferrand *et al.*, unpublished results) do confirm this hypo-thesis and may suggest at least two different origins. Many more cases exist, that when studied will certainly show a mosaic of different histories. The patterns of population differentiation and distribution of genetic diversity described in this paper will help to unravel a number of interesting situations.

Our data may be summarized in a scenario that depicts an evolutionary history of rabbit populations at a global scale (Figure 5). This scenario should be viewed as a hypothesis to be tested in the future, when the introduction of hypervariable markers like microsatellites as well as the sequencing of multiple nuclear genes and mtDNA should help to better understand the evolutionary history of this species.

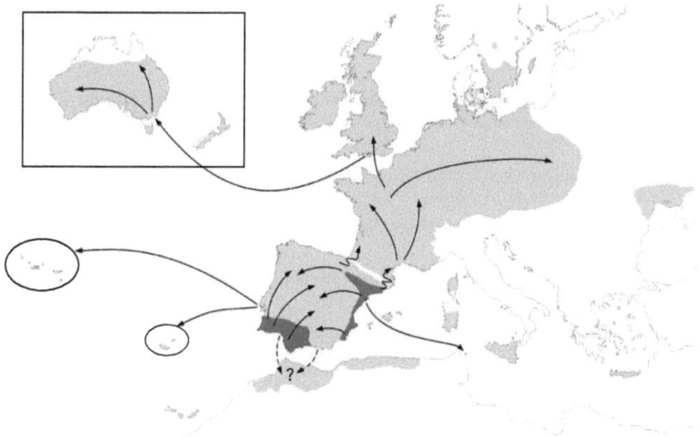

Figure 5*. A possible scenario for the origins and history of major groups of rabbit populations. This scenario summarizes the available paleontological, archaeozoological, historical and genetic data. Dark-gray areas represent the putative refugial areas for the two major groups of populations, and light-gray areas the present-day distribution of the rabbit in Europe, North Africa, some Mediterranean and Atlantic islands, Australia and New Zealand. Straight arrows represent both natural and human-mediated geographical expansions while broken arrows indicate restricted gene flow causing a large reduction in genetic diversity.*

Acknowledgements

This project was partially supported by JNICT (project STRDA/C/CEN/471/92), FCT (FMRH/BSAB/50/98 sabbatical grant to NF), Direcção-Geral de Florestas and CIBIO/UP. We thank P. C. Alves, F. Queirós, J.L. Rosa, A. Loureiro, S. Moreno, R. Villafuerte, E. Castién and M. Monnerot for their help in sampling campaigns or for sending samples, and B. Carvalho and J.C. Machado for support in the laboratory. We are also indebted to Steve Weiss for comments and suggestions in previous drafts of this manuscript.

References

Altland K, Becher P, Banzhoff A (1987) Paraffin oil protected high resolution hybrid isoelectric focusing for the demonstration of substitutions of neutral aminoacids in denatured proteins: the case of four human transthyretin (prealbumin) variants associated with familial amyloidotic polineuropathy. ***Electrophoresis*** **8**, 293-297.

Amorim A, Kompf J, Schunter F, Ritter H (1982) Aminolevulinate dehydratase (E.C. 4.2.1.24): linkage analysis. ***Human Genetics*** **61**, 48-49.

Anderson RM, May RM (1982) Coevolution of hosts and parasites. ***Parasitology*** **85**, 411-426.

Arana A, Zaragoza P, Rodellar C, Amorena B (1989) Blood biochemical polymorphisms as markers for genetic characteristics of wild Spanish and domestic rabbits. ***Genetica*** **79**, 1-9.

Arnaud JM (1987) Os concheiros mesolíticos dos vales do Tejo e Sado: semelhanças e diferenças. ***Arqueologia*** **15**, 53-64.

Avise JC (1994) ***Molecular Markers, Natural History and Evolution***. Chapman & Hall, London.

Avise JC, Hamrick JL (1997) ***Conservation Genetics: Case Histories from Nature***. Chapman & Hall, New York.

Barrett SCH, Richardson BJ (1986) Genetic attributes of invading species. In: ***Ecology of Biological Invasions*** (eds. Groves RH, Burden JJ), pp. 21-33. Cambridge University Press, Cambridge.

Beaty TH, Prenger VL, Virgil DG, Lewis B, Kwiterovich PO, Bachorik PS (1992) A genetic model for control of hypertriglyceridemia and apolipoprotein B levels in the John Hopkins Colony of St. Thomas Hospital rabbits. ***Genetics*** **132**, 1095-1104.

Beaucournu JC (1980) Les ectoparasites du lapin de garenne *Oryctolagus cuniculus*: apports a son histoire. ***Bulletin Mensuel de l'Office National de la Chasse***, (numéro special), 23-35.

Beaucournu JC, Launay H (1990) ***Les puces de France et du Bassin Méditerranéen Occidental***. Fédération Française des Societés de Sciences Naturelles, Paris.

Ben Amor M (1998) ***Contribution à la reconstitution de l'histoire des populations de lapins des îles Kuriat et Zembra: Analyse du polymorphisme de l'ADN mitochondrial et de marqueurs protéiques/isoenzymatiques***. PhD Thesis, University of Tunis II.

Biju-Duval C, Ennafaa H, Dennebouy N, Monnerot N, Mignotte F, Soriguer R, El Gaied A, El Hili A, Mounolou JC (1991) Mitochondrial DNA evolution in lagomorphs: origin of systematic heteroplasmy and organization of diversity in european rabbits. ***Journal of Molecular Evolution*** **33**, 92-102.

Boursot P, Auffray J-C, Britton-Davidian J, Bonhomme F (1993) The evolution of the house mice. ***Annual Review of Ecology and Systematics*** **24**, 119-152.

Branco M, Ferrand N (1998) Genetic polymorphism of rabbit (*Oryctolagus cuniculus*) tissue acid phosphatases (ACP2 and ACP3). ***Comparative Biochemistry and Physiology*** **120B**, 405-409.

Branco M, Ferrand N (2002) Genetic polymorphism of antithrombin III, haptoglobin and haemopexin in the European rabbit (*Oryctolagus cuniculus*). ***Biochemical Genetics*** **40**, 387-393.

Branco M, Ferrand N (2003) Biochemical and population genetics of rabbit (*Oryctolagus cuniculus*) carbonic anhydrases I and II in the Iberian Peninsula and France. ***Biochemical Genetics*** **41**, 391-405

Branco M, Machado JC, Ferrand N (1999) Extensive genetic polymorphism of peptidases A, B, C and D, in wild rabbit (*Oryctolagus cuniculus*) populations from the Iberian Peninsula. ***Biochemical Genetics*** **37**, 237-249.

Branco M, Ferrand N, Monnerot M (2000) Phylogeography of the European rabbit (*Oryctolagus cuniculus*) in the Iberian Peninsula inferred from RFLP analysis of the cytochrome b gene. ***Heredity*** **85**, 307-317.

Brown, AHD, Marshall DR, Weir BS (1981) Current status of the charge state model for protein polymorphism. In: ***Genetic Studies of Drosophila Populations*** (eds. Gibson JB, Oakeshott JG), pp. 15-43. Australian National University Press, Canberra.

Cabrera J (1914) ***Fauna Ibérica***. Museo Nacional de Ciências Naturales de Madrid, Madrid.

Callou C (1995) Modifications de l'aire de répartition du lapin (*Oryctolagus cuniculus*) en France et en Espagne, du Pleistocène à l'époque actuelle. État de la question. ***Anthropozoologica*** **21**, 95-114.

Cavalli-Sforza LL, Piazza A, Menozzi P, Mountain J (1988) Reconstruction of human evolution: bringing together genetic, archaeological and linguistic data. ***Proceedings of the National Academy of Sciences, USA*** **85**, 6002-6006.

Cavalli-Sforza LL, Menozzi P, Piazza A (1994) ***The History and Geography of Human Genes***. Princeton University Press, Princeton.

Cazenave PA, Benammar A, Sogn JA, Kindt TJ (1987) Immunoglobulin genes in the feral rabbit. In: ***The Rabbit in Contemporary Immunological Research*** (ed. Dubiski S), pp. 148-163. Longman Scientific & Technical, John Wiley & Sons, New York.

Chikhi L, Destro-Bisol G, Bertorelle G, Pascali V, Barbujani G (1998) Clines of nuclear DNA markers suggest a largely Neolithic ancestry of the European gene pool. ***Proceedings of the National Academy of Sciences, USA*** **95**, 9053-9058.

Clutton-Brock J (1987) ***A Natural History of Domesticated Animals***. Cambridge University Press, Cambridge.

Clutton-Brock J (1992) Domestication of animals. In: ***The Cambridge Encyclopedia of Human Evolution*** (eds. Jones S, Martin R, Pilbeam D), pp. 380-385. Cambridge University Press, Cambridge.

Comes HP, Abbott RJ (1998) The relative importance of historical events and gene flow on the population structure of a Mediterranean ragwort *Senecio gallicus* (Asteraceae). ***Evolution*** **52**, 355-367.

Cooper SJB, Ibrahim KM, Hewitt GH (1995) Postglacial expansion and genome subdivision in the European grasshopper *Chorthippus parallelus*. ***Molecular Ecology*** **4**, 49-60.

Curtain CC, Wood DH, Sobey WR (1973) The distribution of immunoglobulin allotypes in rabbit populations in Australia and on Macquarie Island. ***Animal Blood Groups and Biochemical Genetics*** **4**, 101-109.

Dobson M (1998) Mammal distributions in the western Mediterranean: the role of human intervention. ***Mammal Review*** **28**, 77-88.

Donard E (1982) ***Recherches sur les léporidés quaternaires (Pleistocène moyen et supérieur, Holocène)***. PhD Thesis, University of Bordeaux I.

Dubiski S (1987) ***The Rabbit in Contemporary Immunological Research***. Longman Scientific & Technical, John Wiley & Sons, New York.

Esteves PJ, Lanning D, Ferrand N, Knight KL, Zhai SK, van der Loo W (2005) The evolution of the immunoglobulin heavy chain variable region (IgV_H) in Leporids: an unusual case of transspecies polymorphism. ***Immunogenetics*** **57**, 874-882.

Farris JS (1972) Estimating phylogenetic trees from distance matrices. ***American Naturalist*** **106**, 645-668.

Felsenstein J (1982) How can we infer geography and history from gene frequencies? ***Journal of Theoretical Biology*** **96**, 9-20.

Ferrand N (1989) Biochemical and genetic studies on rabbit hemoglobin. I. Electrophoretic polymorphism of the β-chain. ***Biochemical Genetics*** **27**, 673-678.

Ferrand N (1990) Biochemical and genetic studies on rabbit hemoglobin. II. Electrophoretic polymorphism of the α-chain. ***Biochemical Genetics*** **28**, 117-122.

Ferrand N (1995) ***Variação genética de proteínas em populações de coelho (*****Oryctolagus cuniculus*****). Análise da diferenciação subespecífica, subestruturação, expansão geográfica e domesticação***. PhD Thesis, University of Oporto.

Ferrand N, Rocha J (1992) Demonstration of serum albumin (ALB) polymorphism in wild rabbit (*Oryctolagus cuniculus*) by means of isoelectric focusing. ***Animal Genetics*** **23**, 275-278.

Ferrand N, Carvalho G, Amorim A (1988) Transferrin (Tf) polymorphism in wild rabbit, *Oryctolagus cuniculus*. Animal ***Genetics*** **19**, 295-300.

Flux JEC (1994) World distribution. In: ***The European Rabbit: The History and Biology of a Successful Colonizer*** (eds. Thompson HV, King CM), pp. 8-21. Oxford Science Publications, Oxford.

Flux JEC, Fullagar PJ (1992) World distribution of the rabbit *Oryctolagus cuniculus* on islands. ***Mammal Review*** **22**, 151-205.

Gibb JA (1990) The European Rabbit *Oryctolagus cuniculus*. In: ***Rabbits, Hares and Pikas*** (eds. Chapman JA, Flux JEC), pp. 116-129. International Union for Conservation of Nature and Natural Resources, Gland.

Hardy C, Vigne JD, Casane D, Dennebouy N, Mounolou JC, Monnerot M (1994) Origin of European rabbit (*Oryctolagus cuniculus*) in a Mediterranean island: zooarchaeology and ancient DNA examination. ***Journal of Evolutionary Biology*** **7**, 217-226.

Hardy C, Callou C, Vigne JD, Casane D, Dennebouy N, Mounolou JC, Monnerot M (1995) Rabbit mitochondrial DNA diversity from prehistoric to modern times. ***Journal of Molecular Evolution* 40**, 227-237.

Harris H, Hopkinson DA (1976) ***Handbook of Enzyme Electrophoresis in Human Genetics***. North-Holland Publishers, Amsterdam.

Hartl GB (1987) Biochemical differentiation between the wild rabbit (*Oryctolagus cuniculus* L.), the domestic rabbit and the brown hare (*Lepus europaeus* Pallas). ***Zeitschrift für Zoologische Systematik und Evolutionsforschung* 24**, 309-316.

Karl SA, Avise J (1992) Balancing selection at allozyme loci in oysters: implications from nuclear RFLPs. ***Science* 256**, 100-102.

Kunkele J, Von-Holst D (1996) Natal dispersal in the European wild rabbit. ***Animal Behaviour* 58**, 1047-1059.

Langman RE (1989) ***The Immune System***. Academic Press, New York.

Launay H, Beaucournu JC (1982) Critères taxinomiques et essai de paléobiogeographie de deux espèces jumelles, *Xenopsylla ramesis* (Rothschild 1904) et *Xenopsylla cunicularis* Smit 1957 (*Siph. Pulicidae*): approche biométrique. ***Annales de la Societé Entomologique de France*** (N.S.) **18**, 43-54.

Leisner V (1983) As diferentes fases do Neolítico em Portugal. ***Arqueologia* 7**, 7-15.

Lentacker A (1986) Preliminary results of the fauna of Cabeço de Amoreira and Cabeço de Arruda (Muge, Portugal). ***Trabalhos de Antropologia e Etnologia* 26**, 9-26.

Lewontin RC (1970) The units of selection. ***Annual Reviews in Ecology and Systematics* 1**, 1-18.

Lewontin RC (1991) Twenty-five years ago in Genetics: electrophoresis in the development of evolutionary genetics: milestone or millstone? ***Genetics* 128**, 657-662.

Lopez-Martinez N (1989) ***Revision sistematica y biostratigrafica de los lagomorphos (Mammalia) del neogeno y cuaternario de España***. Memorias del Museo Paleontológico de la Universidad de Zaragoza, Diputación General de Aragón.

Mage R, Lieberman R, Potter M, Terry WT (1973) The immunoglobulin allotypes. In: ***The Antigens***, vol. 1 (ed. Sela M), pp. 299-376. Academic Press, New York.

Monnerot M, Vigne JD, Biju-Duval C, Casane D, Callou C, Hardy C, Mougel F, Soriguer R, Dennebouy N, Mounolou JC (1994) Rabbit and man: genetic and historic approach. ***Genetics, Selection, Evolution* 26** (Supplement 1), 167-182.

Nei M (1973) Analysis of gene diversity in subdivided populations. ***Proceedings of the National Academy of Sciences, USA* 70**, 3321-3323.

Nei M (1975) ***Molecular Population Genetics and Evolution***. North-Holland, Amsterdam.

Nei M (1978) The theory of genetic distance and evolution of human races. ***Japanese Journal of Human Genetics* 23**, 341-369.

Nei M (1987) ***Molecular Evolutionary Genetics***. Columbia University Press, New York.

Nei M (1995) Genetic support for the out-of-Africa theory of human evolution. ***Proceedings of the National Academy of Sciences, USA* 92**, 6720-6722.

Nei M, Roychoudhury AK (1982) Genetic relationship and evolution of human races. ***Evolutionary Biology* 14**, 1-59.

Nei M, Roychoudhury AK (1993) Evolutionary relationships of human populations on a global scale. ***Molecular Biology and Evolution* 10**, 927-943.

Nei M, Tajima F, Tateno Y (1983) Accuracy of estimated phylogenetic trees from molecular data. ***Journal of Molecular Evolution* 19**, 153-170.

Ota T (1993) ***DISPAN: Genetic Distance and Phylogenetic Analysis*.** Institute of Molecular and Evolutionary Genetics, Pennsylvania State University, University Park.

Pages MV (1980) Essai de reconstitution de l'histoire du lapin de garenne en Europe. ***Bulletin Mensuel de l'Office National de la Chasse*** (numéro special), 13-21.

Pamilo P, Nei M (1988) Relationships between gene trees and species trees. ***Molecular Biology and Evolution* 5**, 568-583.

Peterka M, Hartl GB (1992) Biochemical-genetic variation and differentiation in wild and domestic rabbits. On the significance of genetic distances, dendrograms and the estimation of divergence times in domestication studies. ***Zeitschrift für Zoologische Systematik und Evolutionsforschung* 30**, 129-141.

Queney G, Ferrand N, Marchandeau S, Azevedo M, Mougel F, Branco M, Monnerot M (2000) Absence of a genetic bottleneck in a wild rabbit (*Oryctolagus cuniculus*) population exposed to a severe viral epizootic. ***Molecular Ecology* 9**, 1253-1264.

Ramshaw JAM, Coyne JA, Lewontin RC (1979) The sensitivity of gel electrophoresis as a detector of genetic variation. ***Genetics* 93**, 1019-1037.

Reumer JWF, Sanders EAC (1984) Changes in the vertebrate fauna of Menorca in prehistoric and classical times. ***Zeitschrift für Säugetierkunde* 49**, 321-325.

Richardson BJ, Rogers PM, Hewitt GM (1980) Ecological genetics of the wild rabbit in Australia. II. Protein variation in British, French and Australian rabbits and the geographical distribution of the variation in Australia. ***Australian Journal of Biological Sciences* 33**, 371-383.

Righetti PG (1990) ***Immobilized pH Gradients: Theory and Methodology*.** Elsevier, Amsterdam.

Righetti PG, Gianazza E, Bianchi-Bosisio A, Wajcman H, Cossu G (1989) Electrophoretically silent hemoglobin mutants as revealed by isoelectric focusing in immobilized pH gradients. ***Electrophoresis* 10**, 595-599.

Rogers PM, Arthur CP, Soriguer R (1994) The rabbit in continental Europe. In: ***The European Rabbit: The History and Biology of a Successful Colonizer*** (eds. Thompson HV, King CM), pp. 22-63. Oxford Science Publications, Oxford.

Rzhetsky A, Nei M (1992) Statistical properties of the ordinary least-squares, generalized least-square, and minimum-evolution methods of phylogenetic inference. ***Journal of Molecular Evolution* 35**, 367-375.

Saitou N, Imanishi T (1989) Relative efficiencies of the Fitch-Margoliash, maximum-parsimony, maximum-likelihood, minimum-evolution, and neighbor-joining methods of phylogenetic tree construction in obtaining the correct tree. ***Molecular Biology and Evolution* 6**, 514-525.

Saitou N, Nei M (1987) The neighbor-joining method: a new method for reconstructing phylogenetic trees. ***Molecular Biology and Evolution* 4**, 406-425.

Sharples CM, Fa JE, Bell DJ (1996) Geographical variation in size in the European rabbit *Oryctolagus cuniculus* (Lagomorpha: Leporidae) in Western Europe and North Africa. ***Zoological Journal of the Linnean Society* 117**, 141-158.

Siebert G, Kompf J, Ritter H (1980) Galactose-1-phosphate uridyltransferase (E.C.2.7.7.11): a simple routine method for detecting individuals heterozygous for the silent allele Gt^0. ***Human Genetics* 54**, 273-275.

Singh RS, Rhomberg LR (1987) A comprehensive study of genic variation in natural populations of *Drosophila melanogaster*. II. Estimates of heterozygosity and patterns of geographic differentiation. ***Genetics*** **117**, 255-271.

Sneath PHA, Sokal RR (1973) ***Numerical Taxonomy***. W. H. Freeman, San Francisco.

Su C, Nei M (1999) Fifty million years old polymorphism at an immunoglobulin variable region gene locus in the rabbit evolutionary lineage. ***Proceedings of the National Academy of Sciences, USA*** **96**, 9710-9715.

Sugiura M, Ito Y, Hirano K, Sawaki S (1977) Detection of dipeptidase and tripeptidase activities on polyacrylamide gel and cellulose acetate gel by the reduction of tetrazolium salts. ***Analytical Biochemistry*** **81**, 481-484.

Swofford DL, Selander RB (1989) BIOSYS-1. A computer program for the analysis of allelic variation in population genetics and biochemical systematics. Release 1.7. Illinois Natural History Survey.

Thompson HV, King CM (eds.) 1994) ***The European Rabbit: The History and Biology of a Successful Colonizer***. Oxford Science Publications, Oxford.

van der Loo W (1987) Studies on the adaptive significance of the immunoglobulin alleles (Ig allotypes) in wild rabbit. In: ***The Rabbit in Contemporary Immunological Research*** (ed. Dubiski S), pp. 164-190. Longman Scientific & Technical, John Wiley & Sons, New York.

van der Loo W (1993) Variance analysis of immunoglobulin alleles in natural populations of rabbit (*Oryctolagus cuniculus*): the extensive interallelic divergence at the *b* locus of the immunoglobulin light chain constant region could be the outcome of overdominance-type selection. ***Genetics*** **135**, 171-187.

van der Loo W, Verdoodt B (1992) Patterns of interallelic divergence at the rabbit *b* locus of the immunoglobulin light chain constant region are in agreement with population genetical evidence for overdominant selection. ***Genetics*** **132**, 1105-1117.

van der Loo W, Arthur CP, Drees-Wallage M, Richardson B (1987) Nonrandom allele associations between unlinked protein loci: are the polymorphisms of the immunoglobulin constant regions adaptive? ***Proceedings of the National Academy of Sciences, USA*** **84**, 3075-3079.

van der Loo W, Ferrand N, Soriguer R (1991) Estimation of gene diversity at the *b* locus of the constant region of the immunoglobulin light chain in natural populations of European rabbit (*Oryctolagus cuniculus*) in Portugal, Andalusia and on the Azorean islands. ***Genetics*** **127**, 789-799.

van der Loo W, Mougel F, Sánchez MS, Bouton C, Castien E, Fonseca A, Ferrand N, Soriguer R, Monnerot M (1999) Cytonuclear disequilibria in wild populations of rabbit (*Oryctolagus cuniculus* L.) suggest unequal allele turnover rates at the *b* locus (*IGKC1*). ***Immunogenetics*** **49**, 629-643.

Vieira J (1993) ***Utilização de marcadores genéticos na caracterização e certificação de populações domésticas e selvagens de coelho (*Oryctolagus cuniculus*)***. MSc Thesis, University of Oporto.

Vieira J, Ferrand N (1995) Four new alleles at the mannose-6-phosphate isomerase locus in rabbit. ***Animal Genetics*** **26**, 37-38.

Vigne JD (1988) Données préliminaires sur l'histoire du peuplement mammalian de l'ilot de Zembra (Tunisie). ***Mammalia*** **52**, 567-574.

Villafuerte R (1994) ***Riesgo de predación y estrategias defensivas del conejo,* Oryctolagus cuniculus, *en el Parque Nacional de Doñana***. PhD Thesis, University of Cordoba.

Weisbroth SH, Flatt RE, Kraus AL (1974) ***The Biology of the Laboratory Rabbit***. Academic Press, Orlando.

Williams K, Parer I, Conan B, Burley J, Braysher M (1995) ***Managing Vertebrate Pests: Rabbits***. CSIRO Division of Wildlife and Ecology, Australian Government Publishing Service, Canberra.

Zapolski EJ, Princiotto JV (1980) Binding of iron from nitrilotriacetate analogs by human transferrin. ***Biochemistry*** **19**, 3599-3603.

Zeuner FE (1963) ***A History of Domesticated Animals***. Hutchinson, London.

Zhivolovsky LA, Feldman MW (1995) Microsatellite variability and genetic distances. ***Proceedings of the National Academy of Sciences, USA*** **92**, 11549-11552.

Appendix

Table A1. *Allele frequencies at 20 protein loci for the 13 European rabbit populations.*

		Population												
Locus	Allele	1	2	3	4	5	6	7	8	9	10	11	12	13
ACP3	1	0.88	1.00	1.00	1.00	1.00	0.82	1.00	1.00	0.89	0.98	0.71	0.42	0.40
	2						0.18				0.02	0.29	0.58	0.60
	3	0.12								0.11				
	n	132	21	36	43	57	11	16	28	26	53	29	57	96
ADA	1	0.26	0.21	0.17		0.34	0.83	0.53	0.29	0.21	0.26	0.67	0.50	0.58
	2	0.04	0.10	0.03				0.03		0.12				
	3	0.70	0.62	0.76	1.00	0.66	0.13	0.44	0.71	0.65	0.73	0.31	0.35	0.18
	4		0.07											
	5											0.02	0.15	0.24
	6						0.04			0.02	0.01			
	7			0.04										
	n	89	21	36	50	54	12	16	28	26	53	29	58	77
ALB	1	0.65	0.91	0.51	0.52	1.00	0.62	0.75	0.44	0.44	0.39	0.34	0.83	1.00
	2	0.34	0.09	0.49	0.48		0.38	0.25	0.56	0.56	0.61	0.66	0.17	
	3	0.01												
	n	120	22	36	58	67	13	16	70	26	54	47	57	100
CAI	1	1.00	0.98	1.00	1.00	1.00	1.00	1.00	1.00	1.00	1.00	0.86	0.74	1.00
	2											0.14	0.26	
	3		0.02											
	n	55	21	36	47	49	12	16	29	26	52	29	57	100
CAII	1	0.15		0.03			0.46	0.03	0.38	0.28	0.23	0.45	0.59	0.83
	2	0.03	0.24	0.08				0.06		0.02	0.06			
	3						0.03							
	4							0.03						
	5	0.82	0.76	0.88	1.00	1.00	0.25	0.88	0.62	0.68	0.71	0.55	0.41	0.17
	6			0.01			0.04			0.02				
	n	58	21	36	58	50	12	16	29	25	54	29	59	145
GALT	1	0.70	0.91	0.81	1.00	1.00	0.63	0.59	0.72	0.84	0.93	1.00	1.00	0.99
	2			0.03										
	3	0.26	0.02	0.15			0.37	0.13	0.04		0.03			0.01
	4									0.08				
	5	0.04		0.01				0.13	0.22	0.02	0.04			
	6									0.06				
	7							0.15						
	8								0.02					
	9		0.07											
	n	74	21	36	50	50	12	16	27	26	53	29	44	99

Continued on next page

Table A1. *Continued.*

		Population												
Locus	Allele	1	2	3	4	5	6	7	8	9	10	11	12	13
GC	1	0.92	0.90	0.79	1.00	1.00	1.00	0.91	0.83	0.96	0.82	1.00	1.00	1.00
	2	0.01												
	3	0.06	0.02	0.03				0.06			0.13			
	4	0.01	0.02	0.15						0.04	0.04			
	5			0.03				0.03						
	6								0.11					
	7								0.06					
	8										0.02			
	9		0.05											
	n	83	21	36	50	50	13	16	27	26	57	28	44	100
GPI	1	0.97	1.00	0.95	1.00	1.00	1.00	0.97	0.94	1.00	0.99	1.00	1.00	1.00
	2	0.03							0.06					
	3			0.01				0.03			0.01			
	4			0.04										
	n	37	21	57	50	47	56	16	63	33	37	18	58	47
HBA	1											0.97	0.89	0.76
	2	0.68	0.33	0.79	0.56	0.89	0.63	0.22	0.57	0.84	0.90	0.03	0.11	0.20
	3	0.26	0.67	0.21	0.44	0.11	0.12	0.78	0.43	0.16	0.10			0.04
	4	0.06					0.08							
	5						0.17							
	n	62	21	36	55	46	12	16	28	25	54	29	58	109
HBB	1	0.18		0.08			0.08	0.06			0.06	0.07	0.08	0.49
	2	0.82	1.00	0.92	1.00	1.00	0.92	0.94	1.00	1.00	0.94	0.93	0.92	0.51
	n	68	21	36	55	44	12	16	28	25	54	29	58	143
HBX	1								0.03					
	2										0.03			
	3	0.46	0.72	0.61	0.51	0.32	0.61	0.70	0.69	0.37	0.61	0.31	0.38	0.55
	4	0.13		0.10	0.49	0.68		0.10	0.04	0.28	0.19		0.12	0.02
	5	0.41	0.28	0.29			0.26	0.20	0.23	0.35	0.16	0.28	0.11	0.05
	6						0.13		0.01		0.01	0.39	0.39	0.38
	7											0.02		
	n	46	32	33	57	28	39	15	28	58	36	27	88	110
MPI	1	0.10	0.01	0.04	0.01			0.12	0.11		0.01	0.50	0.50	0.56
	2	0.57	0.91	0.58	0.47	0.50	1.00	0.13	0.81	0.82	0.80	0.50	0.50	0.39
	3													0.05
	4	0.20	0.01		0.23	0.32			0.02		0.05			
	5	0.08	0.02	0.25	0.29	0.18		0.75	0.06		0.12			
	6	0.05	0.04	0.13						0.18				
	7										0.01			
	8		0.01								0.01			
	n	159	49	24	46	49	12	4	27	25	42	25	25	66
NP	1	0.84	0.98	0.76	0.76	0.30	1.00	0.81	0.94	0.96	0.96	1.00	1.00	1.00
	2	0.13	0.02	0.24	0.24	0.70		0.19	0.06	0.04	0.04			
	3	0.03												
	n	257	23	36	47	70	12	16	27	26	53	29	57	87

***Table A1**. Continued.*

		Population												
Locus	**Allele**	**1**	**2**	**3**	**4**	**5**	**6**	**7**	**8**	**9**	**10**	**11**	**12**	**13**
PEPA	1	0.44	0.19	0.43	0.74	0.80	1.00	0.66	0.32	0.42	0.68	1.00	1.00	1.00
	2	0.56	0.81	0.57	0.26	0.20		0.34	0.55	0.58	0.32			
	3								0.13					
	n	62	21	35	23	53	12	16	28	26	53	27	42	48
PEPB	1	0.77	0.80	0.78	0.45	0.72	0.92	0.97	0.95	1.00	0.91	0.93	1.00	1.00
	2	0.23		0.15	0.55	0.28						0.07		
	3		0.04	0.07			0.08	0.03	0.05		0.07			
	4		0.16								0.02			
	n	124	23	36	57	59	12	16	30	26	54	29	54	23
PEPC	1	0.59	0.54	0.52	0.22	0.20	1.00	0.53	0.61	0.44	0.40	0.98	1.00	1.00
	2	0.29	0.15	0.41	0.67	0.68		0.22	0.30	0.15	0.23	0.02		
	3	0.04							0.02		0.06			
	4		0.24		0.11	0.12		0.22	0.02	0.14	0.10			
	5	0.06	0.07	0.07				0.03	0.05	0.27	0.18			
	6	0.02												
	7										0.03			
	n	133	23	35	55	62	12	16	30	26	54	29	54	44
PEPD	1	0.80	0.74	0.83	0.92	0.88	1.00	0.59	0.46	0.58	0.61	1.00	1.00	1.00
	2	0.14	0.17	0.15	0.08	0.12		0.38	0.45	0.36	0.18			
	3	0.06	0.09					0.03	0.07	0.02	0.16			
	4			0.02					0.02	0.04				
	5										0.02			
	6										0.03			
	n	56	23	36	56	56	12	16	29	25	54	29	51	24
PGD	1	0.34	0.71	0.36	0.59	0.44	0.88	0.47	0.54	0.77	0.39	0.98	0.60	0.98
	2												0.27	0.02
	3	0.63	0.29	0.64	0.41	0.56		0.53	0.46	0.23	0.61	0.02	0.13	
	4	0.03												
	5						0.12							
	n	236	21	36	46	70	12	15	26	26	53	29	44	160
TF	1	0.79	0.64	0.60	0.61	0.65	1.00	0.69	0.53	0.73	0.62	1.00	1.00	1.00
	2	0.01												
	3	0.20	0.05	0.40	0.39	0.35		0.31	0.34	0.27	0.29			
	4								0.01					
	5										0.07			
	6								0.08					
	7								0.04		0.02			
	8		0.07											
	9		0.24											
	n	117	21	36	58	48	13	16	65	26	52	25	49	150
SOD	1	1.00	1.00	1.00	1.00	1.00	1.00	1.00	0.82	1.00	1.00	1.00	1.00	1.00
	2								0.18					
	n	85	55	48	57	14	12	16	63	202	176	22	50	238

Chapter 8

Patterns of hemoglobin polymorphism [α-globin (HBA) and β-globin (HBB)] across the contact zone of two distinct phylogeographic lineages of the European rabbit (*Oryctolagus cuniculus*)

Rita Campos*, Madalena Branco, Steven Weiss and Nuno Ferrand
*E-mail: ritacampos@mail.icav.up.pt

Abstract
Two loci, HBA and HBB, were studied for protein polymorphism across the contact zone of the European rabbit (*Oryctolagus cuniculus*). Six alleles were identified in HBA and two in HBB. Three alleles at the HBA locus were found to be restricted to some populations, while the other three revealed more broad geographic structure. An apparent substitution of the three major alleles in HBA along an Iberian southwest-northeast axis is proposed to be related to the two formerly described population units, including a hybrid allele within their contact zone. The two alleles of HBB are present in almost all populations at similar frequencies, obscuring the relationship between the two evolutionary units. The starkly contrasting pattern of allelic distribution among populations at these two loci – within a well-established bi-lineage phylogeographic framework – strongly suggests that non-neutral evolutionary processes are involved at a large scale.

Keywords: hemoglobin, hybrid allele, cytonuclear disequilibrium, contact zone, Iberia

Introduction

The European rabbit, *Oryctolagus cuniculus*, originated in southern Iberia but is currently distributed throughout the world, primarily due to human mediated introductions (Flux 1994; Callou 2003). Its well described history and phylogeographic structure offer an excellent framework within which one can pursue more specific population genetic questions directed toward the further understanding of the evolutionary dynamics of the species.

Within the Iberian Peninsula, two distinct groups of populations were identified using protein (Ferrand 1995; Ferrand & Branco, this volume) and immunoglobulin (van der Loo *et al.* 1999) polymorphisms, and these groups correspond to the designation of two subspecies, *O.c. algirus* (southwest

S. Weiss and N. Ferrand (eds.), Phylogeography of Southern European Refugia, 237–255.

Iberia) and *O.c. cuniculus* (northeast Iberia). Allele frequency variation of protein markers supported an estimate of 250 000-500 000 years divergence between the two population groups and further showed a large contact zone bisecting the Peninsula along a southwest-northeast axis (Branco 2000; Ferrand & Branco, this volume). An RFLP analysis of the whole rabbit mitochondrial DNA unveiled the existence of two highly divergent lineages, referred to as A and B. Lineage A was observed in southern Spain and lineage B in northern Spain, two French localities (Camargue and Versailles) and in domestic breeds (Biju-Duval *et al.* 1991). The same work provided a 2 myr divergence estimate between the two maternal lineages, a much greater estimate than that based on nuclear genes. More recently, RFLP data of the mtDNA cytochrome *b* gene gave a detailed view of the geographic distribution of these mitochondrial lineages (A and B) showing a narrower region of contact than that seen with protein markers (Branco *et al.* 2000). A strong correlation between both sets of genetic markers and the geographic distribution of the two subspecies was shown (Branco 2000), as well as cytonuclear disequilibrium (van der Loo *et al.* 1999). Such genetic structure reflects a long period of isolation in two refugial areas throughout the Pleistocene epoch. During the most recent interglacial, rabbits are thought to have expanded from these two refugia and subsequently formed the secondary contact zone in central Iberia (Branco *et al.* 2002). This scenario is concordant with similar patterns for other organisms, which demonstrate that the Iberian Peninsula was one of the major refuge areas for many southern European species (Webb & Bartlein 1992; Hewitt 1996; Myers *et al.* 2000; see also Gomez & Lunt, this volume).

The domestication of the European rabbit is recent, and may have resulted from a single event. Data from immunoglobulins (van der Loo *et al.* 1999), proteins (Ferrand 1995), mtDNA and microsatellites (Queney *et al.* 2002) indicate that all domestic breeds originate from the genetic pool available in France and only two variants of the B haplogroup were observed. Though a number of studies focused on allozyme variation in wild rabbit populations (e.g. Ferrand *et al.* 1988; Vieira & Ferrand 1995; Branco *et al.* 1998; Branco & Ferrand 2002), only more recent studies have been carried out in the context of the rabbit's evolutionary history (Branco *et al.* 1999; Branco & Ferrand 2003; Ferrand & Branco, this volume). We chose to study rabbit hemoglobin genes (HB), α-globin (HBA) and β-globin (HBB), based on the availability of samples (hemoglobin is the most abundant protein in blood), the extensive knowledge of its evolution and regulation (Hardison 1991, 2001) and the description of polymorphism in domestic rabbit (Ferrand 1989, 1990). The relationship of these two genes is additionally interesting as each contributes two polypeptide chains to the hemoglobin molecule but are located on different chromosomes and in a radically different genomic context: HBA has been assigned to chromosome 6 (Xu & Hardison 1991) and HBB to chromosome 1 (Xu & Hardison 1989).

A considerable amount of research has been carried out on rabbit HB, but almost exclusively focused on domestic animals (see review in Hardison 1991).

The first polymorphism described in these genes was based on the determination of the amino acid sequence. Up to six amino acid changes in HBA (von Ehrenstein 1966) and four in HBB (Bricker & Garrick 1974) have been reported. In both genes, substitutions are arranged in two linked sets (Hunter & Munro 1969; Garrick *et al.* 1974). At the molecular level, the entire nucleotide sequence of both gene clusters is well described (Margot *et al.* 1989; Hardison *et al.* 1991). Although DNA sequencing is now much more routine than in the recent past, it can still be a costly and time consuming process in the framework of a population genetic question. The use of acid gel electrophoresis in the study of genetic polymorphism in rabbit α- and β-globins proved particularly useful in large-scale studies. The application of such analysis in a sample of Portuguese domestic rabbit and some Iberian wild populations led to the recognition of five alleles in HBA (Ferrand 1990, 1995) and two in HBB (Ferrand 1989).

In this study, we extended the analysis of genetic polymorphism of HBA and HBB of wild rabbit populations from their region of origin to the mitochondrial contact zone as well as locations outside Iberia, where rabbits have expanded their range in modern times (north of France). A new PCR-RFLP method to genotype rabbit HBB is also described. The overall distribution pattern of HBA and HBB alleles allows new insights on the rabbit's evolutionary history and the processes involved in the establishment of the present ranges of the two evolutionary units.

Materials and methods

Sampling

We sampled 23 wild rabbit populations and two domestic breeds (Figure 1). The wild populations cover almost the entire natural range of the species in Iberia, and a representative subset of the most recently colonized area outside Iberia (five French populations). Studies on ancient mtDNA showed that rabbit populations from the Pyrenees to northern Europe present less variability than Iberian ones, as well as more evidence of anthropogenic influence, based on the presence of domestic rabbit haplotypes (Hardy *et al.* 1995; Loreille *et al.* 1997). Thus, by studying some French populations we can estimate the existing genetic diversity in northern European rabbit populations. Blood samples were collected and prepared according to previously published protocols (Ferrand 1989, 1990).

Electrophoresis

For determination of HBA and HBB phenotypes, we used acid starch gel electrophoresis followed by a general protein stain (details in Ferrand 1989, 1990).

Allele 5 of HBA is the only variant observed in the native hemoglobin and was detected through conventional electrophoresis in agarose gels (Ferrand 1995).

PCR amplification and RFLP analysis of HBB

The molecular analysis of HBB was done using a PCR-RFLP protocol. DNA was extracted from red blood cells or liver using a standard ammonium acetate method. We amplified a 1597 bp fragment containing the entire HBB gene with a pair of primers designed from the HBB gene cluster (EMBL accession number X07786). DNA was added to a PCR mix, containing 2U of Taq polymerase EcoTaq (Ecogen), 10x buffer, 3mM $MgCl_2$, 5% DMSO, 0.2μM of each primer, HBB1F (5'-AGATACATAGAAGGAAGGCT) and HBB3R (5'-CAGCATATGGCATATGTTGC) and 0.8 mM dNTPs. PCR amplification was performed with an initial denaturing step at 94° C for 3 min, followed by 40 cycles of 1 min at 94° C, 1 min at 54° C and 2.5 min at 72° C. All reactions ended with a final extension step at 72° C for 10 min. For the RFLP analysis, we digested the resulting PCR fragments overnight with the restriction enzyme *Xce*I (Fermentas), according to the manufacturer's instructions. This enzyme cuts the HBB*1 allele only, in the third nucleotide following the first polymorphic position (β^{52} Hist/Asn; Bricker & Garrick 1974; Margot *et al.* 1989). The complete digestion of allele HBB*1 produces two fragments of 1066 and 531bp whereas HBB*2 alleles display a single fragment of 1597bp. Eighty-two wild and two domestic animals were analyzed, accordingly.

Data analysis

Allele frequencies were calculated by direct gene counting. The degree of population differentiation given by Wright's fixation index, θ, (F_{ST}; Weir & Cockerham 1984) was calculated independently for each locus with GenePop web version 3.1c (updated from version 1.2; Raymond & Rousset 1995). A transect running southwest to northeast was drawn and populations within this transect were used to measure the association between nuclear and cytoplasmic genes. For the same individuals, the allele frequencies of HBA and HBB were plotted against the frequencies of the two major mtDNA types (A and B), described in Branco *et al.* (2000). We tested the null hypothesis of no association between the alleles at the nuclear loci and the mitochondrial locus ($D{\approx}0$) calculating the χ^2 for each allele by $[D_T^2/f(A)*f(M)]*2N$, where *A* is the nuclear allele and M the mitochondrial haplotype. Disequilibria are defined as the difference between the frequency of a cytonuclear genotype and the product of the frequencies of the nuclear allele and the mitochondrial haplotypes, within populations [$Ds = f(A,M) - f(A)*f(M)$] and between populations [$cov(A/M) = f(AK)*f(MK) - f(A)*f(M)$ for population *K*; Asmussen & Arnold 1991].

Results

Polymorphism in HBA

Three major alleles were detected in locus HBA: HBA*1, HBA*2 and HBA*3. These alleles correspond to those previously observed in a sample of Portuguese domestic rabbit (Ferrand 1990) and present unequal frequency distributions across Iberia and France. HBA*1 was found in the northeast region of the Iberian Peninsula and in France, whereas HBA*2 predominates in southwest Iberia. In central Iberia, the third allele, HBA*3, is the most common (Table 1, Figures 1 & 2).

***Table 1**. Allelic frequencies at the HBA and HBB loci in the studied wild and domestic rabbit populations. Populations within the contact zone are shaded in light grey, as defined by protein polymorphism, and dark grey as defined by mitochondrial DNA.*

Population	Locus									
	HBA							HBB		
	N	HBA*1	HBA*2	HBA*3	HBA*4	HBA*5	HBA*6	*N*	HBB*1	HBB*2
Domestic										
Portuguese domestic	163	**0.75**	**0.20**	**0.05**	0.00	0.00	0.00	226	**0.50**	**0.50**
English	36	**0.71**	**0.10**	0.00	0.00	0.00	**0.19**	36	**0.71**	**0.29**
Wild continental										
Portimão	27	**0.02**	**0.96**	**0.02**	0.00	0.00	0.00	25	**0.22**	**0.78**
Huelva	54	0.00	**0.91**	**0.09**	0.00	0.00	0.00	54	**0.06**	**0.94**
Doñana	78	0.00	**0.90**	**0.10**	0.00	0.00	0.00	69	**0.01**	**0.99**
Las Lomas	49	0.00	**0.56**	**0.44**	0.00	0.00	0.00	51	**0.01**	**0.99**
Vila Viçosa	27	0.00	**0.86**	**0.14**	0.00	0.00	0.00	36	**0.08**	**0.92**
Infantado	53	0.00	**0.81**	**0.19**	0.00	0.00	0.00	55	**0.02**	**0.98**
Santarém	61	0.00	**0.68**	**0.26**	0.06	0.00	0.00	67	**0.18**	**0.82**
Badajoz	16	0.00	**0.22**	**0.78**	0.00	0.00	0.00	16	**0.06**	**0.94**
Idanha	49	0.00	**0.23**	**0.77**	0.00	0.00	0.00	44	**0.05**	**0.95**
Cabreira	11	0.00	**0.64**	**0.36**	0.00	0.00	0.00	16	**0.06**	**0.94**
Ciudad Real	20	0.00	**0.05**	**0.95**	0.00	0.00	0.00	20	**0.10**	**0.90**
Amoladeras	11	0.00	**0.82**	**0.18**	0.00	0.00	0.00	11	**0.36**	**0.64**
Bragança	27	0.00	**0.50**	**0.50**	0.00	0.00	0.00	26	**0.11**	**0.89**
Toledo	48	**0.04**	**0.23**	**0.73**	0.00	0.00	0.00	57	**0.20**	**0.80**
Alicante	38	**0.12**	**0.50**	**0.38**	0.00	0.00	0.00	39	**0.50**	**0.50**
Tudela	49	**0.10**	**0.47**	**0.29**	**0.04**	**0.10**	0.00	49	**0.17**	**0.83**
Tarragona	21	**0.69**	**0.02**	**0.29**	0.00	0.00	0.00	23	**0.26**	**0.74**
Lérida	26	**0.69**	**0.12**	**0.19**	0.00	0.00	0.00	26	**0.02**	**0.98**
Perpignan	19	**0.66**	**0.34**	0.00	0.00	0.00	0.00	26	0.00	**1.00**
Camargue	29	**0.96**	**0.04**	0.00	0.00	0.00	0.00	29	**0.07**	**0.93**
Carlucet	-	-	-	-	-	-	-	25	**0.03**	**0.97**
Vaulx-en-Velin	-	-	-	-	-	-	-	25	0.00	**1.00**
Versailles	58	**0.88**	0.12	0.00	0.00	0.00	0.00	58	**0.08**	**0.92**

Despite its wide distribution in Iberian populations, outside the putative hybrid zone the frequency of the HBA*3 allele is low. The alleles HBA*4, HBA*5 and the newly described HBA*6 are present in low frequencies and have limited geographic distributions. HBA*4 is present both in Santarém (south of Portugal) and Navarra (north of Spain), and HBA*5 was found only in Navarra. HBA*6 was found both in an insular Tunisian population (P. Esteves, pers. com.) and a European domestic breed (English) (Table 1).

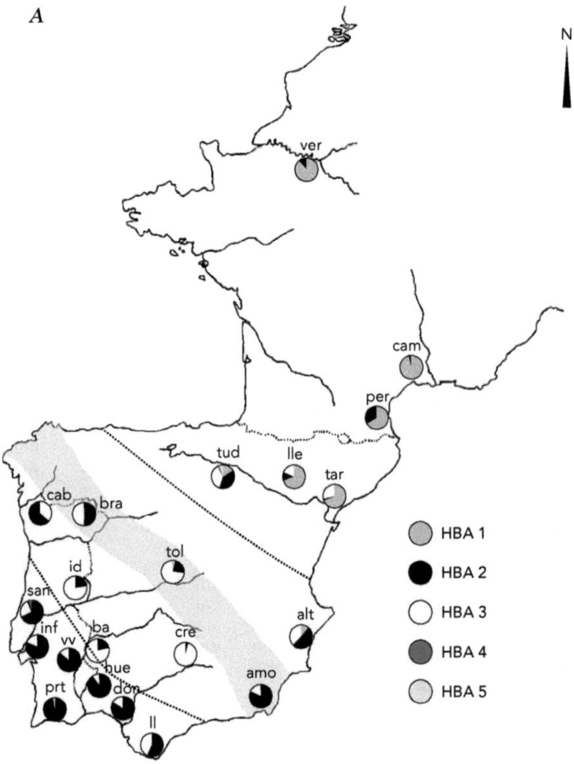

Figure 1. Continued.

Continued on next page

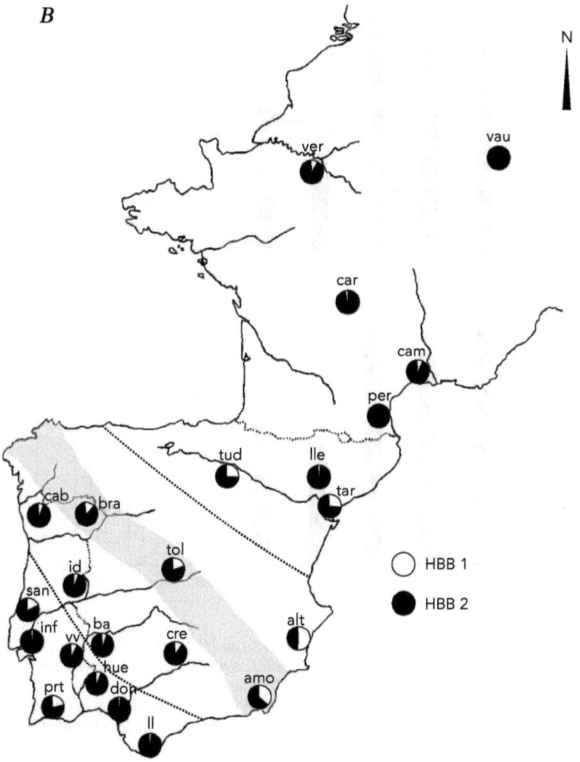

Figure 1. *Geographical distribution of the five HBA (A) and two HBB (B) alleles in the Iberian Peninsula and France. prt: Portimão, inf: Infantado, vv: Vila Viçosa, hue: Huelva, don: Doñana, ll: Las Lomas, san: Santarém, ba: Badajoz, cab: Cabreira, id: Idanha-a-Nova, cre: Ciudad Real, bra: Bragança, tol: Toledo, amo: Amoladeras, alt: Alicante, tud: Tudela, tar: Tarragona, lle: Lérida, pep: Perpignan, tv: Camargue, car: Carlucet, ver: Versailles, vau: Vaulx-en-Velin. Shaded area represents the mitochondrial hybrid zone as reported in Branco* et al. *(2000) and the area within the dotted lines is the contact zone defined by protein polymorphism. Samples from Portimão, Huelva, Las Lomas, Bragança, Toledo, Tudela, Lérida, Perpignan, Camargue and Versailles were used in the graphics depicted in Figure 2.*

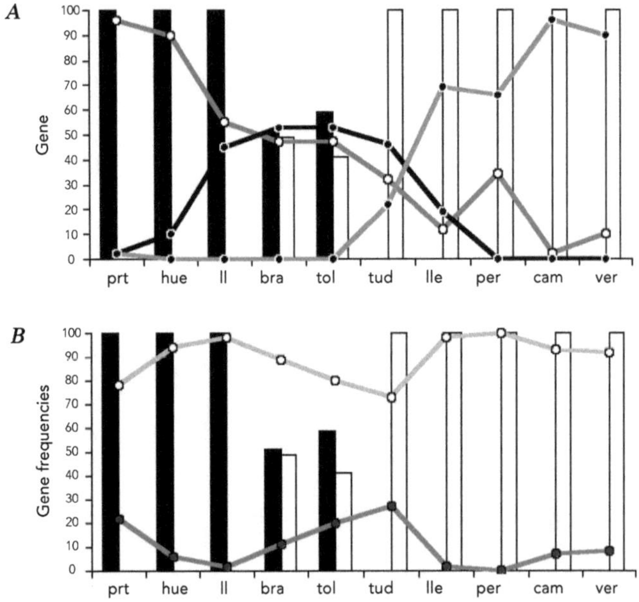

Figure 2. *Frequency association between HBA (**A**) and HBB (**B**) alleles and mtDNA. Gene frequencies of HBA and HBB are described in Table 1; mtDNA lineages A (black bars) and B (white bars) are according to Branco* et al. *(2000). The populations are shown in an ordered transect from south of Portugal to north of France.* ***A****) HBA alleles are HBA*1 (grey line, black circle), HBA*2 (grey line, white circle) and HBA*3 (black line, black circle).* ***B****) HBB alleles are HBB*1 (dark grey line) and HBB*2 (light grey line).*

Mean expected heterozygosities were relatively high, being higher in domestic rabbit (0.43) than in wild populations (0.36) (Table 2). The wild populations with the lowest values were Portimão, Ciudad Real and Versailles, due to the predominance of a single allele (Table 1). The most frequent allele in each population corresponds to the expected area of occurrence for each of the three common alleles: HBA*1 in the north (Versailles, north of France), HBA*2 in the south (Portimão, south of Portugal) and HBA*3 in central Iberia (Ciudad Real, central Spain).

Polymorphism in HBB

Only the two previously described alleles, HBB*1 and HBB*2, (Ferrand 1989) were detected in this study (Table 1 & Figure 1). Except for Perpignan and Vaulx-en-Velin (France), all populations studied thus far display HBB*1 at a low frequency (≈ 0.10) and HBB*2 at a high frequency (≈ 0.90). This explains the low values of expected heterozygosity calculated for all the wild populations (0.19), as opposed to the high values in the domestic samples (0.45) (Table 2).

Table 2. *Expected heterozygosity at HBA and HBB loci in the studied wild and domestic rabbit populations. Populations within the contact zone are shaded in light grey, as defined by protein polymorphism, and dark grey as defined by mitochondrial DNA.*

Population	Locus	
	HBA	HBB
Domestic		
Portuguese domestic	0.40	0.50
English	0.45	0.41
mean	*0.43*	*0.45*
Wild continental		
Portimão	0.08	0.34
Las Lomas	0.49	0.02
Doñana	0.18	0.02
Huelva	0.16	0.11
Vila Viçosa	0.24	0.15
Infantado	0.31	0.04
Santarém	0.47	0.29
Badajoz	0.39	0.11
Idanha	0.35	0.09
Cabreira	0.46	0.11
Ciudad Real	0.09	0.18
Amoladeras	0.29	0.46
Bragança	0.50	0.22
Toledo	0.41	0.32
Alicante	0.59	0.50
Tudela	0.67	0.28
Tarragona	0.44	0.38
Lérida	0.47	0.04
Perpignan	0.45	0.00
Camargue	0.08	0.13
Carlucet	-	0.06
Vaulx-en-Velin	-	0.00
Versailles	0.21	0.15
mean	*0.36*	*0.19*

The introduction of a PCR-RFLP methodology proved to be very efficient for the discrimination of the two electrophoretic variants of HBB. This further permitted the recognition of the two well-defined molecular lineages, corresponding precisely to the HBB*1 and HBB*2 alleles. This finding was confirmed by sequencing some domestic animals, with a previously determined phenotype.

Table 3. *Genetic differentiation within and between rabbit groups. Taxonomic groups are here defined based on protein polymorphism, whereby 'Hybrid' is the differentiation within the hybrid zone, and* O.c. algirus *x* O.c. cuniculus *is the differentiation between these lineages excluding the hybrid zone. For geographic regions SWIP is the southwest Iberian Peninsula; CIP: the central Iberian Peninsula; NEIP: the northeast Iberian Peninsula; and FR: France. For mtDNA defined groups, A + B is the differentiation within the hybrid zone defined by mtDNA, and mtDNA A x mtDNA B is the differentiation between the two lineages excluding the hybrid zone.*

	Locus	
	HBA	HBB
All populations	0.44	0.15 (0.08)*
Taxonomic groups		
Oryctolagus c. algirus	0.11	0.04
Hybrid	0.20	0.23 (0.10)*
Oryctolagus c. cuniculus	0.22	0.12
O.c. algirus x *O.c. cuniculus*	0.54	≈ 0
Geographical regions		
SWIP	0.30	0.03
CIP	0.20	0.20 (0.09)*
NEIP + FR	0.22	0.12
mtDNA defined groups		
mtDNA A	0.35	0.02
mtDNA A+B (hybrid)	0.21	0.09
mtDNA B	0.26	0.25 (0.12)*
mtDNA A x mtDNA B	0.38	0.05

* Without Alicante population.

Table 4. *Cytonuclear disequilibrium in wild rabbit populations along a transect from the southwest of Iberia to north of France.*

alleles	Cov(HBA/mtA)	Ds	D_T	χ_1^2
HBA 1	−0.14145	−0.001	−0.14245	28.691*
HBA 2	0.11845	−0.007	0.11145	14.257*
HBA 3	0.02182	0.008	0.02982	1.903
	Cov(HBB/mtA)	**Ds**	D_T	χ_1^2
HBB 1	0.00436	0.0012	0.00556	0.144
HBB 2	−0.00436	−0.005	−0.00936	0.048

*The values of χ are significant at $p < 0.05$.

Genetic structure

In HBA, 44% of the total genetic differentiation was among rather than within populations (Table 3). This result is true whether considering the global data set or partitioning the populations into geographic (SWIP, CIP and NEIP+FR, see Table 3) or mitochondrial groups (A, A+B and B). However, when considering the two subspecies and the populations inside the large hybrid zone defined by protein variation (Branco 2000), almost all (89%, 78% and 80% in *O.c. algirus*, *O.c. cuniculus* and hybrid populations, respectively) of the genetic diversity was within populations. The F_{ST} value between the two putative subspecies (0.54), although lower than that obtained for mtDNA (0.65; Branco 2000), is higher than that obtained for a set of 16 polymorphic protein loci (0.18; Ferrand 1995). In contrast, for HBB, no trace of differentiation between the subspecies can be detected ($F_{ST} \approx 0$; Table 3). In this locus only 15% of the total diversity was found between populations and this value drops to 8% when the Alicante population is not considered. This population was the only one where a strong deviation from Hardy-Weinberg equilibrium was detected, as well as a similar frequency of each allele (Table 1).

Cytonuclear systems

We found no association between HBB alleles and mtDNA types, but there was a strong non-random association between HBA alleles and the two mtDNA lineages across the Iberian Peninsula (Figure 2 & Table 4). The HBA*1 allele occurs almost exclusively in populations that exhibit the B mtDNA lineage and exhibits a very strong negative correlation with lineage A; Table 4. HBA*2 is present in populations with the A mtDNA lineage and HBA*3 is characteristic of central Iberian populations, where both mtDNA types are found (no significant association between this allele and either lineage was found, Table 4). The remaining three alleles were not studied for cytonuclear associations due to their very limited distribution.

Discussion

Geographical partitioning of alleles at HBA locus

The overall distribution pattern of HBA alleles reflects the differentiation between southern and northern populations (Table 1). The most relevant contribution to this observation is the presence or absence of HBA*1 allele and the frequency of HBA*2 allele. While southern populations characteristically present HBA*2, this allele is almost absent in northern populations, where HBA*1 predominates (Figure 1, Table 1). This association is probably the primary cause of the marked population differentiation ($F_{ST} = 0.44$), in contrast to an almost absence of differentiation within each group of populations (F_{ST} = 0.11 in *O.c. algirus*, F_{ST} = 0.20 in *O.c. cuniculus* and F_{ST} = 0.22 in the hybrid populations), as each can be identified through a characteristic allele. The detection of HBA*1 in Portimão (south of Portugal) is an exception to this pattern. However, because the sample came from a region undergoing frequent restocking to ameliorate the effects of hunting pressure, it is probable that admixture from domestic rabbits or other allochthonous animals has occurred. The association between HBA alleles and the two putative subspecies is particularly evident when compared with mitochondrial DNA data. Figure 2 shows a very strong association between the HBA* 1 allele and the mtDNA lineage B, characteristic of the northeast group, *O.c. cuniculus*, and HBA* 2 with lineage A, prevalent in the southwest group, *O.c. algirus*.

Combining data from nuclear and cytoplasmic markers (cytonuclear systems) is a very informative methodology for studying the evolutionary history of natural populations (Asmussen *et al.* 1987). Special attention has been given to the detection of a non-random association between a nuclear and a cytoplasmic gene in hybrid zones (Asmussen *et al.* 1989; Arnold 1993). An association between mtDNA lineages and alleles at an immunoglobulin locus has already been observed in rabbits (van der Loo *et al.* 1999). By measuring cytonuclear disequilibrium within a hybrid zone it is possible to access information on the level of gene flow and direction of introgression between the hybridizing taxa (Arnold 1993). Hybrid alleles have been broadly identified in secondary contact zones of numerous species (Woodruff 1989; Bradley *et al.* 1993; Schilthuizen & Gittenberger 1994; Arntzen 2001) and have been hypothesized to confer fitness advantage over the parental type alleles. In our case, the distribution of HBA*3 coincides with the putative hybrid zone defined in terms of mitochondrial haplotypes (Branco *et al.* 2000), as well as proteins (Branco 2000), where there is a near absence of intrapopulation divergence ($F_{ST} \approx 0.09$, without Alicante population; Table 3). Thus, we propose that the HBA*3 allele is a hybrid allele. Some studies show that hybrizymes often derive from parental type alleles through a single nucleotide substitution (Bradley *et al.* 1993; Hoffman & Brown 1995; Schilthuizen *et al.* 1999). The complete sequencing of at least the three major HBA alleles detected by electrophoresis should clarify the question of whether HBA*3 is a hybrid allele.

The identification of private and regional alleles in HBA (HBA*4, HBA*5 and HBA*6; Table 1) is in accordance with what is known in other markers, such as proteins (Ferrand 1995; Vieira & Ferrand 1995; Branco *et al.* 1998; Branco & Ferrand 1998; Branco *et al.* 1999;), mtDNA (Branco *et al.* 2000), microsatellites (Queney *et al.* 2001) or immunoglobulins (van der Loo *et al.* 1999). This is probably a consequence of the strong local structure that characterizes rabbit populations. Rare alleles are also frequently associated with populations that have maintained relatively stable effective population sizes through time (Chakraborty *et al.* 1980). The fact that most rare alleles are found in populations of southwestern Iberia has been interpreted as favorable evidence that the origin of the species was within this area (further supported by archaeological remains of *Oryctolagus* sp., Lopez-Martinez 1989) and that large and stable effective population sizes have been maintained (Ferrand 1995; Branco *et al.* 1999, 2000). The observation of stationary (homoplasic) size distributions of microsatellite loci between the two rabbit lineages (Queney *et al.* 2001) was also deemed concordant with the theoretical predictions of such a pattern, given large effective populations sizes (Nauta & Weissing 1996).

Tracing the origins of HBA alleles

The origin of two allopatric alleles within a locus is evident from mtDNA (Branco *et al.* 2000, 2002) and Y-chromosome (Geraldes *et al.* 2005; Geraldes & Ferrand in press) studies. Both markers support a similar phylogeographic scenario, where the prolonged isolation of two population groups led to the formation of two geographically distinct units in *O. cuniculus*. The survey of 16 protein-coding loci highlighted the existence of the same signature through a high number of rare and characteristic alleles (Ferrand 1995). In HBA, the heterogeneous distribution of the most common alleles, HBA*1 and HBA*2, provides evidence for their origins. The fact that HBA*1 is nearly fixed in domestic rabbits suggests an origin within *O.c. cuniculus* during the time when these animals were separated from the southern group, *O.c. algirus*. The present distribution of this allele most likely reflects the spread of *O.c. cuniculus* populations across Iberia and northern Europe. This hypothesis predicts an alternative scenario for HBA*2 allele, assigning its origin within *O.c. algirus*. Therefore, the simultaneous migration of rabbits and their alleles can be superimposed on the actual distribution of both subspecies in Iberia and Europe. Under this assumption, HBA*3 probably descended from either HBA*1 or HBA*2. Its predominance in the hybrid zone suggests that HBA*3 may confer some fitness advantage over each of the former alleles. The results from electrophoretic polymorphism highlight an association between HBA*2 and HBA*3 (Table 1, Figure 2), giving some weight to the presumption that HBA*3 derives from a mutation in HBA*2.

HBA*4, HBA*5 and HBA*6 are probably more recent alleles, also resulting from a mutational event in the ancestral HBA*1 or HBA*2 alleles. Because HBA*4 is identified both in one *O.c. algirus* and one *O.c. cuniculus* population, it seems reasonable to concur that there is cryptic genetic variation and that there are at least two alleles displacing the same electrophoretic pattern. The amino acid changes involving the major alleles result from noncharged substitutions (von Ehrenstein 1966), so the electrophoretic system has probably reached its detection limit at this locus.

Homogeneous distribution of HBB alleles among all rabbit populations

Unexpectedly, the two HBB alleles identified in domestic rabbit (Ferrand 1989) were also observed in all wild populations, except Perpignan and Vaulx-en-Velin, and at similar frequencies (Table 1 & Figure 1). Consequently, this locus exhibits extremely low values of expected heterozygosity (Table 2) and population differentiation ($F_{ST} = 0.14$). The exception to this pattern is Alicante, where similar frequencies of both alleles are observed. Ferrand (1995) observed that *O.c. algirus* populations consistently show higher expected heterozygosities compared to *O.c. cuniculus* populations. Furthermore, this lack of population structure of HBB does not correspond to what was expected under the scenario of historical differentiation and recent secondary contact (Ferrand 1995; Branco *et al.* 2000). On the contrary, such results suggest a coalescence time for HBB alleles prior to the separation of *O.c. algirus* and *O.c. cuniculus*.

Is HBB under selection?

From allozyme data alone, we can postulate that HBB*1 and HBB*2 alleles originated within the (ancestral) *Oryctolagus cuniculus* group, in southwestern Iberia, and then spread throughout the entire range of rabbit populations. This implies an estimated time of origin for the two alleles of more than 2 myr, or prior to the Pleistocene epoch. Such a deep coalescence allows the prediction of recombination within and between the two alleles. Amino acid sequence determination of HBB alleles in domestic rabbit revealed the existence of two well differentiated linked sets of four amino acid substitutions (Galizzi 1970; Garrick *et al.* 1974). These authors postulated that the maintenance of these substitutions could be attributed either to selection or to linkage disequilibrium (LD; Garrick *et al.* 1974). A recent work on the extent of LD blocks in humans showed that the range of LD varies along the genome (Reich *et al.* 2001). Moreover, LD tends to be greater in more recent populations and in populations that were subject to severe decreases in population size than in older and more stable populations (Tishkoff *et al.* 1996; Reich *et al.* 2001). Therefore, by looking at the wild animal genome, we would expect to see an 'evolutionary route', that is, the gradual appearance of mutations leading to the establishment of those four amino acid changes. However, the introduction of a PCR-RFLP approach revealed that the linked set of amino acids appears to be present in all analyzed animals, both domestic and wild. The sequencing of both electrophoretic alleles confirmed this hypothesis and further allowed the identification of 5 fixed differences between them. Although the sequences were obtained from domestic animals only, this result prompts us to postulate another evolutionary scenario for the appearance of HBB alleles. Under this alternate hypothesis, the two alleles differentiated during the isolation of the two groups. The ubiquitous presence of both alleles in similar frequencies in all populations studied so far can be a result of natural selection. The excess of one allele, as seen for the HBB*2 in our study, was found to be a general feature of selective sweeps (Wiehe 1998). The effect that a 'selective sweep' (the fixation of an advantageous mutation; Maynard Smith & Haigh 1974) has on the genealogy of the variants of the locus under selection is often designated as 'star-like'. This happens because after the sweep new mutations start to arise, but scattered as low-frequency variants (Hudson 1990). There is now a wide range of statistical tests available to assess deviations from the neutral model using nucleotide sequence data (e.g. Hudson *et al.* 1987; Tajima 1989). By sequencing the HBB gene from a representative set of wild rabbits it will be possible to test whether there is a skew in the spectrum of allele frequencies towards low frequency variants.

Final remarks

The application of acid starch gel electrophoresis to the analysis of genetic variation within HBA and HBB loci proved to be a very efficient approach. Although these systems are probably at their limit of detection of extra-variation, we can assume that the relevant alleles in each locus were detected. By studying the genetic variation of two unlinked but closely related loci across the distribution area of the rabbit, we were able to describe: i) a putative hybrid allele in HBA, ii) the inability to distinguish the two separate groups of rabbit populations through the analysis of HBB electrophoretic polymorphism, and iii) the differences in the pattern of geographical distribution of alleles in the two loci, that could be attributed to divergent forces shaping the actual genetic variation in each gene. The data presented here will serve as a guide for further and more refined investigations into the evolutionary history of the rabbit, and further emphasize the advantage of pursuing such questions within systems with a well-described phylogeographic framework.

Acknowledgements

This work was partially financed by project grants POCTI/BSE/34151/1999 and POCTI/BSE/40280/2001 from FCT (Fundação para a Ciência e Tecnologia). R. Campos is financed by PhD grant SFRH/BD/1409/2000 and M. Branco by post-doctoral grant SFRH/BPD/5610/2001, both from FCT.

References

Arnold J (1993) Cytonuclear disequilibria in hybrid zones. *Annual Review of Ecology and Systematics* **24**, 521-554.

Arntzen JW (2001) Genetic variation in the Italian crested newts, *Triturus carnifex*, and the origin of a non-native population north of the Alps. ***Biodiversity and Conservation* 10**, 971-987.

Asmussen MA, Arnold J (1991) The effects of admixture and population subdivision on cytonuclear disequilibria. ***Theoretical Population Biology* 39**, 273-300.

Asmussen, MA, Arnold J, Avise JC (1987) Definition and properties of disequilibrium statistics for associations between nuclear and cytoplasmic genotypes. ***Genetics* 115**, 755-768.

Asmussen MA, Arnold J, Avise JC (1989) The effects of assortative mating and migration on cytonuclear association in hybrid zones. ***Genetics* 122**, 923-934.

Biju-Duval C, Ennafaa H, Dennebouy N, Monnerot M, Mignotte F, Soriguer RC, El Gaaied A, El Hili A, Mounolou J-C (1991) Mitochondrial DNA evolution in lagomorphs: origin of systematic heteroplasmy and organization of diversity in European rabbits. ***Journal of Molecular Evolution* 33**, 92-102.

Bradley RD, Bull JJ, Johson AD, Hillis DM (1993). Origin of a novel allele in a mammalian hybrid zone. ***Proceedings of the National Academy of Sciences, USA* 90**, 8939-8941.

Branco M (2000) ***Estrutura genética das populações de coelho europeu (*Oryctolagus cuniculus*) na Península Ibérica. Isolamento, diferenciação de duas unidades evolutivas, expansão geográfica e contacto secundário.*** PhD Thesis, University of Oporto.

Branco M, Ferrand N (1998) Genetic polimorphism of the rabbit (*Oryctolagus cuniculus*) tissue acid phosphatases (ACP2 and ACP3). ***Comparative Biochemistry and Physiology B* 120**, 405-409.

Branco M, Ferrand N (2002) Genetic polymorphism of antithrombin III, haptoglobin, and haemopexin in wild and domestic European rabbits. ***Biochemical Genetics* 40**, 387-393.

Branco M, Ferrand N (2003) Biochemical and population genetics of the rabbit, *Oryctolagus cuniculus*: carbonic anhydrases Iand II, from the Iberian Peninsula and France. ***Biochemical Genetics* 41**, 391-404.

Branco M, Lopes G, Ferrand N (1998) Genetic polymorphism of properdin factor B (BF) in domestic rabbit. ***Animal Genetics* 29**, 135

Branco M, Machado JC, Ferrand N (1999) Extensive genetic polymorphism of peptidases A, B, C and D in wild rabbit (*Oryctolagus cuniculus*) populations from the Iberian Peninsula. ***Biochemical Genetics* 37**, 237-249.

Branco M, Ferrand N, Monnerot M (2000) Phylogeography of the European rabbit (*Oryctolagus cuniculus*) in the Iberian Peninsula inferred from RFLP analysis of the cytochrome b gene. ***Heredity* 85**, 307-317.

Branco M, Ferrand N, Monnerot M, Templeton AR (2002) Postglacial dispersal of the European rabbit (*Oryctolagus cuniculus*) on the Iberian Peninsula reconstructed from nested clade and mismatch analyses of mitochondrial DNA genetic variation. ***Evolution* 56**, 792-803.

Bricker J, Garrick M (1974) An isoleucine-valine substitution in the β-chain of rabbit hemoglobin. ***Biochimica et Biophysica Acta* 351**, 437-441.

Callou C (2003) De la garenne au clapier: étude archéozoologique du lapin en Europe occidental. ***Memoires du Muséum National d'Histoire Naturelle* 189**, 1-356.

Chakraborty R, Fuerst PA, Nei M (1980) Statistical studies on protein polymorphism in natural populations. III. Distribution of allele frequencies and the number of alleles per locus. ***Genetics* 94**, 1039-1063.

Ferrand N (1989) Biochemical and genetic studies on rabbit hemoglobin. I. Electrophoretic polymorphism of the β-chain. ***Biochemical Genetics*** **27**, 673-678.

Ferrand N (1990) Biochemical and genetic studies on rabbit hemoglobin. II. Electrophoretic polymorphism of the α-chain. ***Biochemical Genetics*** **28**, 117-122.

Ferrand N (1995) ***Variação genética de proteínas em populações de coelho*** **(Oryctolagus cuniculus)*****. Análise da diferenciação subespecífica, subestruturação, expansão geográfica e domesticação.*** PhD Thesis, University of Oporto.

Ferrand N, Carvalho G, Amorim A (1988) Transferrin (Tf) polymorphism in wild rabbit, (*Oryctolagus cuniculus*). ***Animal Genetics*** **19**, 295-300.

Flux JEC (1994) The rabbit in continental Europe. In: ***The European Rabbit: The History Biology of a Successful Colonizer*** (eds. Thompson HV, King CM), pp. 22-63. Oxford Science Publications, Oxford.

Galizzi A (1970) The amino acid sequence of the β-chains of rabbit hemoglobin. ***European Journal of Biochemistry*** **17**, 49-55.

Garrick M, Bricker J, Garrick L (1974) An electrophoretically silent polymorphism for the beta chains of rabbit hemoglobin and associated polyribosome patterns. ***Genetics*** **76**, 99-108.

Geraldes A, Ferrand N (in press) A 7-bp insertion 3' to the HGM box suggests the duplication and concerted evolution of the rabbit SRY gene. ***Genetics Selection Evolution***.

Geraldes A, Rogel-Gaillard C, Ferrand N (2005) High levels of nucleotide diversity in the European rabbit (*Oryctolagus cuniculus*) SRY gene. ***Animal Genetics*** **36**, 349-351.

Hardison R (1991) Evolution of globin gene families. In: ***Evolution at the Molecular Level*** (eds. Selander RK, Clark AG, Whittam TS), pp 272-289. Sinauer Associates, Sunderland.

Hardison R (2001) New views of evolution and regulation of vertebrate α-like globin gene clusters from an orphaned gene in marsupials. ***Proceedings of the National Academy of Sciences, USA*** **98**, 1327-1329.

Hardison R, Krane D, Vandenbergh D, Cheng J-F, Mansberger J, Taddie J, Schwartz S, Huang X, Miller W (1991) Sequence and comparative analysis of the rabbit globin gene cluster reveals a rapid mode of evolution in a G+C-rich region of mammalian genomes. ***Journal of Molecular Biology*** **222**, 233-249.

Hardy C, Callou C, Vigne JD, Casane D, Dennebouy N, Mounolou JC, Monnerot M (1995) Rabbit mitochondrial DNA diversity from prehistoric to modern times. ***Journal of Molecular Evolution*** **40**, 227-237.

Hewitt GM (1996) Some genetic consequences of ice ages, and their role in divergence and speciation. ***Biological Journal of the Linnean Society*** **58**, 247-276.

Hoffman SM, Brown WM (1995) The molecular mechanism underlying the 'rare allele phenomenon' in a subspecific hybrid zone of the California field mouse, *Peromyscus californicus*. ***Journal of Molecular Evolution*** **41**, 1165-1169.

Hudson RR (1990) Gene genealogies and the coalescent process in models with selection and recombination. In: ***Oxford Surveys in Evolutionary Biology*** (eds. Futuyma D, Antonovics J), pp. 1-44. Oxford University Press, Oxford.

Hudson RR, Kreitman M, Aguade M (1987) A test of neutral molecular evolution based on nucleotide data. ***Genetics*** **116**, 153-159.

Hunter T, Munro A (1969) Allelic variants in the aminoacid sequence of the α-chain of rabbit hemoglobin. ***Nature*** **223**, 1270-1272.

Lopez-Martinez N (1989) ***Revisión sistemática y biostratigráfica de los lagomorphos (Mammalia) del neogeno y cuaternario de España.*** Memorias del Museo Paleontologico de la Universidade de Zaragoza, n°3. Disputación General de Aragon, Zaragoza.

Loreille O, Mounolou JC, Monnerot M (1997) History of the rabbit and ancient DNA. ***Comptes-rendus des Séances de la Société de Biologie* 191**, 537-544.

Margot JB, Demers GW, Hardison RC (1989) Complete nucleotide sequence of the rabbit β-like globin gene cluster. Analysis of intergenic sequences and comparison with human β-like globin gene cluster. ***Journal of Molecular Biology* 205**, 15-40.

Maynard Smith J, Haigh J (1974) The hitch-hiking effect of a favourable gene. ***Genetical Research* 23**, 23-35.

Myers N, Mittermeier RA, Mittermeier CG, da Fonseca GAB, Kent J (2000) Biodiversity hotspots for conservation priorities. ***Nature* 403**, 853-858.

Nauta MJ, Weissing FJ (1996) Constraints on allele size at microsatellite loci: implications for genetic differentiation ***Genetics* 143**, 1021-1032.

Queney G, Ferrand N, Weiss S, Mougel F, Monnerot M (2001) Stationary distributions of microsatellite loci between divergent population groups of the european rabbit (*Oryctolagus cuniculus*). ***Molecular Biology and Evolution* 18**, 2169-2178.

Queney G, Vachot A-M, Brun J-M, Dennebouy N, Mulsat P, Monnerot M (2002) Different levels of human intervention in domestic rabbits: effects on genetic diversity. ***Journal of Heredity* 93**, 205-209.

Raymond M, Rousset F (1995) GENEPOP (version 1.2): population genetics software for exact tests and ecumenicism. ***Journal of Heredity* 86**, 248-249.

Reich DE, Cargill M, Bolk S, Ireland J, Sabeti PC, Richter DJ, Lavery T, Kouyoumjian R, Farhadian SF, Ward R, Lander ES (2001) Linkage disequilibrium in the human genome. ***Nature* 411**, 199-204.

Schilthuizen M, Gittenberger E (1994) Parallel evolution of an sAat-'hybrizyme' in hybrid zones in *Albinaria hippolyti* (Boettger). ***Heredity* 73**, 244-248.

Schilthuizen M, Hoekstra RF, Gittenberger E (1999) Selective increase of a rare haplotype in a land snail hybrid zone. ***Proceedings of the Royal Society of London, B* 266**, 2181-2185.

Tajima F (1989) Statistical method for testing the neutral mutation hypothesis by DNA polymorphism. ***Genetics* 123**, 585-595.

Tishkoff SA, Dietzsch E, Speed W, Pakstis AJ, Kidd JR, Cheung K, Bonne-Tamir B, Santachiara AS, Moral P, Krings M, Paabo S, Watson E, Risch N, Jenkins T, Kidd KK (1996) Global patterns of linkage disequilibrium at the CD4 locus and modern human origins. ***Science* 271**, 1380-1387.

van der Loo W, Mougel F, Sánchez MS, Bouton C, Castien E, Fonseca A, Ferrand N, Soriguer R, Monnerot M (1999) Cytonuclear disequilibria in wild populations of rabbit (*Oryctolagus cuniculus L.*) suggest unequal turnover rates at the b locus (IGKC1). ***Immunogenetics* 49**, 629-643.

Vieira J, Ferrand N (1995) Four new alleles at the mannose-6-phosphate isomerase locus in rabbit. ***Animal Genetics* 26**, 37-38.

von Ehrenstein, G (1966) Translational variations in the amino acid sequence of the α-chain of rabbit hemoglobin. ***Cold Spring Harbor Symposium on Quantitative Biology* 31**, 705-714.

Webb T, Bartlein PJ (1992) Global changes during the last 3 million years: climatic controls and biotic response. ***Annual Review of Ecology and Systematics* 23**, 141-173.

Weir BS and Cockerham CC (1984) Estimating F-statistics for the analysis of population structure. ***Evolution* 38**, 1358-1370.

Wiehe T (1998) The effect of selective sweeps on the variance of the allele distribution of a linked multiallele locus: hitchhiking of microsatellites. ***Theoretical Population Biology* 53**, 272-283.

Woodruff DS (1989) Genetic anomalies associated with *Cerion* hybrid zones: the origin and maintenance of new electrophoretic variants called hybrizymes. ***Biological Journal of the Linnean Society of London* 36**, 281-294.

Xu J, Hardison R (1989) Localization of the β-like globin gene cluster and the genes for parathyroid hormone and c-Harvey-ras 1 to region q14----q21 of rabbit chromosome 1 by *in situ* hybridization. ***Cytogenetics and Cell Genetics* 52**, 157-161.

Xu J, Hardison R (1991) Localization of the α-like globin gene cluster to region q12 of rabbit chromosome 6 by *in situ* hybridization. ***Genomics* 9**, 362-365.

Chapter 9

Spatial genetic structure of an explicit glacial refugium of maritime pine (*Pinus pinaster* Aiton) in southeastern Spain

Santiago C. González-Martínez*, Aranzazu Gómez, José S. Carrión, Dolores Agúndez, Ricardo Alía and Luis Gil
*E-mail: santiago@inia.es

Abstract
The Iberian Peninsula has been described as a glacial refugium for numerous organisms. In particular, there is evidence both from pollen records and genetic studies that shows the existence of Mediterranean conifers (*Pinus halepensis* Miller, *Pinus pinaster* Aiton, and *Pinus pinea* L) in southeastern Spain during the last glacial stage. Data from eight polymorphic allozyme markers were used to study the spatial genetic structure of 11 native populations of maritime pine, *P. pinaster*, in this region. Models of isolation by distance were adjusted to different groups of populations to test specific hypotheses about the role of mountain ranges in shaping the spatial genetic structure of maritime pine in southeastern Spain. In addition, pairwise gene interchange was analyzed using migration matrix models and maximum likelihood methods to make joint estimates of dispersal rates and population sizes. A complex pattern in the distribution of gene diversity was found, involving historical isolation due to geographical variables for particular populations. The role of mountain ranges in glacial refugia i) reducing the risk of a population bottleneck by altitudinal migration in response to climatic change, and ii) acting as geographical barriers to gene flow, is discussed.

Keywords: genetic structure, glacial refugia, likelihood, Mediterranean basin, conifers

Introduction

Over the last decade, phylogeographic studies have revealed high genetic diversity and richness in southern Europe for most temperate species, whereas low genetic variation has been usually found in northern populations. The richness of southern populations is assumed to be the result of persistence of populations and accumulation of variation over several glacial stages (Hewitt 1996, 1999; Bennett 1997). Postglacial colonization followed concordant south to north migration routes in many species from Europe and the

S. Weiss and N. Ferrand (eds.), Phylogeography of Southern European Refugia, 257–269.

Pacific Northwest of the United States (Hewitt 2001). Fossil records in central and northern Europe have shown that, when typical migration rates for mesothermophilous species are invoked, there would not have been enough time for some species to arrive from their putative southern refugia (i.e. Reid's paradox; Clark *et al.* 1998). Rapid, long-distance dispersal of small groups ahead of the main dispersing populations has been suggested (Hewitt 1999), but increasing evidence supports the existence of cryptic northern refugia in sheltered areas with suitable microhabitats (Willis *et al.* 2000; Steward & Lister 2001). Thus, current populations in central and northern Europe might be a result of the interaction between local or regional survival and continental-scale migration.

Hewitt (2001) suggested that in southern refugia, genomes are greatly subdivided geographically due to survival in disjunct locations without large geographical displacement. Coalescent simulations in grasshoppers have shown that glaciations promoted divergence among populations due to i) drift associated with colonization of previous glaciated areas and ii) differentiation among multiple allopatric glacial refugia (Knowles 2001). Analysis of spatial genetic structure within putative glacial refugia can provide relevant information about the distribution of gene diversity in stable populations and its causes. This is particularly true in long-lived organisms like forest trees (Petit & Vendramin, this volume), where no genetic structure is usually found in recently colonized regions, partly because there has been an insufficient number of generations for the accumulation of variation and corresponding divergence between populations.

Major refugial areas have been identified in the Iberian Peninsula for a diversity of organisms (Santucci *et al.* 1998; Taberlet *et al.* 1998; Comes & Abbot 1998; Willis & Whittaker 2000; Gómez & Lunt, this volume). Recent studies show a very high amount of genetic variation in forest tree populations of southern Iberia. Ferris *et al.* (1998) found three major cpDNA types in European white oaks (*Quercus robur* and *Q. petraea*) and suggested a postglacial migration pathway from the Iberian Peninsula based on the distribution of one of them. More recently, a consortium of 16 laboratories have studied chloroplast DNA variation in European white oaks (*Quercus robur*, *Q. petraea*, *Q. pubescens*, *Q. frainetto*, *Q. faginea*, *Q. pyrenaica*, *Q. canariensis* and *Q. macranthera*). Four of the six chloroplast lineages recognized so far in European white oaks were represented in the Iberian Peninsula and there was strong evidence for at least two major refugia in Spain (Olalde *et al.* 2002; Petit *et al.* 2002). Sinclair *et al.* (1999), in a wide-range study of mitochondrial variants of Scots pine, *Pinus sylvestris* L, observed within-population genetic variation in Iberian populations, whereas elsewhere in Europe populations were fixed for one mitotype. Moreover, a population in southern Spain (Baza) showed a private mitotype. The Baza population is located close to Sierra Nevada, a region with a great level of endemism.

Mediterranean pines such as Aleppo pine, *Pinus halepensis*, and maritime or cluster pine, *P. pinaster*, have also shown high levels of diversity in southern Spain. These levels could be related to persistence during several glacial cycles (Salvador *et al.* 2000; Gómez *et al.* 2001, 2005; González-Martínez *et al.* 2001). Gene variation in Aleppo pine is clinally distributed from north to south, showing maximum levels of diversity in the extreme edges of its distribution (Agúndez *et al.* 1999; Gómez *et al.* 2005). For maritime pine, chloroplast variation studies, including populations from most of the native range of the species showed three main refugial areas: the Atlantic coast of Portugal, southwestern Iberia, and Pantellaria and Sardinia in Italy (Ribeiro 2001 and references within; G.G. Vendramin pers. comm.). Burban & Petit (2003) found three maternal lineages in maritime pine using mtDNA-RFLP analysis (named *western*, *eastern* and *Moroccan*). The *western* mitotype might be associated with the Iberian glacial refugia and the eastern with those described for the Italian islands of Sardinia and Pantellaria.

Maritime pine exhibits high genetic diversity in Spain. While Vendramin *et al.* (1998) found 34 different haplotypes in ten populations from Portugal, France, Italy and northern Africa, up to 69 haplotypes were recently found in seven Spanish populations using the same cpSSR markers (our unpublished results). Moreover, an allozyme study of 32 Iberian populations showed that populations from southern Spain displayed the highest allelic richness in the Iberian Peninsula, including 82% of the total number of alleles (González-Martínez 2001). Population differentiation is relatively high in maritime pine from the Iberian Peninsula ($G_{ST} = 0.077$; Salvador *et al.* 2000) and weak, yet significant, fine-scale structure due to restricted gene dispersal has been found in a classical locality from central Spain (González-Martínez *et al.* 2002).

The primary aim of this work is to analyze the spatial genetic structure of maritime pine within one putative refugial area of southeastern Spain. The region under study is a physiographically complex, mountainous territory, so it represents an excellent model system to study the effect of mountain ranges as barriers to interpopulational gene flow. Finally, spatial analysis of gene diversity in maritime pine provides relevant information for genetic conservation of forest resources.

Materials and methods

Plant Material

Seeds were collected from 11 populations covering the native range of *P. pinaster* in southern Spain (Figure 1 & Table 1). Six provenance regions (i.e. native locations of the species used as breeding units) have been delimited in southern Spain based on ecological and historical data (Alía *et al.* 1996). To test the provenance division as units for management and conservation practices, we sampled at least one population from each of them. One of the provenances, 'Sierra de Segura-Alcaraz', is one of the most important areas of maritime pine in southern Spain, covering more than 70 000 hectares. Seedlots from this provenance produce around 100-300 kg of seeds per year, and are used in plantations all around Spain. In this case, four populations were sampled to estimate within-population variation in the 'Sierra de Segura-Alcaraz' breeding unit. Two other populations included in this study ('Gaucín' - MA2, and 'Sierra de Oria' - AL) are considered geographically marginal. In each stand, two to three cones were collected from 80 trees, at least 50 m apart from each other. The material analyzed was either 70-80 female gametophytes per population (four populations) or gametophytes and embryos of 35-40 seeds per population (seven populations). The present study included four from the 12 populations analyzed by Salvador *et al.* (2000) and five from those included in the study made by González-Martínez *et al.* (2001).

Table 1. *Geographical location and provenance of origin of the 11 studied* Pinus pinaster *populations.*

Population	Code	Provenance	Latitude	Longitude	Altitude
Riopar	AB	Sierra de Segura-Alcaraz	38° 28′ 05″ N	2° 27′ 31″ W	1200
Cazorla	J1	Sierra de Segura-Alcaraz	37° 55′ 05″ N	2° 55′ 11″ W	1100-1200
Siles	J2	Sierra de Segura-Alcaraz	38° 21′ 30″ N	2° 34′ 30″ W	800-1500
Segura	J3	Sierra de Segura-Alcaraz	38° 17′ 03″ N	2° 26′ 12″ W	800-1000
Estepona	MA1	Sierra Bermeja	36° 31′ 05″ N	5° 07′ 11″ W	500
Gaucín	MA2	Serranía de Ronda	36° 32′ 10″ N	5° 17′ 56″ W	500-700
Cómpeta	MA3	Sierra almijara-Nevada	36° 51′ 44″ N	3° 53′ 33″ W	1000-1500
Jubrique	MA4	Sierra Bermeja	36° 33′ 05″ N	5° 10′ 41″ W	1000
La Peza	GR	Sierra Almijara-Nevada	37° 16′ 26″ N	3° 22′ 10″ W	1400
Sierra de Oria	AL	Oria	37° 30′ 49″ N	2° 20′ 11″ W	1300
Moratalla	MU	Moratalla	38° 05′ 57″ N	2° 11′ 37″ W	1000-1200

Molecular Markers

Seeds were stored at 4°C in a dry environment until enzyme extraction. For enzyme extraction, seeds were soaked on filter paper at room temperature for 24 h and germinated in a Petri dish in an incubator. Enzymes were extracted from 3-4 mm long radicles, following Conkle *et al.* (1982). Different studies dealing with allozyme variation in maritime pine from the Iberian Peninsula have shown that only a limited number of markers are significantly polymorphic, even when more than 20 allozyme loci have been successfully analyzed in this species (Castro 1989; Salvador 1997). Thus, only loci known to be polymorphic in the Iberian Peninsula at the 95% level were used. Horizontal starchgel electrophoresis of seven allozymes encoded by eight loci was performed. The enzyme systems and the scored loci were: isocitrate dehydrogenase (IDH; EC 1.1.1.42), malate dehydrogenase (MDH2 and MDH3;

EC 1.1.1.37), phosphoglucose isomerase (PGI2; EC 5.3.1.9), acid phosphatase (ACPH; EC 3.1.3.2), glutamate dehydrogenase (GDH; EC 1.4.1.3), glutamate-oxalacetate transaminase (GOT2; EC 2.6.1.1), and leucine aminopeptidase (LAP; EC 3.4.11.1). Genetic interpretation of enzyme systems and staining methods can be found in Castro (1989) and Salvador (1997). All loci included in this work showed to be neutral under the conditions of the Ewens-Watterson neutrality test (see Hartl & Clark 1997, pp. 298-300).

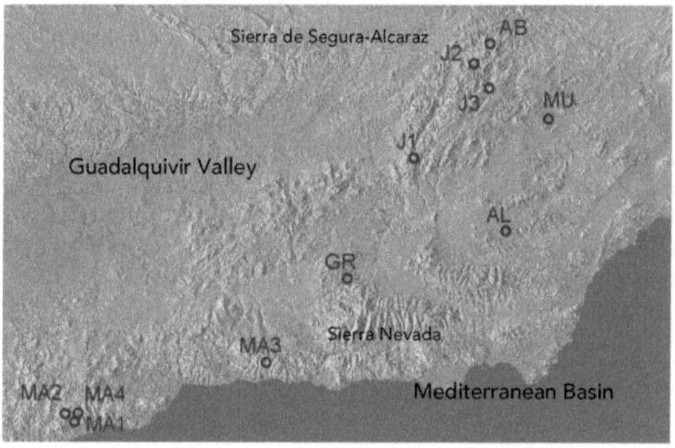

Figure 1. *Physiography of southeastern Iberia and location of* P. pinaster *populations.*

Data Analysis

Spatial genetic structure was analyzed using the isolation by distance approach of Rousset (1997). In two-dimensional habitats, this method involves the regression of $F_{ST}/(1-F_{ST})$ estimates for pairs of populations on the logarithm of distance. Rousset (1997) showed that under isolation by distance, the variables considered show a linear relationship, the slope of which is inversely correlated with the product of effective population density and the second moment of parental axial distance ($4D\pi\sigma^2$). The parameter σ^2 is a measure of the speed at which two lineages descending from a common ancestor depart from each other in space (Rousset 2001). The absence of a pattern of isolation by distance (null slope of the regression, $D\sigma^2$ infinite) was tested by an exact permutation procedure using Genepop version 3.3 (M. Raymond & F. Rousset, ISEM, Université de Montpellier 2, France). We computed the correlation between pairwise $F_{ST}/(1-F_{ST})$ and the logarithm of distance in two cases: i) within the 'Sierra de Segura-Alcaraz' provenance; and ii) using all the populations. The first analysis was done in order to study the spatial genetic structure among populations of the same seed collection region. In the second case, we used two types of distances: straight geographic distances and the length of the shortest pathway below 800 meters above sea level (m a.s.l.). Nowadays populations of maritime pine are usually distributed at medium altitudes. During the Ice Ages, displacements of this species across some of the high mountains that separate populations in this region may have been difficult. Our hypothesis is that any spatial structure would be more easily detected using distances along low altitude pathways.

Gene interchange between pairs of populations in a subset of six populations (one from each provenance region) was analyzed using migration matrix models and maximum likelihood methods. The estimation procedure used an expansion of the coalescent theory that included migration. Sampling of genealogies has been done using a Markov chain Monte Carlo (MCMC) approach and the Metropolis-Hastings algorithm (see description in Chib & Greenberg 1995). Two runs with 10 short chains (200 trees used out of 4000 sampled) and 2 long chains (2000 trees used out of 40 000 sampled) were computed. As the correlation between the two runs was high (>80%), we present only the average values. *Migrate* version 1.1. software (Beerli 1997-2001) has been used to perform this analysis (see Beerli & Felsenstein 2001 for a detailed description).

Results

Maritime pine showed a marginally significant spatial structure but only when the length of the shortest pathway below 800 meters above sea level was considered (b = 0.00816; $p = 0.09$). In this case, the slope of the correlation between $F_{ST}/(1-F_{ST})$ and the logarithm of the distance provided an indirect estimation of $4D\pi\sigma^2$ for dispersal in two dimensions ($4D\pi\sigma^2 = 122.55$). No genetic spatial structure was found within the 'Sierra de Segura-Alcaraz' provenance, nor when straight geographic distances between all pairs of populations were used. The correlation graph between $F_{ST}/(1-F_{ST})$ and the logarithm of the distance is shown in Figure 2. Most outliers in the graph included the 'Sierra de Oria' marginal population. The close genetic similarity between two groups of populations, 'Sierra de Segura-Alcaraz' and 'Sierra Bermeja/ Ronda' is also remarkable. These populations, while lying several kilometers apart, are connected by the Guadalquivir valley.

Average pairwise gene interchange estimated using *Migrate* software was low ($Nm = 1.6029$). Some populations included in different provenance regions, but located close to each other (e.g. MA2 and MA4), were practically isolated ($Nm < 1$; Table 2). Moreover, altitudes as low as 1000 m a.s.l.

(at present, maritime pine can grow at 1800 m a.s.l.) were enough to effectively isolate the populations AL (Sierra de Oria) and J1 (Cazorla), even when these two populations are separated by less than 50 km. The 'Sierra de Oria' population had a higher number of immigrants than emigrants, as shown by a likelihood ratio test performed with the three closest populations: MU, J1 and MA3 ($p < 0.000$; see also Figure 3). Asymmetric gene flow is expected at the edge of the distribution range of a species due to expansion/retreat processes.

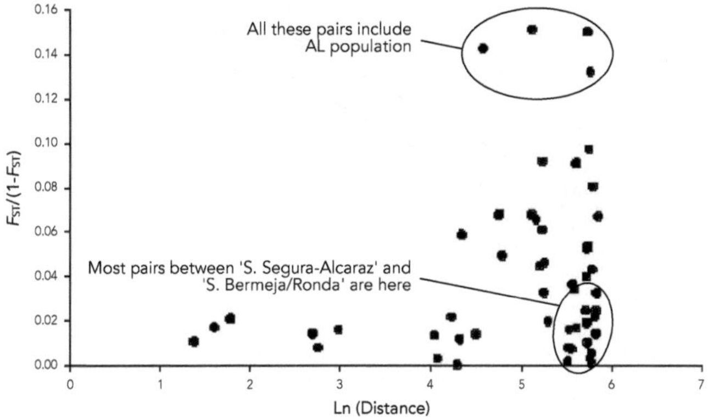

Figure 2. *Correlation graph between $F_{ST}/(1\text{-}F_{ST})$ and the logarithm of the distance. All pairs of populations and the length of the shortest pathway below 800 meters above sea level were used.*

Table 2. *Pairwise gene flow (*Nm*) among six* P. pinaster *populations (one from each provenance region) in southeastern Spain.*

POP (x)	*Nm* (x receiving population)						
	J1,x	MA3,x	MA2,x	MA4,x	MU,x	AL,x	*Average*
J1		1.6482	0.8896	0.7716	0.8380	1.5401	*1.1375*
MA3	0.9082		1.1647	1.2346	0.5501	1.0674	*0.9850*
MA2	2.7209	3.7466		0.9551	0.2753	0.8005	*1.6997*
MA4	0.3352	0.2188	0.8776		0.6887	1.5348	*0.7310*
MU	1.9738	2.8169	3.8128	1.3741		2.1699	*2.4295*
AL	1.2184	2.8912	2.5738	1.6951	4.7951		*2.6347*
Average	*1.4313*	*2.2643*	*1.8637*	*1.2061*	*1.4294*	*1.4225*	

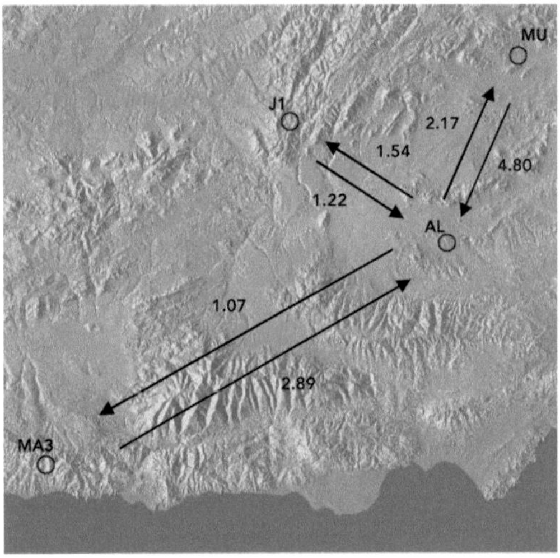

Figure 3. *Pairwise gene interchange (*Nm*) between AL (Sierra de Oria) marginal population and, i) J1 and MA3, two close populations but separated from AL by different mountain ranges and ii) MU, a population with a direct gene flow pathway.*

Discussion

Temperate forest trees have contrasting spatial patterns of genetic structure at the natural range scale. Some show clinal variation of allele frequencies (e.g. *Quercus petraea*, Zanetto & Kremer 1995; *Pinus halepensis*, Agúndez *et al.* 1999; *Pinus sylvestris*, Prus-Glowacki & Stephan 1994; *Picea abies*, Bucci & Vendramin 2000), but no gene diversity structure has been found in others (e.g. *Castanea sativa*, Fineschi *et al.* 2000; *Prunus spinosa*, Mohanty *et al.* 2000; *Sorbus torminalis*, Demesure *et al.* 2000). Within a given species, it is also common to find genetic structure differences at a regional scale. Maritime pine, in particular, presented a geographical pattern within Mediterranean populations (Salvador *et al.* 2000), but no structure in the Atlantic populations (Portugal, Ribeiro *et al.* 2001; southwestern France, Mariette *et al.* 2001). The transfer of seeds and the high gene flow among regions have probably erased the original spatial structure in Portugal (Ribeiro 2001).

The high mountain ranges that separate populations of maritime pine in southern Spain, have probably played an important role in shaping the present distribution of gene variation. Southern maritime pine populations showed a significant spatial structure, but only when the length of the shortest pathway at low altitude (< 800 m a.s.l.) was considered. Fine-scale estimations of *Nm* using matrix migration models for pairs of populations showed that pairwise gene interchange was very low, even at short distances, when mountains higher than 1000 m a.s.l. separated populations (e.g. AL and J1). The ability of mountains to serve as effective barriers to gene flow is well documented. Hewitt (2001) pointed out the importance of barriers such as the Pyrenees or the Alps in shaping the actual patterns of genetic diversity in several European animal and plant species. At a regional scale, mountains i) promote more stable population dynamics due to displacements in altitude in response to climatic changes (Comes & Kadereit 1998), and ii) increase the differentiation between populations as a consequence of the isolation of populations, even during the mildest periods of the glaciations. In maritime pine, a global F_{ST} value of 0.085 was found in the Iberian Peninsula. This value is 0.042 when the populations from southern Spain were removed from the analysis (data recalculated from González-Martínez *et al.* 2001).

The oscillating climates of the last glacial stage must have profoundly influenced the altitudinal location of plant species in southern Spain. Horizontal transfer of chloroplast DNA types among species of the *Armeria* complex (Plumbaginaceae) in Sierra Nevada showed a scenario that must have involved populations ascending or descending mountains (Gutiérrez-Larena *et al.* 2002). Our results indicate that the expansion during interglacials in maritime pine must have taken place by low altitude pathways, through territories presently covered by meso- and thermo-Mediterranean woods such as *Quercus faginea*, *Q. ilex rotundifolia*, *Q. suber*, *Q. coccifera*, *Pistacia lentiscus*, *Juniperus phoenicea*, *J. oxycedrus* and *Olea europaea*. An interesting case study is provided by the recolonization of oaks (*Quercus robur*, *Q. petraea* and *Q. pubescens*) across the Swiss Alps. Mátyás & Sperisen (2001) suggested that oak species coming from a glacial refugium in Italy were able to cross the Alps in the area of the Brenner pass (1371 m a.s.l.). In maritime pine, populations along the Guadalquivir valley showed high genetic similarity ('Sierra de Segura-Alcaraz' and 'Sierra Bermeja/Ronda' provenances). This valley may have acted as a corridor for different gene pools of the species during the mildest periods of the most recent glaciation.

Overall, this picture agrees with the paleoecological information on the species. The pollen records of Cañada de la Cruz, Siles, and Villaverde, situated across elevational and latitudinal gradients have been correlated to produce a picture of Upper Pleistocene and Holocene environmental history in the Segura region (Carrión 2002). In particular, the Siles lake paleoecological record (2° 30′, 38° 24′ N, 1320 m a.s.l.) showed that *P. pinaster*, together with a number of temperate and Mediterranean mesothermophilous trees and shrubs, persisted in these mountains during the last glacial times.

Apart from the relevance of this persistence, which had already been shown in the southern Iberian System (Carrión & van Geel 1999) although not as prominently, it is worth highlighting that maritime pine has been extremely sensitive to climatic changes, with rapid (century-scale) altitudinal patterns of displacement, not only during the arrival of late-glacial amelioration, but even throughout the Holocene (Carrión *et al.* 2001b). Furthermore, the Siles record provides support to the view that *Pinus pinaster* could survive in southern European mountains at relatively elevated locations during the last glacial stage. This hypothesis was put forward by Bennett *et al.* (1991), who contended that tree survival would have been especially important in those mountain ranges that, like the Balkans, allowed rapid altitudinal displacements of tree populations in response to climatic pulses. As in the Balkans, the Segura mountains probably permitted latitudinal movements of tree populations owing to their approximate north-south orientation. Interestingly, *P. pinaster* is absent from the Holocene Villaverde pollen record (Carrión *et al.* 2001a), which suggests a more recent distribution in the northern platforms of Sierra de Alcaraz, in contrast with the Segura range, resulting from a recent expansion and/or introduction by humans.

Some practical implications can be drawn from the present spatial structure analysis in southern Spain. First, the lack of genetic structure within 'Sierra de Segura-Alcaraz' provenance makes seed collection from multiple stands unnecessary in this region, and thus seed collection concentrated in a few highly productive stands is recommended. Second, plantations with plant material from 'Sierra de Segura-Alcaraz' are recommended only locally and in populations connected by the Guadalquivir valley but not in the south-easternmost range of the species, where historical isolation has produced highly differentiated populations. Third, southern Spain is a focal area for *in situ* conservation of maritime pine genetic resources. Stands selected for conservation purposes should cover a wide range of locations, irrespective of the geographical distance between them. The location of genetic reserves has to be considered carefully and genetic, ecological and demographic information should be combined to define conservation priorities.

Acknowledgements

The study was funded by the cooperation project DGCN–INIA CC00-0035 and the INIA project SC97-118. Thanks to Diana Barba for her technical assistance in the lab and to Fernando del Caño and Alfonso Piñero for the field collections. Laura Salvador from the University of Huelva gently provided data from AB, MA1, GR and AL populations. Thanks are extended to Maria M. Ribeiro who discussed a draft of the manuscript and to Patricia C. Grant who revised the English.

References

Agúndez D, Degen B, von Wuehlisch G, Alía R (1999) Multilocus analysis of *Pinus halepensis* Mill. from Spain: genetic diversity and clinal variation. ***Silvae Genetica*** **48**, 173-178.

Alía R, Martín S, de Miguel J, Galera R, Agúndez D, Gordo J, Catalán G, Gil L (1996) ***Las regiones de procedencia de* Pinus pinaster *Ait.*** OA de Parques Nacionales, DGCONA, Madrid.

Beerli P (1997-2001) ***Migrate: documentation and program, part of LAMARC. Version 1.1.*** Revised April 30, 2001. http://evolution.genetics.washington.edu/lamarc.html.

Beerli P, Felsenstein J (2001) Maximum likelihood estimation of a migration matrix and effective population sizes in n subpopulations using a coalescent approach. ***Proceedings of the National Academy of Sciences, USA*** **98**, 4563-4568.

Bennett KD (1997) ***Evolution and Ecology: The Pace of Life.*** Cambridge University Press, Cambridge.

Bennett KD, Tzedakis PC Willis KJ (1991) Quaternary refugia of north European trees. ***Journal of Biogeography*** **18**, 103-115.

Bucci G, Vendramin GG (2000) Delineation of genetics zones in the European Norway spruce natural range: preliminary evidence. ***Molecular Ecology*** **9**, 923-934.

Burban C, Petit RJ (2003) Phylogeography of maritime pine inferred with organelle markers having contrasted inheritance. ***Molecular Ecology*** **12**, 1487-1495.

Carrión JS (2002) Patterns and processes of late Quaternary environmental change in a montane region of southwestern Europe. ***Quaternary Science Reviews*** **21**, 2047-2066.

Carrión JS, van Geel B (1999) Fine-resolution Upper Weichselian and Holocene palynological record from Navarrés (Valencia, Spain) and a discussion about factors of Mediterranean forest succession. ***Review of Palaeobotany and Palynology*** **106**, 209-236.

Carrión JS, Andrade A, Bennett KD, Navarro C, Munuera M (2001a) Crossing forest thresholds: inertia and collapse in a Holocene sequence from south-central Spain. ***The Holocene*** **11**, 635-653.

Carrión JS, Munuera M, Dupré M, Andrade A (2001b) Abrupt vegetation changes in the Segura Mountains of southern Spain throughout the Holocene. ***Journal of Ecology*** **89**, 783-797.

Castro LFT (1989) ***Isoenzimas do* Pinus pinaster *Ait. numa perspectiva de aplicação ao melhoramento genético da espécie.*** PhD Thesis, Universidade de Trás os Montes e Alto Douro.

Chib S, Greenberg E (1995) Understanding the Metropolis-Hastings algorithm. ***The American Statistician*** **49**, 327-335.

Clark JS, Fastie C, Hurtt G, Jackson ST, Johnson C, King JA, Lewis M, Lynch J, Pacala S, Prentice C, Schupp EW, Webb III T, Wyckoff P (1998) Reid's paradox of rapid plant migration: dispersal theory and interpretation of paleoecological records. ***BioScience*** **48**, 13-24.

Comes HP, Abbott RJ (1998) The relative importance of historical events and gene flow on the population structure of a Mediterranean ragwort, *Senecio gallicus* (Asteraceae). ***Evolution*** **52**, 355-367.

Comes HP, Kadereit JW (1998) The effect of Quaternary climatic changes on plant distribution and evolution. ***Trends in Plant Science*** **3**, 432-438.

Conkle MT, Hodgkiss PD, Nunnally LB, Hunter SC (1982) ***Starch gel electrophoresis of conifer seeds: a laboratory manual***. United States Department of Agriculture, Forest Service, General Technical Report PSW-64.

Demesure B, Le Guerroué B, Lucchi G, Prat D, Petit RJ (2000) Genetic variability of a scattered temperate forest tree: *Sorbus torminalis* L. (Crantz). ***Annals of Forest Science*** **57**, 63-71.

Ferris C, King RA, Vainola R, Hewitt GM (1998) Chloroplast DNA recognizes three refugial sources of European oaks and suggests independent eastern and western immigrations to Finland. ***Heredity*** **80**, 584-593.

Fineschi S, Taurchini D, Villani F, Vendramin GG (2000) Chloroplast DNA polymorphism reveals little geographical structure in *Castanea sativa* Mill. (Fagaceae) throughout southern European countries. ***Molecular Ecology*** **9**, 1495-1503.

Gómez A, Alía R, Bueno MA (2001) Genetic diversity of *Pinus halepensis* Mill. populations detected by RAPD loci. ***Annals of Forest Science*** **58**, 869-875.

Gómez A, Vendramin GG, González-Martínez SC, Alía R (2005) Genetic diversity and differentiation of two Mediterranean pines (*P. halepensis* Mill. and *P. pinaster* Ait.) along a latitudinal cline using cpSSR markers. ***Diversity and Distributions*** **11**, 257-263.

González-Martínez SC (2001) ***Estructura poblacional y flujo genético de* Pinus pinaster *Aiton en el noroeste de la Península Ibérica***. PhD Thesis, Universidad Politécnica de Madrid.

González-Martínez SC, Agúndez D, Alía R, Salvador L, Gil L (2001) Geographical variation of gene diversity of *Pinus pinaster* Ait. in the Iberian Peninsula. In: ***Genetic Response of Forest Systems to Changing Environmental Conditions*** (eds. Müller-Starck G, Schubert R), pp. 161-171. Kluwer Academic Publishers, Dordrecht, Boston, London.

González-Martínez SC, Gerber S, Cervera MT, Martínez-Zapater JM, Gil L, Alía R (2002) Seed gene flow and fine-scale structure in a Mediterranean pine (*Pinus pinaster* Ait.) using nuclear microsatellite markers. ***Theoretical and Applied Genetics*** **104**, 1290-1297.

Gutiérrez-Larena B, Fuertes-Aguilar J, Nieto-Feliner G (2002) Glacial-induced altitudinal migrations in *Armeria* (Plumbaginaceae) inferred from patterns of chloroplast DNA haplotype sharing. ***Molecular Ecology*** **11**, 1965-1974.

Hartl DL, Clark AG (1997) ***Principles of Population Genetics***, 3rd edition. Sinauer Associates, Sunderland.

Hewitt GM (1996) Some genetic consequences of ice ages, and their role in divergence and speciation. ***Biological Journal of the Linnean Society*** **58**, 247-276.

Hewitt GM (1999) Post-glacial recolonization of European biota. ***Biological Journal of the Linnean Society*** **68**, 87-112.

Hewitt GM (2001) Speciation, hybrid zones and phylogeography – or seeing genes in space and time. ***Molecular Ecology*** **10**, 537-549.

Knowles L (2001) Did the Pleistocene glaciations promote divergence? Tests of explicit refugial models in montane grasshoppers. ***Molecular Ecology*** **10**, 691-701.

Mariette S, Chagné D, Lézier C, Pastuszka P, Raffin A, Plomion C, Kremer A (2001) Genetic diversity within and among *Pinus pinaster* populations: comparison between AFLP and microsatellite markers. ***Heredity*** **86**, 469-479.

Mátyás G, Sperisen C (2001) Chloroplast DNA polymorphisms provide evidence for postglacial re-colonisation of oaks (*Quercus* spp.) across the Swiss Alps. ***Theoretical and Applied Genetics*** **102**, 12-20.

Mohanty A, Martín JP, Aguinagalde I (2000) Chloroplast DNA diversity within and among populations of the allotetraploid *Prunus spinosa* L. ***Theoretical and Applied Genetics* 100**, 1304-1310.

Olalde M, Herrán A, Espinel S, Goicoechea PG (2002) White oaks phylogeography in the Iberian Peninsula. ***Forest Ecology and Management* 156**, 89-102.

Petit RJ, Brewer S, Bordács S, Burg K, Cheddadi R, Coart E, Cottrell J, Czaikl UM, van Dam B, Deans JD, Espinel S, Fineschi S, Finkeldey R, Glaz I, Goicoechea PG, Jensen JS, König AO, Lowe AJ, Madsen SF, Mátyás G, Munro RC, Popescu F, Slade D, Tabbener H, deVries SGM, Ziegenhagen B, de Beaulieu J-L, Kremer A (2002) Identification of refugia and post-glacial colonisation routes of European white oaks based on chloroplast DNA and fossil pollen evidence. ***Forest Ecology and Management* 156**, 49-74.

Prus-Glowacki W, Stephan BR (1994) Genetic variation of *Pinus sylvestris* from Spain in relation to other European populations. ***Silvae Genetica* 43**, 7-14.

Ribeiro MM (2001) ***Genetics of* Pinus pinaster *Aiton with cytoplasmic and nuclear markers*.** PhD Thesis, Swedish University of Agricultural Sciences.

Ribeiro MM, Plomion C, Petit RJ, Vendramin GG, Szmidt AE (2001) Variation in chloroplast single-sequence repeats in Portuguese maritime pine (*Pinus pinaster* Ait.). ***Theoretical and Applied Genetics* 102**, 97-103.

Rousset F (1997) Genetic differentiation and estimation of gene flow from F-statistics under isolation by distance. ***Genetics* 145**, 1219-1228.

Rousset F (2001) Inferences from spatial population genetics. In: ***Handbook of Statistical Genetics*** (eds. Balding D, Bishop M, Cannings C), pp. 239-269. John Wiley & Sons, New York.

Salvador L (1997) ***Estudio de la variabilidad genética de* Pinus pinaster *en España usando marcadores proteicos e isoenzimáticos*.** PhD Thesis, Universidad Politécnica de Madrid.

Salvador L, Alía R, Agúndez D, Gil L (2000) Genetic variation and migration pathways of maritime pine (*Pinus pinaster* Ait.) in the Iberian Peninsula. ***Theoretical and Applied Genetics* 100**, 89-95.

Santucci F, Emerson B, Hewitt G (1998) Mitochondrial DNA phylogeography of European hedgehogs. ***Molecular Ecology* 7**, 1163-1172.

Sinclair WT, Morman JD, Ennos RA (1999) The postglacial history of Scots pine (*Pinus sylvestris* L.) in western Europe: evidence from mitochondrial DNA variation. ***Molecular Ecology* 8**, 83-88.

Steward JR, Lister AM (2001) Cryptic northern refugia and the origins of the modern biota. ***Trends in Ecology and Evolution* 11**, 608-613.

Taberlet P, Fumagalli L, Wust-Saucy A, Cosson J (1998) Comparative phylogeography and postglacial colonization routes in Europe. ***Molecular Ecology* 8**, 1923-1934.

Vendramin GG, Anzidei M, Madaghiele A, Bucci G (1998) Distribution of genetic diversity in *Pinus pinaster* Ait. as revealed by chloroplast microsatellites. ***Theoretical and Applied Genetics* 97**, 456-463.

Willis KJ, Whittaker RJ (2000) The refugial debate. ***Science* 287**, 1406-1407.

Willis KJ, Rudner E, Sümegi P (2000) The full-glacial forests of central and southeastern Europe. ***Quaternary Research* 53**, 203-213.

Zanetto A, Kremer A (1995) Geographical structure of gene diversity in *Quercus petraea* (Matt.) Liebl. I. Monolocus patterns of variation. ***Heredity* 75**, 506-517.

Part IV

First contributions of islands and other peninsulas to the phylogeography of southern Europe

Chapter 10

Endemism in Sardinia

Andrea Grill*, Paolo Casula, Roberta Lecis and Steph Menken
*E-mail: grill@science.uva.nl

Abstract
The Tyrrhenian islands are known for their highly relictual fauna and flora and are one of the ten Mediterranean hotspots of plant diversity and endemism. There is little detailed information available on species' biogeography, and new species are still being discovered. This chapter is the first to put together information from several groups of organisms endemic to Sardinia (viz. plants, butterflies, amphibians, lizards, and beetles), with a particular focus on butterflies and amphibians. Reviewing recent literature, we describe distributional patterns and point to centers of endemicity, which we compare with the location and extent of existing protected areas in Sardinia, in order to assess their usefulness in protecting endemic species. Further, we discuss the geological history of the Mediterranean basin and relate geophysical events to molecular-based estimates of species' divergence times to investigate when and how Sardinian endemics came to the island and describe scenarios of speciation that might have resulted from vicariance, dispersion, and human transportation. The divergence time estimates we summarize here support that the cladogenetic events leading to the Sardinian lineages of various taxa have occurred after the separation of the Sardo-Corsican microplate from the continental landmass and after the rotation of the Corso-Sardinian plate. Furthermore, there is evidence that the split of many Sardinian taxa has occurred after the marine regressions (± 5 mya). Areas of high endemism generally coincide with mountainous areas. The main centers of endemism in Sardinia are already included in a network of natural parks but these areas have not yet been officially accepted as protected areas by the Sardinian authorities. Giving them official status would be a step towards safeguarding the unique nature of Sardinia.

Keywords: endemism, diversity, butterflies, amphibians, lizards, beetles, plants, evolution, conservation

Introduction

Numerous examples in recent literature show that endemism has high conservation priority (Munguira 1995; Hurtrez-Boussès 1996; Schnittler & Ludwig 1996; Gruttke *et al.* 1999; Médail & Quézel 1999; Cook & MacDonald 2001; Grill *et al.* 2002). Endemic species are often 'specialists' that depend on particular and often localized resources, making them especially

S. Weiss and N. Ferrand (eds.), Phylogeography of Southern European Refugia, 273–296.

vulnerable to changes in climate, land use, or habitat management (Munguira 1995). Consequently, increased human pressure will induce greater losses in endemic species than in more widespread biota (Oberdorff *et al.* 1999). This has resulted in increased interest in detecting areas rich in endemics, which are usually termed biodiversity hotspots (Morrone & Crisci 1995; Martín *et al.* 2000; Myers *et al.* 2000). Countries or regions harboring endemic species carry a particular conservation responsibility, as the disappearance of the species from those areas would mean their global extinction. Contrary to general belief, endemic species and regions of high endemism are poorly known even in western Europe (Deharveng 1996). For butterflies, there have been recent efforts to overcome this lack of knowledge, and areas of high endemism and/or species richness in Europe have been defined and pointed out to policy and decision makers as 'Prime Butterfly Areas' (Balletto *et al.* 2003). In the Mediterranean basin, efforts to identify centers of endemicity and species richness revealed that it is one of the world's most important regions of plant diversity, harboring an astounding 20% of the world's total floristic richness in only 2% of the earth's surface (Médail & Quézel 1997, 1999; Médail & Verlaque 1997; Verlaque *et al.* 1997). These studies clearly underline the importance of the Mediterranean region in global conservation.

But the importance of areas of high endemicity goes beyond the straight forward aims of conservation biology. Species with restricted distributions and taxa representing unique lineages within a species flock can be interesting model organisms to investigate the processes of differentiation, speciation, and coevolution, and often provide the key for answers to broader scale biogeographic questions. The study of islands where many endemic species or subspecies evolved has provided fundamental insights into the relationships between geographical patterns and biological processes for famous evolutionary biologists like Darwin, Wallace, Mayr, and Wilson (Drake *et al.* 2002 and references therein).

Islands often hold a large proportion of endemic species (Gómez-Campo 1985; Hurtrez-Boussès 1996; Whittaker 1997). Indeed, the Macaronesian (Madeira and the Canary Islands; see Vargas, this volume) and the Tyrrhenian islands (Balearic Islands, Corsica, and Sardinia) are among the most important hotspots of endemism within Europe (Médail & Quézel 1999). Notably, the Tyrrhenian islands together with the Maritime-Ligurian Alps are known for their high relictual endemism (Médail & Quézel 1999). Island endemics are the most vulnerable of all endemic taxa, as 1) islands are unlikely to offer refuges during ecological change, and 2) island populations are usually limited in size. Generally, all endemic taxa are potentially threatened by hybridization, competition, predation or disease when interacting with introduced taxa. An island is mostly referred to as a stretch of land surrounded by a mass of water isolating it from other land areas. By widening this definition to include areas isolated by habitat unsuitable for the taxon under consideration (Hudson 1998), analogies can be drawn for land patches or refuges that have become isolated through Quaternary Ice-age events, causing populations to differentiate in allopatry. Pleistocene glaciations probably induced insular speciation *sensu lato*. Hence, real islands like the Mediterranean island of Sardinia, provide 'laboratories' *in natura* for the study of evolutionary questions (Caccone *et al.* 1994; Salomon 2001).

Objectives

Endemic species in Europe have received little comprehensive study (Médail & Quézel 1999), and this is especially true for those found in Sardinia. There is little detailed information on the distribution and biogeography of Sardinian endemics, and new species are still being discovered (Rota 1992; Gentili *et al.* 1998; Sabella *et al.* 1998; Selvi 1998; Mossa *et al.* 1999; Bacchetta *et al.* 2000). Here we review published data from several groups of plants, butterflies, amphibians, lizards and beetles, endemic to Sardinia in order to define centers of endemicity and see if these overlap with the existing protected areas in Sardinia. As butterflies and amphibians are the focus of our own research we particularly concentrate on them. Furthermore, we explore the factors that may have promoted Sardinian species to diverge from their continental ancestors or sister species.

1. Distributional patterns of endemic species

The Tyrrhenian islands belong to the ten Mediterranean hotspots of plant diversity and endemism defined by Médail & Quézel (1997), where plant richness is >2000 species per 15 000 km^2 and at least 10% of the species are narrow endemics. This high richness is primarily due to paleogeographical and historical factors (Verlaque *et al.* 1997). In the following section we give some examples of distributional and ecological patterns of endemicity in Sardinia (summarized in Table 1) and indicate the main centers of endemism on the island. When we speak of endemics, we consider two different groups: species exclusively endemic to Sardinia and species endemic to Sardinia and Corsica or additional Tyrrhenian islands.

Plants

Echium anchusoides (Boraginaceae) was only recently described (Bacchetta *et al.* 2000) and is endemic to the main siliceous massifs of Sardinia, situated in the mountainous zones of the island. The same is true for the Sardinian oak *Quercus ichnusa* (Fagaceae) (Mossa *et al.* 1999), the Sardo-Corsican thyme, *Thymus herba-barona* (Lamiaceae), the shrub *Santolina insularis* (Compositae), and the perennials *Glechoma sardoa*, and *Lamium corsicum* (Lamiaceae). *Glechoma sardoa* and *L. corsicum* are both endemic to Sardinia and Corsica (Brotzu 1998). In contrast, the Sardo-Corsican endemic plants *Vinca sardoa* (Apocynaceae) and *Ornithogalum biflorum* (Hyacinthaceae) can be found at all altitudes in a variety of habitats, including road margins and the edges of cultivated fields (Brotzu 1998; Sacchetti *et al.* 1999). The perennial herb *Anchusa crispa* (Boraginaceae) occurs in open herbaceous vegetation on low-lying sand dunes (Quilini *et al.* 2001).

Butterflies

Endemic Lepidoptera are usually found at altitudes above 500 meters (Cobolli *et al.* 1995; Biermann 1998; Kleinekuhle 1999). Only three of the 14 Sardo-Corsican endemics are observed at equal frequencies in coastal (sea level) and mountainous habitats (Kleinekuhle 1999; Grill 2002). One of the 14 species, *Hipparchia aristaeus aristaeus*, which is usually known from mountainous areas, has also been recorded on the coast (Cobolli *et al.* 1995). These coastal localities, however, are not its main distributional center. The distribution areas of Sardinian endemics are often strictly related to the composition of the underlying substrate. *Lysandra coridon gennargenti*, for example, strictly follows the distributional pattern of its host plant, *Hippocrepis comosa* (Fabaceae), which is typically found on calcareous grounds and as a consequence, the butterfly is restricted to calcareous outcrops in the 'Barbagia di Seulo' and the 'Supramonte di Orgosolo' mountains. In Sardinia *H. comosa* is most probably the only food plant used by *L. coridon gennargenti*, as the plants used by the continental European populations of *L. coridon*, viz. *Coronilla* sp. and *Astragalus glaucus* (Fabaceae), do not occur in Sardinia. *Hippocrepis comosa* is rather common on other Tyrrhenian islands and the Mediterranean mainland. Individuals of *H. comosa* on Sardinia, however, are much more delicate than those found on the continent with island populations restricted to mountainous areas whereas on the Italian mainland they are found from 0-2900 m a.s.l. (Pignatti 1982). Recognizing the Sardinian form as a distinct endemic taxon could therefore be appropriate.

The Sardinian blue, *Pseudophilotes barbagiae*, is exclusively dependent on *Thymus herba-barona*, the above mentioned Sardo-Corsican endemic thyme species, which grows between 1000 and 2000 m a.s.l (Pignatti 1982); its distribution is thus restricted to a few slopes in the Barbagiae and the Gennargentu mountains, and Mount Limbara. *L. coridon gennargenti* and *P. barbagiae* are among the rarest butterfly species of Europe (Grill 2002). The Sardinian meadow brown, *Maniola nurag*, has its distributional centers around the three main mountainous areas of the island. The endemic hesperid, *Spialia sertorius therapne* has been observed on the Gennargentu, Limbara, and Sette Fratelli mountains (Cobolli *et al.* 1995).

Table 1. *Distributional patterns of Sardinian and Sardo-Corsican endemics. The geographic extent of species' distribution is shown as well as the altitudes where they occur. S = Sardinia, C = Corsica, T = other Tyrrhenian islands, GS = Gennargentu-Supramonte Massif, SF = Sette Fratelli, LI = Mount Limbara, WC = West-Central Sardinia, CO = Coastal areas.*

		Region						Altitude		
Distribution area		**GS**	**SF**	**LI**	**WC**	**CO**	**other**	**<500 m**	**>500 m**	**>1000 m**
Plants										
Echium anchusoides	S	+							+	
Quercus ichnusa	S	+							+	
Thymus herba-barona	S	+								+
Santolina insularis	S	+							+	
Glechoma sardoa	SC	+							+	
Vinca sardoa	SC						+	+	+	
Ornithogalum biflorum	SC						+	+	+	
Anchusa crispa	SC					+		+		
Lamium corsicum	SC	+						+	+	
Butterflies										
Papilio hospiton	SC	+	+	+			+	+	+	
Aglais urticae ichnusa	SC	+	+	+	+				+	
Argynnis elisa	SC	+		+					+	
Argynnis paphia immaculata	SC	+		+					+	
Euchloe insularis	SC	+	+	+	+	+	+	+	+	
Coenonympha corinna	SC	+	+	+	+	+	+	+	+	
Hipparchia aristaeus aristaeus	SC	+				+		+	+	
Hipparchia neomiris	SC	+				+		+	+	
Maniola nurag	S	+	+	+					+	
Lasiomata megera paramegera	SC	+	+	+	+		+	+	+	
Lysandra coridon gennargenti	S	+							+	
Plebejus idas bellieri	SC	+		+					+	
Pseudophilotes barbagiae	S	+								+
Spialia sertorius therapne	S	+	+	+	+				+	
Lizards										
Podarcis tiliguerta	SC	+	+	+	+	+	+	+	+	
Archeolacerta bedriagae	SC	+		+	+		+	+	+	
Salamanders										
Euproctus platycephalus	S	+	+	+				+	+	
Speleomantes supramontis	S	+							+	+
Speleomantes genei	S	+	+					+	+	+
Speleomantes flavus	S	+							+	+
Frogs										
Hyla sarda	SCT				+		+	+		
Discoglossus sardus	SCT				+		+	+		
Cave beetles										
Ovobathysciola grafittii	S				+		+	+		
Ovobathysciola majori	S	+					+	+		
Ovobathysciola gestroi	S	+							+	
Patriziella sardoa	S	+						+		
Patriziella nuragica	S					+		+		
Speonomus lostiai	S				+					
Mammals										
Apodemus sylvaticus	SCT	+	+	+	+	+		+	+	

Amphibians

The Sardinian mountain newt, *Euproctus platycephalus* (Salamandridae) appears to occur predominantly in the three main mountain ranges: Gennargentu-Supramonte in central Sardinia, Sette Fratelli in the southeast, and Mount Limbara in the north (Lecis & Norris 2003). In this respect it is similar to most endemic butterfly species. The distribution of this endemic newt covers approximately the eastern side of the island, with very few unconfirmed records in the western areas. The genus *Speleomantes* is represented by four species of cave salamanders, which occur in limestone caves and humid rocky substrates: *S. supramontis* in Supramonte, *S. flavus* in Monte Albo, and *S. genei* and *S. imperialis* in the southeast and in the southwest. The Sardinian tree frog *Hyla sarda*, endemic to Corsica, Sardinia, and the Tuscany archipelago, inhabits lowlands and temporary waters all over the island and is locally quite abundant, but probably declining in numbers (Colomo 1999). The Sardinian painted frog, *Discoglossus sardus*, classified as vulnerable, is usually found in stagnant or slow moving waters and is described as widespread (Colomo 1999). Its distribution area covers Corsica and Proquerolles, Port Cros, and the French Hyères archipelago.

Lizards

Archaeolacerta bedriagae (Lacertidae), the Bedriaga's rock lizard, a Sardo-Corsican endemic, seems to occur in areas of Limbara, Marghine, Monte Albo, and Gennargentu, generally in the north and center of Sardinia (Colomo 1999). The insular endemic lacertid lizards, *Algyroides fitzingeri* and *Podarcis tiliguerta*, are described as widespread and common at different altitudes, from sea level to the mountains in both Corsica and Sardinia (Delaugerre & Cheylan 1992; Arnold 2003). Two subspecies, *Podarcis tiliguerta ranzii* and *Podarcis tiliguerta toro*, are both restricted to one little circum-Sardinian island in the north (Molarotto near Olbia) and in the southwest (Toro near Sant'Antioco).

Beetles

The two Sardinian genera of obligate cave-dwelling beetles, *Ovobathysciola* and *Patriziella*, are obviously dependent on cave environments (Caccone & Sbordoni 2001). Recent observations indicate that *Ovobathysciola majori* and *Patriziella sardoa* inhabit numerous caves from sea level to 1000 m elevation in the karst areas of the Supramonte massif (northeast Sardinia) whereas *Ovobathysciola gestroi* is found in the Gennargentu massif. *Ovobathysciola graffitii* and *Patriziella nuragica* have only been found in northwestern Sardinia, near Sassari. *Speonomus lostiai* inhabits a few caves in west-central Sardinia (Caccone & Sbordoni 2001).

2. Speciation molecular divergence, and geological history

Insular speciation usually results from the differentiation between populations settled on an island, and the continental population from which they were isolated (Jacquard 1974; Hudson 1998; Salomon 2001). Diamond (1977) describes three successive stages that can be considered as prerequisites for insular speciation: colonization, settlement, and genetic divergence. Speciation as a consequence of geographical isolation is termed allopatric or geographical speciation. Sympatric speciation results from isolating mechanisms without the involvement of a physical barrier to gene flow. Parapatric speciation takes place when two divergent species have disjunct geographical distributions but there is a contact zone between them, where gene flow is possible.

Estimates of separation times can vary greatly among different genes, and even portions of one particular gene. There is, however, a general consensus that if rates are compared between closely related species for the same DNA region, sequences are very likely to display a clock-like behavior (Caccone & Sbordoni 2001). The existence of well-dated geological events, as is the case for the islands in this study, is a great advantage when trying to calibrate molecular rates in species whose distributions have probably been shaped as a result of these events.

The geological and geophysical history

The geological evolution of the Mediterranean region is characterized by the relatively rapid opening of several back-arc basins, generally floored by oceanic crust, within the framework of the Africa-Eurasia collision and Alpine orogenesis (Speranza *et al.* 2002; Blondel & Aronson 1999 and references therein). In the western Mediterranean, the Liguro-Provencal basin, a triangular sea located between the Provencal-Catalan coasts and the Corsica-Sardinia block, opened during the Oligo-Miocene. Basin spreading and the simultaneous eastward migration of the Alpine belt and Corsica-Sardinia-Calabria blocks were probably driven by the eastward retreat of a Ionian/Adriatic slab passively sinking into the mantle (Malinverno & Ryan 1986). Since the middle-late Miocene, further roll-back of the same slab caused spreading of the Tyrrhenian Sea, the southeastward drift of the Calabrian block, and the orogenesis of the Apennines. The Liguro-Provencal spreading took place simultaneously with the eastward drift of the Corsica-Sardinia block, which rotated at least 30° counterclockwise (e.g. Van der Voo 1993, Speranza *et al.* 2002 and references therein). Paleomagnetic investigations carried out in the 1970's on Oligo-Miocene volcanics of Sardinia suggested that the island was separated from the continental landmass about 33 mya, turned by 35° clockwise up to 21-20.5 mya, and then rotated 30° counterclockwise in a few million years (De Jong *et al.* 1969, 1973; Edel 1979, 1980). Since then, the end of the rotation, fixed at 19 myr by Montigny *et al.* (1981) has been subject to controversy (Edel *et al.* 2001; Speranza *et al.* 2002 and references therein). New paleomagnetic and Ar/Ar results support a begin of the rotation around 21-20.5 mya and an end of the rotation at 18-17.5 mya (Deino *et al.* 2001; Edel *et al.* 2001). But there is growing evidence that the rotation did not end before 16 mya and started after 19 mya (Speranza *et al.* 2002). The question of coupling or decoupling between Corsica and Sardinia during drifting was resolved by Vigliotti *et al.* (1990), who showed that after the Permian the two islands rotated as one block. During a period from 18.3 to17.5 mya the marine transgression occurred (Edel *et al.* 2001). At the same time a NE-SW shortening, interpreted as resulting from the collision of the Sardo-Corsican block with Apulia, affected parts of the island. Speranza *et al.* (2002) propose that at 16-19 mya, the lithosphere of a 'paleo-Ionian' oceanic corridor east of Sardinia sunk in the mantle causing a trench retreat and the Liguro-Provencal spreading. Faster subduction beneath Sardinia than beneath Corsica, due to the heterogeneous nature of the subducting plate, has been put forward as a plausible reason to explain the triangular geometry of the Liguro-Provencal basin and the counterclockwise rotation of Sardo-Corsica. About five million years ago, the Mediterranean Sea was almost entirely desiccated, creating connections of dry land between Sardinia and northern Italy and southern France. Sea level oscillations creating land bridges between Sardinia and Corsica continued from Miocene until well into the Pleistocene (5.7 – 0.23 mya) (Arias *et al.* 1980). In the Quaternary, Sardinia could have been in contact with the mainland via Elba as the sea level was up to 120 meters lower than today. During the last glacial maximum, 20 000 years ago, Sardinia was connected with Corsica.

Lizards and amphibians

Lanza (1983) hypothesized that the split among the Sardinian lizards, *Algyroides fitzingeri*, *Archaeolacerta bedriagae*, and *Podarcis tiliguerta* from mainland relatives is related to Premiocenic or Messinian age, while *P. sicula cettii*, could have diverged during the Pleistocene (Lanza 1983). Oliviero (1998) gives preliminary estimates of divergence times based on DNA sequences for *P. tiliguerta* (13 mya) and *P. sicula cetti* (7 mya), suggesting a Messinian age for both species. Ancient taxa such as *Euproctus platycephalus* and *Speleomantes genei* probably originated from ancestors already present on the Sardo-Corsican microplate prior to its detachment from the continent (Lanza 1983). Other ancient taxa such as three of the four Sardinian *Speleomantes* (*S. flavus, S. imperialis, S. supramontis*) are more closely related to the continental *Speleomantes ambrosii* (Lanza 1983, 1995), indicating that the ancestor of those three species arrived in Sardinia about 5 mya from the Apennines. Jackmann *et al.* (1997), however, infer a close relationship of *S.*

flavus, *S. supramontis*, and *S. genei*, whereas the continental *S. italicus* would be less closely related, and give evidence for a monophyletic origin of the Sardinian *Speleomantes* group.

Caccone and coworkers (1994, 1997) used the split between the Pyrenees and the Corso-Sardinian plate, and the separation of Corsica from Sardinia to calibrate mtDNA (12S and 16S ribosomal RNA and cytochrome *b*) evolutionary rates in newts of the genus *Euproctus*, which comprises three species (with distributions restricted to Corsica, Sardinia, and the Pyrenees). These genetic investigations confirmed records of morphological, anatomical and karyological studies (Bucci-Innocenti *et al.* 1978; Delaguerre & Cheylan 1992): Corsican and Sardinian newts are more closely related to each other than to the Pyrenean newt (Table 2 & 3). This is a sound consequence of a previous speciation event, dated around 29 mya, while the two insular species probably started diverging in Sardinia between 9 and 15 mya (Caccone *et al.* 1994) (Figure 1).

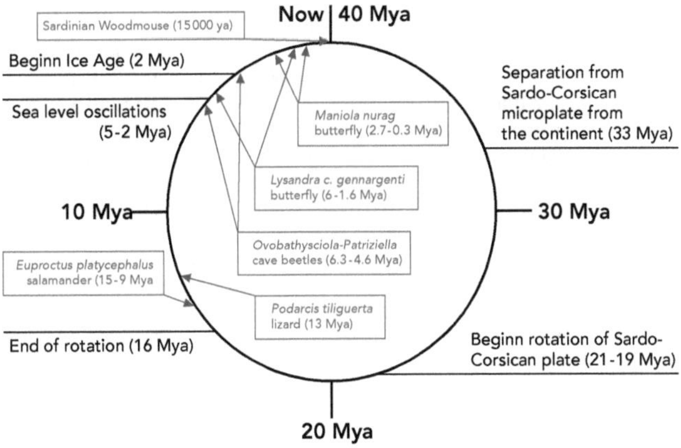

Figure 1. *Time scales of geophysical events in relation to speciation time estimates inferred from molecular data.*

Butterflies and beetles

Recently, Caccone & Sbordoni (2001) used COI divergence rates to estimate the time of isolation of cave beetles of the genera *Ovobathysciola*, *Patriziella*, and *Speonomus*. They estimated divergence times of 16-10 myr among the three *Ovobathysciola* species (*O. grafittii*, *O. majori*, *O. gestroi*), 6.3-4.6 myr for the split between *O. grafittii* and the two *Patriziella* species, and 4.5 to 3.7 myr for the split between *P. sardoa* and *P. nuragica*, all endemic to Sardinia (Table 2 & 3). The species assemblage of cave-dwelling beetles in Sardinia might be explained as the result of vicariance (Caccone & Sbordoni 2001). This assemblage probably began its diversification in a first step due to the separation of Sardinia from the continental landmass, and in a second step after the dramatic changes brought about by sea level oscillations (up to 1000 meters) in the middle and late Miocene (16-5 mya), which separated northern Sardinia from central and southern Sardinia (Steininger & Rögl 1984). These changes might have enhanced the isolation of the ancestral forest-dwelling populations of cave beetles, which then retreated to the wet habitats in the karstic caves of Sardinia. Also the climate changes in the Pliocene (5-2 mya), when the climate switched to cooler and drier conditions, forcing warm and humid subtropical forests to gradually change into a savanna-like vegetation (LaGreca 1998), could have forced some populations into the humid environment of caves where they became isolated from their ancestral forest dwelling populations (Caccone & Sbordoni 2001).

In Lepidoptera, the first expansion events vigorously influencing the European fauna occurred in the Pliocene, when Near Eastern and Balkan species started to invade Europe (10-1 mya. Before the Pliocene, the European fauna was predominantly tropical (Leestmanns 1965; Kleinekuhle 1999). The migration route probably went from central Asia, to the Near East, and Greece, which at this point was still connected to Africa, and from there via Sardinia and Corsica to Tuscany (Leestmanns 1965; Kleinekuhle 1999). During the last glacial maximum, Sardinia was much less affected than Corsica, so that thermophilic species survived in Sardinia, while cold-adapted species could persist in Corsica (Kleinekuhle 1999). This might have enhanced differentiation of a number of endemic species, which during the postglacial warming retreated to higher altitudes. Similarly, it has been suggested that the endemic taxa of the genus *Erebia* in the Alpine region and Scandinavia have arisen due to differentiation within glacial refugia (Roos & Arnscheid 1979; Kleinekuhle 1999). Sardinian examples of endemics, which most likely evolved as a result of Ice-age events are *Aglais urticae ichnusa* (Nymphalidae) and *Hipparchia aristaeus aristaeus* (Satyridae), both endemic to Sardo-Corsica, and related to *Aglais urtica* and *Hipparchia semele* (Satyridae) (Kleinekuhle1999). The latter two species are long-distance dispersers, but probably orientate their routes on determined features of the landscape and appear to be reluctant to cross large sea areas, explaining why they never reinvaded Sardinia or Corsica. Other Sardinian species most probably originate from central Asia. According to Leestmanns (1965), *Argynnis elisa* (Nymphalidae), *Hipparchia neomiris*, *Coenonympha corinna* (Satyridae), and *Papilio hospiton* (Nymphalidae) originated from the Asiatic species *Argynnis clara*, *Hipparchia digna*, Asiatic taxa of the genus

Coenonympha and *Papilio sikimensis* respectively. More recent studies however, conclude a closer relationship between *Papilio machaon saharae* and *Papilio hospiton* (Pierron 1990, 1992).

Table 2. *Examples of species pairs: genetic differences between closely related species/subspecies based on allozyme markers and mtDNA.*

Species	*Podarcis muralis* (Italian mainland)	*Lysandra coridon appenina* (Central Italy)	*Lysandra coridon caelestissima* (Central Spain)	*Maniola jurtina* (Austria, France, Spain)	*Ovobathysciola grafittii* (Sardinia)	*Anillochlamys bueni* (Pyrenees, Spain)	*Speonomus delarouzei* (Pyrenees, Spain)	*Speonomus bruckii* (Pyrenees, Spain)	*Speonomus hygrophilus* (Pyrenees, Spain)	*Apodemus sylvaticus* (France)	*Apodemus sylvaticus* (Belgium)
Lizards											
Podarcis tiliguerta (Sardinia, Corsica) Capula 1996	0.181- 0.318*										
Butterflies											
Lysandra coridon gennargenti (Sardinia) Marchi *et al.* 1996		0.434*	0.337*								
Maniola nurag (Sardinia) Grill *et al.* in prep.				0.065- 0.089*							
Cave beetles											
Ovobathysciola grafittii (Sardinia)						0.828**					
Ovobathysciola majori (Sardinia)					0.327**	0.617**					
Ovobathysciola gestroi (Sardinia)					0.390**	0.597**					
Patriziella sardoa (Sardinia)					0.146**	0.655**					
Patriziella nuragica (Sardinia)					0.160**	0.755**					
Speonomus lostiai Caccone & Sbordoni 2001							0.772**	0.710**	0.860**		
Woodmouse											
Apodemus sylvaticus (Sardinia) Michaux *et al.* 1996										0.001- 0.002*	0.003- 0.007*

* Nei's genetic distances based on allozyme markers.
** Based on mtDNA.

Table 3. *Divergence time estimates of closely related species in millions of years (myr) or years (y).*

Species	*Podarcis* spp. (continental mainland)	*Euproctus montanus* (Corsica)	*Lysandra coridon appenina* (Italy)	*Lysandra coridon caelestissima* (Spain)	*Maniola jurtina* (Austria, France, Spain)	*Ovobathysciola grafittii* (Sardinia)	*Patriziella sardoa* (Sardinia)	*Apodemus sylvaticus* (France, Belgium)
Lizards								
Podarcis tiliguerta (Sardinia, Corsica)	13 myr							
Podarcis sicula settii (Sardinia)	7 myr							
Salamanders								
Euproctus platycephalus		9-15 myr						
Butterflies								
Lysandra coridon gennargenti (Sardinia)			6-1.6 myr	6-1.6 myr				
Maniola nurag (Sardinia)					2.7-0.3 myr			
Cave beetles								
Ovobathysciola majori (Sardinia)						16-10 myr		
Ovobathysciola gestroi (Sardinia)						6.3-4.6 myr		
Patriziella nuragica (Sardinia)							4.5-3.7 myr	
Woodmouse								
Apodemus sylvaticus (Sardinia)								15 000 y

For the Sardinian Blue butterfly, *Lysandra coridon gennargenti*, there is genetic and morphological evidence that it is specifically different from the continental *Lysandra coridon* (Marchi *et al.* 1996; Jutzeler *et al.* 2003). Marchi *et al.* (1996) suggest an allopatric speciation event. They found evidence for an absence of gene flow with the continental populations, indicated by the presence of alternative fixed alleles at several enzymatic loci (*Aat*, *Gpi* and *Pgm*) and significant differences in allele frequencies at other loci, distinguishing the Sardinian population from *L. c. apennina* and *L. caelestissima*. The genetic differentiation of *L. c. gennargenti*, measured using Wright's F_{ST}-

values (0.129-0.923) and Nei's genetic distances (0.337-0.434) indicate that the Sardinian populations evolved as an independent lineage, facilitated by isolation and the strict dependence of the butterflies on specific biotopes. Marchi *et al.* (1996) also found a reduction of genetic variation (Polymorphism = 17.6 %, Heterozygosity = 0.024) with respect to the continental populations (P > 52 %, H greater than or equal to 0.170). Values of Nei's genetic distances (Table 2) between the Sardinian subspecies and the populations of continental Italy, are higher than those found between geographically isolated populations of *L. coridon* from continental Italy, and comparable to or even higher than distance levels found between other endemic taxa that are considered separate species. Our own data on genetic differentiation between *Maniola jurtina* and *Maniola nurag* based on allozyme markers (Grill *et al.*, unpublished work), for example, shows smaller genetic distances (0.065-0.089) although these two butterflies are considered to be different species (Table 2). This suggests that they probably diverged in more recent times than *L. c. gennargenti.* Based on sequence data from two regions of the mtDNA (cytochrome oxidase subunit I and cytochrome *b*) and the assumption of a mutation rate of 1.1% to 1.2%/myr in arthropod mtDNA (Brower 1994; Gaunt & Miles 2002), which translates to about 2.3% sequence divergence/myr, divergence time between *M. nurag* and *M. jurtina* was estimated to be 1.1 to 1.2 myr (Grill *et al.,* unpublished work). However, these numbers are only indicative of the degree of differentiation, and there is no general rule for the relationship between genetic distance and taxonomic status (Menken & Ulenberg 1987; Orr 2001).

Maniola nurag as an example for ecologically induced speciation?

Maniola jurtina, the meadow brown butterfly, has been shown to be closely related to *M. nurag* in allozyme-genetic analyses (Thomson, 1987; Grill *et al.*, unpublished data). The two species are phenotypically similar but nevertheless can usually be distinguished by their wing patterns. However, there is some overlap for individuals flying late in the season, and in exceptional cases, genital preparation might be the only way for determination. Both species fly in Sardinia but have only minor overlap in distribution area and flight period. *M. nurag* emerges a couple of weeks after *M. jurtina*, and has only been found above 500 m a.s.l. whereas *M. jurtina* is most abundant at sea level but can occasionally be observed up to 1000 m a.s.l. (Grill 2001). Adults of *M. nurag* are on the wing from May to September depending on altitude and local weather conditions, *M. jurtina* flies in Sardinia from late April to June. At lower altitudes *nurag* females aestivate during the hottest part of the summer (Tolman & Lewington 1997; Kleinekuhle 1999; Grill 2001). A similar aestivation behavior has been observed in southern populations of the pan-European species *M. jurtina* (Scali & Masetti 1973). *M. nurag* is probably better adapted to the particular conditions in the Sardinian mountains, with extremely dry and hot conditions during Mediterranean summers, and large temperature oscillations between day and night. Body size is smaller than in *M. jurtina*, the body is more compact and darker, and the upper side of both fore- and hindwings are brighter in the endemic species (Grill *et al.* 2004). UV-photographs of wing patterns do not reveal any differences between the two species. In both, the eyespot pupil is bright and visible, as generally observed in satyrids (S. Bryant, pers. comm.). Differentiation between the two species could be related to larval food-plant choice. The larvae of *M. jurtina* feed on a wide range of grass species including *Poa pratensis*, *Festuca rubra*, *F. arundinacea*, *Agrostis stolonifera*, *A. canina*, *Bromus erectus*, *Brachypodium pinnatum*, *Holcus lanatus*, *Avenula pubescens*, and *Anthoxanthum odoratum* (Tolman & Lewington 1997). The island endemic is probably more specialized in its diet, perhaps feeding on grasses that flower relatively late in spring, thus offering oviposition sites that are still green when most other vegetation is already dry.

Ecological and evolutionary isolation in *Euproctus platycephalus*

The three species in the genus *Euproctus*, *E. platycephalus* (Sardinia), *E. asper* (Pyrenees), and *E. montanus* (Corsica) share various morphological, reproductive, and ecological traits, such as the presence of a sixth toe on the male hind legs, the mating behavior (males actively search for females and hold them, curving body and tail in order to manipulate their spermatophores into the female cloaca in an almost real amplexus), and the typical habitat (although the Sardinian *E. platycephalus* seems to occur at lower altitudes than the other two). All three species live in streams, springs, pools or small lakes in mountainous areas. However, the Pyrenean, Corsican and Sardinian mountain ranges differ in geology and climate, so that apparently similar sites might actually be very different as a result of differing environmental conditions. Pyrenean and Corsican mountains are on average higher than Sardinian ridges, and a large part of the area is still covered by *Pinus* and *Quercus* forests. In Sardinia, centuries of deforestation, stock breeding, and fires have gradually changed landscape and microclimate, especially in the center and south of the island (mountains of Gennargentu and Sette Fratelli), where the largest part of the land is currently covered by Mediterranean '*macchia*'. In the Gennargentu mountain system, typical landscape consists of bare or bushy slopes with *Alnus glutinosa* creating gallery forests along water courses. Lecis & Norris (2004b) found evidence for a lack of gene flow between the *E. platycephalus* populations from the three Sardinian mountain ranges, Limbara, Gennargentu, and Sette Fratelli, although the same mtDNA clades are present in all of them.

Mammals

Sadly enough, in Sardinia as well as in Corsica most indigenous land mammals have disappeared. Human activities brought about the extinction of most of the autochthonous mammalian fauna and the gradual introduction of more than 25 taxa, which form the present wild and domestic fauna. Such a complete turnover has also been recorded on other Mediterranean islands. These extinctions include *Prolagus sardus* (Lagomorpha, Leporidae), known from subfossil remains found on Corsica, Sardinia, and adjacent small islands (Vigne 1992). *Prolagus* could possibly have reached Sardinia during the desiccation of the Mediterranean during the Miocene (Schüle 1993). Its origin seems to be in Mongolia from where its ancestors reached Corsica and then Sardinia. Skeletal remains indicate that *Prolagus* was still present on Corsica and Sardinia less than 2000 years ago (Vigne 1992). The final report of a living population was made in 1774 by F. Cetti, who observed, *"giant rats whose burrows are so abundant that one might think the surface of the soil had been recently turned over by pigs"* on the island of Tavolara of northeastern Sardinia. It probably attained a length of 200-250 mm but must have undergone rapid evolutionary changes following the arrival of humans on Corsica and Sardinia about 9000 years ago (Vigne 1992). These modifications include an increase in the size of the skull but a reduction of the post cranial skeleton. In Neolithic times *Prolagus* was an important part of human diet in Sardinia, testified by the great amount of skeletons found in human-inhabited caves, like the Grotta di Corbeddu near Oliena. While it apparently survived longer than other extinct, mice-like, insectivorous mammals of the Mediterranean islands (*Nesiotites*, *Tyrrhenicola*, *Rhagamys*), its final extinction seems to have been caused by human predation. Another species that probably became extinct due to human influence is the giant deer *Megaceros*. All extant wild ungulates on the Mediterranean islands are feral domestic animals, or continental game introduced during the Neolithic or later, and none of them have Pleistocene ancestors (Schüle 1993).

The largest mammal on the island is *Cervus elaphus corsicanus*, the Sardinian form of the European deer. It is smaller, darker and more delicately built than continental deer and restricted to three main regions, *viz.* Capoterra, Sette Fratelli forest, and the World Wildlife Fund park of Monte Arcosu. The Sardinian deer is protected by regional legislation and is a target species in the Italian Natura 2000 network. Most of the other mammals presently living on the island have been introduced by man, albeit perhaps hundreds of years ago (Blondel & Vigne 1994; Michaux *et al.* 1996). The Tyrrhenian form of the wood mouse, *Apodemus sylvaticus milleri*, also has an anthropogenic origin. Allozyme data suggest that all the Tyrrhenian wood mice and those of peninsular Italy have a common origin but differ from the northwestern subspecies, *A. sylvaticus sylvaticus*. The Tyrrhenian mice are well isolated from those living on the western edge of the Alpine chain, including the eastern

Pyrenean beech forest. They invaded the islands via the route of Etruria to Elba and Corsica. This hypothesis is in agreement with archaeological evidence of relations between island and mainland populations of Neolithic humans (Klein Hoffmeiyer *et al.* 1986; Michaux *et al.* 1996). According to this theory, wood mice would have colonized the islands as 'lifters' on human boats. The island-specific alleles of Corsican and Elban *Apodemus* are completely absent in Sardinian mice. This indicates that Sardinia was invaded directly from Italy without the detour across Elba and Corsica (Michaux *et al.* 1996).

3. Reflections on conservation issues in butterflies and salamanders

Island species and particularly endemics, are intrinsically more vulnerable to extinction than more widespread species. Habitat destruction, and/or competition with newly introduced species may have severe effects on islands biodiversity. Low genetic variability, resulting from inbreeding or genetic drift has often been reported to decrease species' fitness, and consequently make them more vulnerable (Keller *et al.* 2002). Hybridization and consequent genetic assimilation might be additional threats.

A well known example where natural hybridization is frequent are the butterflies *Papilio machaon* and *Papilio hospiton* (Aubert *et al.* 1996, 1997). Laboratory crosses show that hybrids are not sterile. However, genetic assimilation does not seem to be a threat for *P. hospiton* as developmental perturbations impair the viability of further hybrid progenies.

A recent assessment of endemic Sardinian butterflies suggests that the lycaenids, *Pseudophilotes barbagiae* and *Lysandra coridon gennargenti*, are the only two butterfly species classified as 'vulnerable' according to the IUCN threat categories (Grill *et al.* 2002). However, the two more conspicuous but probably less threatened species *Papilio hospiton* and *Argynnis elisa*, are listed in Appendix II of the Bern Convention, which, since 1988 legislates for the protection of invertebrates at a European level. *Papilio hospiton* is also listed in Annex II of the European Habitats Directive (Council Directive 92/43/EEC).

Maniola nurag could become vulnerable as a result of its complicated life-history. Oviposition only takes place after a female diapause during the hottest part of the summer. Female butterflies need large amounts of nectar before laying their eggs, so the timing of oviposition is probably related to the availability of thistles as high quality nectar resources (D. Jutzeler, pers. comm.). As a consequence, female survival over the summer diapause is a crucial factor for the viability of *M. nurag* populations. This particular ecological characteristic makes the species very susceptible to human-induced changes of their habitat (as shown for *M. jurtina*) (Scali 1971) and might become a particular concern if the climate warms. Under a warmer climate scenario, imagos would emerge earlier in spring but resume activity later in autumn. The consequence would be a prolonged aestivation phase that would increase the risk of female death before oviposition. Another effect could be that with increasing temperature, individuals from intermediate altitudes move higher up the mountain slopes. But due to the island situation these areas would only serve as a limited refuge. Although *M. jurtina* performs the same female diapause in its southern European populations, this species is much less vulnerable than *M. nurag* on a global scale. First, *M. jurtina* is much more widespread and abundant, and second, it does not aestivate in the northern part of its range (Scali 1971; Tolman & Lewington 1997).

As for amphibians, the Sardinian newt is the only species in the genus *Euproctus* for which there is high conservation concern: it is classified by the IUCN as a critically endangered species due to deterioration of its freshwater stream habitat (IUCN 2000). Recent studies have pointed out a contraction in the mountainous range inhabited by the species, as newt populations are no longer present in some localities that some years ago were still occupied (Lecis & Norris 2003). *E. platycephalus* is fully aquatic, and consequently the severe droughts in Sardinia during the last decade caused a strong decline of populations and numbers of individuals. Other threats to this endemic urodele might be the introduction of brown trout, *Salmo trutta* , illegal fishing methods, water pollution and other anthropogenic disturbances (Lecis & Norris 2004a). Long term detailed field surveys would be required to investigate the actual extent of population decline. In the genus *Speleomantes*, all four Sardinian species are rare, *S. flavus* is classified as vulnerable, *S. flavus*, *S. genei*, *S. imperialis*, *S. supramontis*, are considered to be at a lower risk.

4. Conclusions and perspectives

Although this overview is far from being complete, we think it points to some general distributional and ecological patterns of Sardinian endemics and how they have evolved. Divergence time estimates from various sources suggest that the cladogenetic splits leading to the Sardinian lineages have occurred well after the separation of the Sardo-Corsican plate from the continent. Many taxa seem to be younger than the marine regressions in Miocene (5 mya), and may even have arisen during the severe climatic changes of the latest Ice Age (Figure 2).

In Sardinia, areas of high endemism generally coincide with mountains (Table 1, Figure 2). For butterflies, areas of maximum endemism also coincide with areas of maximum species richness (Biermann 1998; Kleinekuhle

1999). This is probably due to the high proportion of endemics among the entire biotic community of Sardinia (25% for butterflies) (Kleinekuhle 1999). Butterfly species richness reaches a maximum in the Gennargentu massive and decreases from east to west and towards the lowlands, and reaches its minimum at the coast. The patterns of endemicity in Sardinia seem to be in agreement with what has been shown for the Iberian Peninsula (Martín *et al.* 2000) and the general European pattern (Balletto 1995). But as indicated in those studies, it is expected that each taxonomic group follows a different pattern related to its individual ecological characteristics or dispersal ability. As yet, estimates of total numbers of species endemic to Sardinia are only available for a few taxonomic groups. Consequently, these last conclusions remain preliminary, and it might as well be that Sardinian mountain massifs generally have higher species richness and endemic species just follow this pattern.

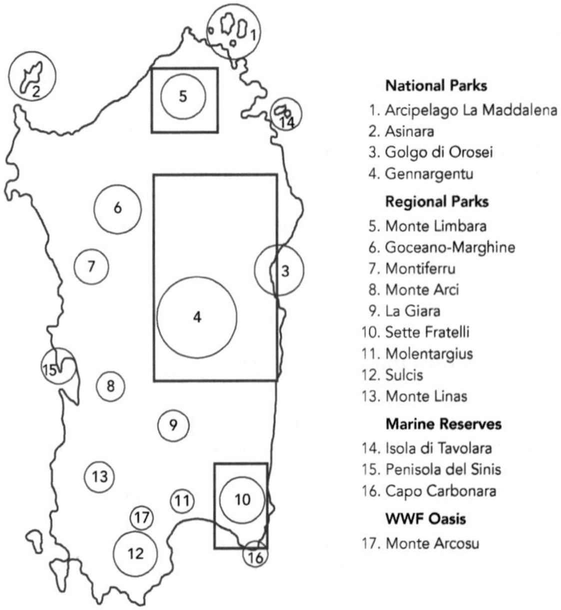

Figure 2. *Existing network of protected areas in Sardinia. Centers of endemicity are indicated with rectangles.*

Considering that most developmental efforts of the tourism industry take place in coastal areas, the concentration of endemic species at higher altitudes might help protect them from the negative effects of increased human pressure. Nevertheless, giving those areas extra protection status is surely not superfluous. There are several protected areas in Sardinia (such as national and regional parks, WWF Oasis, marine reserves, and sites designated as 'Relevant Natural Areas'), however, many political and economical problems need to be resolved before the protection of these areas can be implemented. Given the high number of endemic species in Sardinia, it is necessary that designated reserves and parks do not only exist on paper but conservation and management of the island's unique habitats and species are implemented in practice. This requires increasing the awareness of local people, promotion of field surveys and publication of updated atlases.

Three main centers of endemism, namely the Limbara, Gennargentu, and Sette Fratelli mountains are already included in a network of natural parks that have been proposed to become protected areas but have not yet been officially accepted as such by the Sardinian authorities (Figure 2). Giving them equal status as the already established marine reserves in Villasimius, Asinara, and the Maddalena islands would be a further step to safeguard the unique nature of Sardinia.

Acknowledgements

We thank Johann Genser, Geology Department, Univ. Salzburg, Austria, for his advice regarding the geological part of this chapter, Gabriel Nève and Emmanual Cosson, Laboratoire Systématique Evolutive, CASE 5, Univ. de Provence, Marseille, France, and Steven Weiss, Univ. Graz, Austria, for valuable comments on the manuscript. The Austrian Academy of Science provided a DOC grant financing large parts of this research.

References

Arias C, Azzararoli A, Bigazzi G, Bonadonna FP (1980) Magnetostratigraphy and Pliocene-Pleistocene boundary in Italy. ***Quarternary Research*** **13**, 65-74.

Arnold N (2003) ***Amphibians and Reptiles of Europe***. Princeton University Press, Princeton.

Aubert J, Descimon H, Michel F (1996) Population biology and conservation of the Corsican swallowtail butterfly *Papilio hospiton* Géné. ***Biological Conservation*** **78**, 247-255.

Aubert J, Barascud B, Descimon H, Michel F (1997) Ecology and genetics of interspecific hybridization in the swallowtails, *Papilio hospiton* Géné and *Papilio machaon* L., in Corsica (Lepidoptera: Papilionidae). ***Biological Journal of the Linnean Society*** **60**, 467-492.

Bacchetta G, Brullo S, Selvi F (2000) *Echium anchusoides* (Boraghinaceae), a new species from Sardinia (Italy). ***Nordic Journal of Botany*** **20**, 271-278.

Balletto E (1995) Endemism, areas of endemism, biodiversity and butterfly conservation in the Euro-Mediterranean area. ***Bolletino del Museo Regionale di Scienze Naturali Torino* 13**, 445-491.

Balletto E, Bonelli S, Cassulo L, Meregalli M, Tontini L, Grill, A (2003) Italy. In: ***Prime Butterfly Areas in Europe: Priority Sites for Conservation*** (eds. van Swaay CA, Warren MS), pp. 352-353. National Reference Centre for Agriculture, Nature and Fisheries, Ministry of Agriculture, Nature and Fisheries, The Netherlands.

Biermann H (1998) Tabellarische Übersicht über die Tagfalter der tyrrhenischen Inseln und des angrenzenden Festlandes. ***Atalanta* 28**, 277-287.

Blondel J, Aronson J (1999) ***Biology and Wildlife of the Mediterranean Region***. Oxford University Press, Oxford.

Blondel J, Vigne JD (1994) Space, time and man as determinants of diversity of birds and mammals in the Mediterranean region. In: ***Species Diversity in Ecological Communities*** (eds. Ricklefs RE, Schutler D), pp. 135-146. University of Chicago Press, Chicago.

Brotzu R (1998) ***Fiori spontanei della Sardegna***. Il Maestrale. Sassari.

Brower AVZ (1994) Rapid morphological radiation and convergence among races of the butterfly *Heliconius erato* inferred froom patterns of mitochondrial DNA evolution. ***Proceedings of the National Academy of Sciences, USA* 91**, 6491-6495.

Bucci-Innocenti S, Lascialfari D, Ragghianti M, Mancino G (1978) The karyotypes of the Corsican and Sardinian mountain salamanders, *Euproctus montanus* and *Euproctus platycephalus* (Urodela: Salamandridae). ***Experientia* 34**, 1001-1004.

Caccone A, Sbordoni V (2001) Mitochondrial biogeography of cave life: a study using mitochondrial DNA from Bathysciine beetles. ***Evolution* 55**, 122-130.

Caccone A, Milinkovitch MC, Sbordoni V, Powell JR (1994) Molecular biogeography: calibrating mitochondrial rDNA evolutinoary rates in mountain newts (*Euproctus*) using the Corsica-Sardinian microplate disjunction. ***Journal of Evolutionary Biology* 7**, 227-245.

Caccone A, Milinkovitch MC, Sbordoni V, Powell JR (1997) Mitochondrial DNA rates and biogeography in European newts (genus *Euproctus*). ***Systematic Biology* 46**, 126-144.

Capula M (1996) Evolutionary genetics of the insular lacertid lizard *Podarcis tiliguerta*: genetic structure and population heterogeneity in a geographically fragmented species. ***Heredity* 77**, 518-529.

Cobolli M, Lucarelli M, Sbordoni V (1995) Le farfalle diurne delle piccole isole circumsarde. ***Biogeographia* 8**, 569-582.

Colomo S (1999) Anfibi. In: ***Guida pratica alla flora e alla fauna della Sardegna***. Edizione Archivio Fotografico Sardo, Nuoro.

Cook JA, MacDonald SO (2001) Should endemism be a focus of conservation efforts along the North Pacific Coast of North America? ***Biological Conservation* 97**, 119-126.

Deharveng L (1996) Soil Collembola diversity, endemism and reforestation: case study in the Pyrenees (France). ***Conservation Biology* 10**, 74-84.

Deino A, Gattacceca J, Rizzo R, Montanari A (2001) $^{40}Ar/^{39}Ar$ dating and paleomagnetism of the Miocene volcanic succession of Monte Furru (western Sardinia): implications for the rotation history of the Corsica-Sardinia microplate. ***Geophysical Research Letters* 28**, 3373-3376.

De Jong KA, Manzoni M, Zijderveld JDA (1969) Paleomagnetism of the Alghero trachyandesites (Sardinia). ***Nature* 224**, 67-69.

De Jong KA, Manzoni M, Stavenga T, Van Dijk F, Van der Voo R, Zijderveld JDA (1973) Paleomagnetic evidence for the rotation of Sardinia during the early Miocene. *Nature* **243**, 281-283.

Delaugerre M, Cheylan M (1992) *Atlas de répartition des batraciens et reptiles de Corse.* Parc Natural Regional de Corse, Ajaccio.

Diamond J (1977) Continental and insular speciation in Pacific island birds. *Systematic Zoology* **26**, 236-268.

Drake R, Mulder PHC, Towns DR, Daugherty CH (2002) The biology of insularity: an introduction. *Journal of Biogeography* **29**, 563-569.

Edel JB (1979) Paleomagnetic study of the Tertiary volcanics of Sardinia. *Journal of Geophysics* **45**, 259-280.

Edel JB (1980) *Etude paléomagnetique en Sardaigne. Conséquences pour la géodynamique de la Méditerranée occidentale.* PhD Thesis, Strasbourg I.

Edel JB., Dubois D, Marchant R, Hernandez J, Cosca M (2001) La rotation miocène inférieur du bloc corso-sarde. Nouvelles constraintes paléomagnetiques sur la fin du mouvement. *Bulletin Societé Géologique France* **172**, 275-283.

IUCN (2000) *The 2000 IUCN Red List of Threatened Species*. Species Information. http://www.redlist.org

Gentili S, Mottura A, Rook L (1998) The Italian fossil primate record: recent findings and their geological context. *Geobios* **31**, 675-686.

Gómez-Campo C (1985) The conservation of the Mediterranean plants: principles and problems. In: *Plant Conservation in the Mediterranean Area* (ed. Gómez-Campo C), pp. 3-8. Dr W Junk Publishers Dordrecht, Netherlands.

Grill A (2001) Conservation parameters of the endemic Sardinian butterfly *Maniola nurag*. *Naturschutz- und Landschaftsplanung* 7, 227-232.

Grill A, Crnjar R, Casula P, Menken SBJ (2002) Applying the IUCN threat categories to island endemics: Sardinian butterflies (Italy). *Journal for Nature Conservation* **10**, 51-60.

Grill A, De Vos R, Van Arkel J (2004) The shape of endemics: notes on male and female genitalia in the genus Maniola (Schrank, 1801), (Lepidoptera, Nymphalidae, Satyrinae). *Contributions to Zoology* **73, 293-303.**

Gruttke H, Ludwig G, Binot-Hafke M, Rieken U (1999) Perspektiven bundesweiter Roter Listen – Ergebnisse eines Symposiums des Bundesamtes für Naturschutz. *Natur und Landschaft* **74**, 281-284.

Hudson RR (1998) Island models and the coalescent process. *Molecular Ecology* 7, 413-418.

Hurtrez-Boussès S (1996) Genetic differentiation among natural populations of the rare Corsican endemic *Brassica insularis* MORIS: Implications for conservation guidelines. *Biological Conservation* 76, 25-30.

Jackman RT, Applebaum G, Wake DB (1997) Phylogenetic relationships of bolitoglossine salamanders: a demonstration of the effects of combining morphological and molecular data sets. *Molecular Biolology and Evolution* **14**, 883-891.

Jacquard A (1974) *Génétique des populations humaines*. 1st edition. Presse Universitaire de France, Paris.

Jutzeler D, Casula P, Gascoigne-Pees M, Grill A, Leigheb G (2003) Confirmation du statut spécifique de *Polyommatus gennargenti* (Leigheb, 1987) de Sardaigne comparé à *Polyommatus coridon* (Poda, 1761) de la région de Schaffhousen (CH) par élevage parallèle (Lepidoptera: Lycaenidae) 1ère partie. ***Linneana Belgica*** **19**, 109-118.

Keller LF, Grant PR, Grant BR, Petren K (2002) Environmental conditions affect the magnitude of inbreeding depression in survival of Darwin's finches. ***Evolution*** **56**, 1229-1239.

Klein Hoffmeiyer G, Martini F, Sanges M, Sondaar PY, Ubjega A (1986). La fine del Pleistocene nella grotta Corbeddu in Sardegna: fossili umani, aspetti paleontologici e cultura materiale. ***Rivista di Science Preistoriche***. Rome.

Kleinekuhle J (1999) Die Tagfalter (Rhopalocera) Sardiniens aus biogeografischer Sicht. ***Oedippus*** **16**, 1-60.

LaGreca M (1998) La situazione paleoclimatica nel Quarternario. ***Biogeographia*** **19**, 1-29.

Lanza B (1983) Ipotesi sulle origini del popolamento erpetologico della Sardegna. ***Lavori Società Italiana Biogeografica (N.S.)*** **8**, 723-744.

Lanza B, Caputo V, Nascetti G, Bullini L (1995) Morphologic and genetic studies of the European plethodontid salamanders: taxonomic inferences (genus *Hydromantes*). ***Museo Regionale di Scienze Naturali Torino*** Monograph XVI.

Lecis R, Norris K (2003) Geographical distribution of the Sardinian brook salamander *Euproctus platycephalus* and implications for its conservation. ***The Herpetological Journal*** **13**, 121-124.

Lecis R, Norris K (2004a) Habitat correlates of distribution and local population decline of the endemic Sardinian newt *Euproctus platycephalus*. ***Biological Conservation*** **115**, 303-317.

Lecis R, Norris K (2004b) Population genetic diversity of the endemic Sardinian newt *Euproctus platycephalus*. ***Biological Conservation*** **119**, 263-270.

Leestmanns R (1965) Etude biogeographique sur les Lepidopteres Diurnes de la Corse. ***Alexanor*** **4**, 17-24, 89-96, 113-120, 179-189.

Malinverno A, Ryan WBF (1986) Extension in the Tyrrhenian Sea and shortening in the Apennines as result of arc migration driven by sinking of the lithosphere. ***Tectonics*** **5**, 227-245.

Marchi A, Addis G, Hermosa VE, Crnjar R (1996) Genetic divergence and evolution of *Polyommatus coridon gennargenti* (Lepidoptera: Lycaenidae) in Sardinia, ***Heredity*** **77**, 16-22.

Martín J, García-Barros E, Gurrea P, Luciañez MJ, Munguira ML, Sanz MJ, Simón JC (2000) High endemism areas in the Iberian Peninsula. ***Belgian Journal of Entomology*** **2**, 47-57.

Médail F, Quézel P (1997) Hot-spot analysis for conservation of plant biodiversity in the Mediterranean Basin. ***Annals of the Missouri Botanical Garden*** **84**, 112-127.

Médail F, Quézel P (1999) Biodiversity hotspots in the Mediterranean Basin: setting global conservation priorities. ***Conservation Biology*** **13**, 1510-1513.

Médail F, Verlaque R (1997) Ecological characteristics and rarity of endemic plants from southeast France and Corsica: implications for biodiversity conservation. ***Biological Conservation*** **80**, 269-271.

Menken SBJ, Ulenberg S (1987) Biochemical Characters in Agricultural Entomology. ***Agricultural Zoology Reviews*** **2**, 306-359.

Michaux JR, Filippucci MG, Libois RM, Fons R, Matagne RF (1996) Biogeography and taxonomy of *Ademus sylvaticus* (the woodmouse) in the Tyrrhenian region: enzymatic variations and mitochondrial DNA restriction pattern analysis. ***Heredity*** **76**, 267-277.

Montigny R, Edel JB, Thuizat R (1981) Oligo-Miocene rotation of Sardinia: K-Ar ages and paleomagnetic data of Tertiary volcanics. ***Earth Planetary Science Letters*** **54**, 261-271.

Morrone JJ, Crisci JV (1995) Historical biogeography: introduction to methods. ***Annual Review of Ecology and Systematics*** **26**, 373-401.

Mossa L, Bacchetta G, Brullo S (1999) *Quercus ichnusae* (Fagaceae), a new species from Sardinia. ***Israel Journal of Plant Sciences*** **47**, 199-207.

Munguira ML (1995) Conservation of butterfly habitats and diversity in European Mediterranean countries. In: ***Ecology and Conservation of Butterflies*** (ed. Pullin AS), pp. 277-287. Chapman & Hall, London.

Myers N, Mittermeier RA, Mittermeier CG, da Fonseca GAB, Kent J (2000). Biodiversity hotspots for conservation priorities. ***Nature*** **403**, 853 - 858.

Oberdorff T, Lek S, Guegan JF (1999) Patterns of endemism in riverine fish of the Northern Hemisphere. ***Ecology Letters*** **2**, 75-81.

Oliviero M, Bologna M, Monciotti A, Amnesi F, Mariottini P (1998) Molecular phylogenetics of the Italian *Podarcis* lizards (Reptilia, Lacertidae). ***Italian Journal of Zoology*** **65**, 315-324.

Orr, A (2001) The genetics of species differences. ***Trends in Ecology & Evolution*** **16**, 343-35.

Pierron M (1990) Contribution a la conaissance del *Papilio machaon saharae* Oberthür, differences avec *Papilio machaon machaon* et hybrides experimentales. ***Alexanor*** **16**, 331-340.

Pierron M (1992) Polyphagie de *Papilio hospiton* (Géné). Influence du trophisme Hybridations expérimentales. ***Alexanor*** **17**, 387-390.

Pignatti S (1982) ***Flora d'Italia***, 3 vols. Edagricole. Bologna.

Quilini A, Debussche M, Thompson, JD (2001) Evidence for the local outbreeding depression in the Mediterranean island endemic *Anchusa crispa* Viv. (Boraginaceae). ***Heredity*** **87**, 190-197.

Regione Sardegna (2000) ***Il problema idrico in Sardegna***. Presidenza della Giunta Regionale, Assesorati Regionali Ente Autonoma del Flumendosa, Ente Sardo Acquedotti e Fognature. Relazione. Cagliari.

Roos P, Arnscheid W (1979) Aspekte der Ökologie und Zoogeographie der europäischen Erebien. ***Atalanta*** **10**, 298-308.

Rota E (1992) New data on the earthworm fauna of Gallura (Northeast Sardinia). ***Soil Biology and Biochemistry*** **24**, 1383-1388.

Sabella G, Brachat V, Buckle C (1998) Revision of Tychus of dalmatinus group, with description of new taxa (Coleoptera: Pselaphidae). ***Annales de la Societé Entomologique de France*** **34**, 217-244.

Sacchetti G, Ballero M, Serafini M, Romagnoli C, Bruni A, Poli F (1999) Laticifer tissue distribution and alkaloid location in *Vinca sardoa* (Stearn) Pign. (Apocynaceae), an endemic plant of Sardinia (Italy). ***Phyton-annales botanicae*** **39**, 265-275.

Salomon M (2001) Evolutionary biography and speciation: essay on a synthesis. ***Journal of Biogeography*** **28**, 13-27.

Scali V (1971) Immaginal diapause and gonadal maturation of *Maniola jurtina* (Lepidoptera: Satyridae) from Tuscany. ***Journal of Animal Ecology* 40**, 435-440.

Scali V, Masetti M (1973) The population structure of *Maniola jurtina* (Lepidoptera: Satyridae): the sex-ratio control. ***Journal of Animal Ecology* 42**, 773-778.

Selvi F (1998) Floral biometrics of the *Anchusa undulata* L. group (Boraginaceae) from the central-eastern Mediterranean. ***Botanical Journal of the Linnean Society* 128**, 251-270.

Schnittler M, Ludwig G (1996) Zur Methodik der Erstellung Roter Listen. ***Schriftenreihe für Vegetationskunde* 28**, 709-739.

Schüle W (1993) Mammals, vegetation and the initial human settlement of the Mediterranean islands – a paleoecological approach. ***Journal of Biogeography* 20**, 399-412.

Speranza F, Villa IM, Sagnotti L, Florindo F, Cosentino D, Cipollari P, Mattei M (2002) Age of the Corsica-Sardinian rotation and Liguro-Provencal Basin spreading: new paleomagnetic and Ar/Ar evidence. ***Tectonophysics* 347**, 231-251.

Steininger FF, Rögl F (1984) Paleogeography and palinspastic reconstruction of the Neogene of the Mediterranean and Paratethys. In: ***The Geological Evolution of the Eastern Mediterranean*** (eds. Dixon, JE, Robertson AHF), pp. 659-668. Geological Society, London

Thomson G (1987) ***Genetic variation in the genus* Maniola**. PhD Thesis, London.

Tolman T, Lewington R (1997) ***Butterflies of Britain and Europe***. Harper Collins, London.

Van der Voo R (1993) ***Paleomagnetism of the Atlantic, Thethys and Iapetus Oceans***. Cambridge University Press, Cambridge.

Verlaque R, Médail F, Quézel P, Babinot JF (1997) Endémism végétal et paléogéographique dans le bassin méditerranéen. ***Geobios Mémoire Spécial* 21**, 159-166.

Vigliotti L, Alvarez W, McWilliams M (1990). No relative motion detected between Corsica and Sardinia. ***Earth Planetary Science Letters* 98**, 313-318.

Vigne JD (1992) Zooarcheology and the biogeographical history of the mammals of Corsica and Sardinia since the last ice age. ***Mammal Review* 22**, 87-96.

Whittaker RJ (1997) ***Island Biogeography: Ecology, Evolution and Conservation***. Oxford University Press, Oxford

Chapter 11

Are Macaronesian islands refugia of relict plant lineages?: a molecular survey

Pablo Vargas
E-mail: vargas@ma-rjb.csic.es

Abstract
From a historical biogeographical perspective, Macaronesia has been considered as an island system where relict plants formerly distributed in Europe and northern Africa found appropriate ecological conditions for survival. In this context, and using previously published data on the Macaronesian flora, three basic relictualism concepts (geographic, taxonomic, and lineage relicts) are evaluated. A significant number of the 88 critically endangered species on the Canary Islands may stem from ongoing range reductions on each island (geographic relicts). Macrofossil evidence supports the occurrence of subtropical angiosperms in Europe in the Tertiary and subsequent extinction of genera and species on the continent, whereas they were sheltered in Macaronesia (taxonomic relicts). Using parsimony-based reconstructions, thirty angiosperm and two gymnosperm phylogenies are used to determine the number of stem-based versus crown-based lineages. In 18 plant groups,16 unequivocal stem-based lineages (lineage relicts) were established in Macaronesia earlier than closely-related lineages differentiated on the mainland. In contrast, 13 lineages display a relatively more recent differentiation in Macaronesia as shown by unequivocal sister-group relationships with respect to their continental siblings. When introducing a time scale, some of these 13 crown-based lineages are found to have evolved in Macaronesia since the Tertiary. Relict status of ancient versus stem-based lineages is discussed and a fourth concept of lineage relictualism (ancient lineage) considered in terms of absolute timing. All stem-based lineages plus some crownbased lineages predate the Quaternary and are thus considered to be ancient lineages. Association between endozoochory and multiple colonizations is observed. Eight plant groups, out of a total of 29 properly sampled, underwent multiple colonizations, of which four tree groups (*Hedera*, *Ilex*, *Juniperus* sect. *Juniperus*, *Olea*) display endozoochorous syndromes. Endozoochory may have been crucial to recurrent long-distance dispersal of fleshy-fruited plants to Macaronesia and the establishment of ancient lineages in the Tertiary.

Keywords: Macaronesia, vascular plants, molecular phylogenetics, taxonomic relicts, geographic relicts, lineage relicts, ancient relicts, ancient dispersal

Introduction

Oceanic islands represent a 'natural laboratory' for the study of evolutionary processes and population dynamics in a known spatio-temporal framework.

S. Weiss and N. Ferrand (eds.), Phylogeography of Southern European Refugia, 297–314.

Macaronesia consists of five volcanic archipelagos (Azores, Madeira, Selvagens, Canaries, Cape Verde) separated from the mainland by distances between circa 100 km (the Canarian island of Lanzarote to Africa) and circa 1500 km (the Azorean island of Santa María to Europe). Distinctive floral elements define the five archipelagos as a biogeographic unit (Médail & Quézel 1999), which is clearly recognized as the Macaronesian region, even though some authors consider it as a subregion of the Mediterranean region (Rivas-Martínez *et al.* 1993). Individual islands vary widely in their age, ranging from circa 21 million years for the Canary Islands (Fuerteventura) to 300 000 years for Azores (Pico) (Rothe 1996). Around 3100 species of flowering plants have been identified in Macaronesia (Hansen & Sundig 1993). This high diversity is a consequence of the pronounced ecological heterogeneity of the region, stemming from large differences in altitude (0-3710 m), latitude 40° N -15° N, annual precipitation means (270-1500 mm), and soil composition (Hobohm 2000). The above characteristics result in a high proportion of endemics, estimated independently by Humphries (1979) and Sundig (1979): Canary Islands (25.5/28%), Cape Verde Islands (15/14%), Madeira Islands (8.16/11%), Azores (5.2/5%) and Selvagem Islands (2.2/1%).

Only a few plant groups have successfully colonized and persisted on these islands after long-distance dispersal. At early stages, successful establishment may have occurred straightforward in a free-to-colonize land (preemption concept), followed by increasing competition and decreasing niche availability over time (Emerson 2002). Isolation from the continents, long distances between archipelagos, relatively young island age, and a limited land surface (circa 14 500 km^2) may have hindered Macaronesia's acquisition and distribution of a greater number of species from the mainland (McArthur & Wilson 1967). In contrast, the diverse abiotic conditions have promoted within-island speciation and endemism in habitats of the five extant vegetation zones (Bramwell 1975): sub-Alpine scrub, pine savanna, evergreen forest, transition zone, and semidesert. Over the last geological periods, a buffered climate has been generated in the five archipelagos due to the influence of the Atlantic Ocean (Fernandopullé 1976). This climate has bestowed Macaronesia with suitable characteristics for sheltering subtropical biota, in contrast to mainland dryness and cold during the late Tertiary and the Quaternary glaciations (Hewitt 2000).

Taking into account the above characteristics we may ask whether Macaronesia has played a significant role as a refugium for plant groups that went extinct on the continents after Tertiary and Quaternary climatic deterioration. In other words, how many plant groups can be considered relicts of Macaronesia?

To address the above questions, I have considered three existing concepts that can be used to classify relicts (Cronk 1992, 1997; Carlquist 1995): geographic, taxonomic, and lineage relicts. The three concepts are repostulated in the context of oceanic islands as follows.

1) *Geographic relict*: the surviving populations of an endemic species on oceanic islands resulting from range reduction of once more widely distributed populations.

2) *Taxonomic relict*: a group of one or more taxa diminished dramatically in number on the continent, but still occurring on islands.

3) *Lineage relict*: stem-based evolutionary sequence of ancestor-descendent species or populations (lineages) exclusive to islands.

This latter concept (lineage relict) is evaluated for the first time for Macaronesian plant groups using molecular evidence. A fourth concept (ancient lineage) is defined in terms of absolute times based on paleoclimatologic periods.

Materials and methods

Accumulation of previous data offers the opportunity of indirectly assessing geographic, taxonomic, and lineage relictualism in Macaronesia. Island characteristics on pollen remains do not allow a conventional survey of geographic and taxonomic relicts. Nevertheless, phylogenetic approaches are used in this study to evaluate the relict status of particular taxa from a considerable number (34) of molecular phylogenies.

The application of the geographic relict concept in Macaronesia is difficult because of the absence of pollen records deposited in stable sites. Accordingly, we are not able to properly evaluate the range reduction of island species that once were more widely distributed. A different approach based on the threatened status of species in the Canarian flora is used to infer range reduction. Critically endangered species present in a range < 100 km^2 and/or represented by two or three known populations with < 250 individuals (IUCN criteria, www.iucn.org) are evaluated to quantify population extinction and, hence, range reduction on islands. I hypothesize that more recently formed species represent a low proportion of threatened species, whereas recurrent extinction of established populations on oceanic islands (MacArthur & Wilson 1967) is interpreted as the major cause of species decline and thus involves a high proportion of the critically endangered species. A comprehensive list of Canarian endangered species (VV. AA. 2000) has been used for this approach.

The taxonomic relict concept is applied according to Cronk (1997): "*a species whose taxonomic isolation is due to* ex situ *extinction (of ancestral taxa and continental taxa descended from these) rather than* in situ *evolution (rapid evolution of traits associated with an island)*". In this study, *ex situ* extinction is referred to continental elimination and island persistence. As in previous publications (Saporta 1889; Depape 1922), I consider well-known macrofossils of genera and species extinct on the mainland that, surprisingly, form part of Macaronesia's present-day flora. Extant Macaronesian species of former European floristic elements are summarized by Bramwell (1976) and Sundig (1979).

Parsimony analyses are used to evaluate i) natural groups of Macaronesian species and continental siblings to infer the number of introductions; ii) relictualism as inferred by stem-based lineages, i.e. by means of sister-group relationships of Macaronesian lineages in phylogenetic trees with respect to their continental siblings; and iii) shifts of ancestral syndromes related to long-distance dispersal to islands. Only a reliable sample and well-supported clades of monophyletic groups are considered to determine the number of lineages involved in different colonization events (Emerson 2002). The lineage relict concept is accepted herein and applied to assess the relevance of Macaronesia in harboring early branching lineages. Based on molecular phylogenies, tree-based inference is analyzed in which stem-based and crown-based clades refer to early and subsequent differentiation, respectively (Hennig 1966). Relictualism inferred by assessment of stem-based groups has been previously discussed in a phylogenetic context (Doyle & Donoghue 1993; Carlquist 1995; Baldwin *et al.* 1998). As every single lineage on oceanic (volcanic) islands originated from mainland lineages, we expect to describe three major patterns of island colonization (Figure 1): A) a single crown-base lineage as a consequence of recent establishment in Macaronesia relative to the continental lineages; B) a single stem-based lineage established early on the islands, whereas further speciation of siblings on the mainland occurs; and C) both crown-based and stem-based lineages resulting from multiple dipersal and establishment at different times.

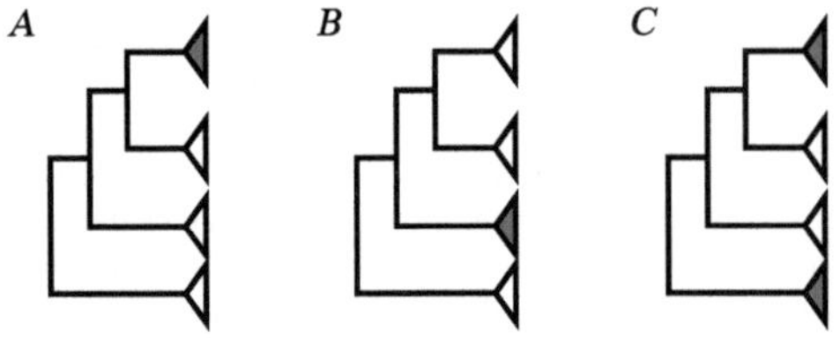

***Figure 1**. The three major cases of phylogenetic relationships of natural groups including Macaronesian taxa. Groups of one or more taxa are represented by full (Macaronesia) and empty (mainland) triangles. **A)** Crown-based lineage as a result of a more recent establishment and evolution in Macaronesia relative to mainland; **B)** Stem-based lineage for a Macaronesian group,with speciation of sibling taxa occurring on the mainland.**C)** Both crown-based and stem-based lineages are the result of multiple dispersal events and establishment at different times (see text for discussion about relative times).*

Table 1. *Plant groups, families, approximate number of Macaronesian species, and literature references of the 34 molecular phylogenetic reconstructions used in the present study.*

Plant group	Family	Taxa number	References
Aeonium alliance	Crassulaceae	c. 70	Mes *et al. 1996;* Van Ham & Hart 1998; Mort *et al.* 2001
Arbutus	Ericaceae	1	Hileman *et al.* 2001
Argyranthemum	Compositae	c. 25	Francisco-Ortega *et al.* 1995, 1997
Armeria	Plumbaginaceae	1	Fuertes & Nieto 2003
Asteriscus	Compositae	c.10	Francisco-Ortega *et al.* 1999b; Goertzen *et al.* 2003
Bellis	Compositae	1	Fiz *et al.* 2002
Bencomia alliance	Rosaceae	c. 9	Helfgott *et al.* 2000
Chamaecytisus	Leguminosae	2	Badr *et al.* 1994; Käss & Wink 1995
Cheirolophus	Compositae	c.15	Susanna *et al.* 1999
Crambe	Cruciferae	c. 11	Francisco-Ortega *et al.* 1999a
Echium	Boraginaceae	c. 33	Böhle *et al.* 1996
Euphorbia	Euphorbiaceae	c. 16	Molero *et al.* 2002
Geranium	Geraniaceae	4	Vargas *et al.*, unpublished
Gonospermum	Compositae	c. 10	Francisco-Ortega *et al.* 2001a
Hedera	Araliaceae	3	Valcárcel *et al.* 2003; Vargas *et al.* 1999a
Ilex	Aquifoliaceae	2	Cuénoud *et al.* 2000
Ixanthus	Gentianaceae	1	Thiv *et al.* 1999
Juniperus	Cupressaceae	2	Martínez & Vargas 2002
Lavatera	Malvaceae	2	Ray 1995; Fuertes *et al.* 2002
Limonium	Plumbaginaceae	c. 20	Lledó *et al.* 1998
Lolium	Graminae	2	Charmet *et al.* 1997
Olea	Oleaceae	2	Hess *et al.* 2000
Pericallis	Compositae	14	Panero *et al.* 1999
Pinus	Pinaceae	1	Krupkin *et al.* 1996
Pulicaria	Compositae	2	Francisco-Ortega *et al.* 2001b
Sambucus	Caprifoliacae	5	Eriksson & Donoghue 1997
Saxifraga	Saxifragaceae	2	Vargas *et al.* 1999b
Sedum	Crassulaceae	9	Van Ham & Hart 1998
Sideritis	Labiatae	24	Barber *et al.* 2002
Sinapidendron	Cruciferae	6	Warwick & Black 1993
Solanum	Solanaceae	c. 31	Bohs & Olmstead 1997
Sonchus alliance	Compositae	c. 40	Kim *et al.* 1996
Teline	Compositae	*c.* 12	Käs & Wink 1995; Percy & Cronk 2002
Tolpis	Compositae	13	Moore *et al.* 2002; Park *et al.* 2001

Table 1 summarizes plant groups studied using parsimony-based analyses from which the relative placement of insular species is inferred. Ancestry of dispersal syndromes is evaluated by means of sister-group reconstructions and morphological characteristics.

Irrespective of the actual mode of dispersal, inference of ancestral characters (plesiomorphies) serves to quantify whether early dispersal syndromes were favored in the successful colonization of islands. The finding of no character shifts simplifies interpretation of a most likely syndrome of island founders, which would have developed dispersal characteristics analogous to those of extant relatives. Classification of these syndromes into five major dispersal-mechanism types (endozoochory, epizoochory, anemochory, hydrochory, unassisted) follows Ridley (1930) and van der Pijl (1979).

Results and discussion

Geographic relicts

The number of threatened species (88) on the Canary Islands varies from island to island, as compiled by Bañares and collaborators (VV. AA. 2000). The largest islands with high numbers of species have more critically endangered species (Figure 2). The ratio of total species to endangered species for each island is as follows: Tenerife (1400/25), Gran Canaria (1300/24), La Gomera (850/12), La Palma (850/8), Fuerteventura (600/9), Lanzarote (650/7), and El Hierro (600/3).

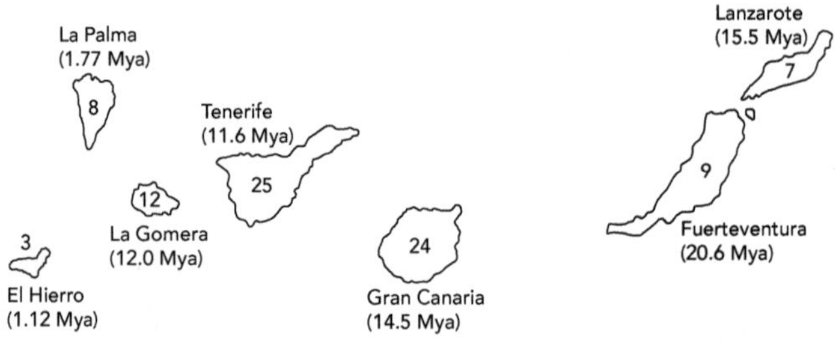

Figure 2. *Number of critically endangered species (88) in the seven Canary Islands, of which a significant number are interpreted as geographic relicts following range reduction. Island ages as in Carracedo (2002).*

Limited distribution of species in oceanic islands may be the result of two evolutionary processes: i) first stages of range expansion after recent speciation events; ii) last extinction stages of ancient species due to population demise. Accordingly, the 88 critically endangered species may represent either range reduction or early stages of range expansion. Analyses of character evolution of plant morphologies are misleading when trying to determine incipient versus decline stages with respect to continental levels of morphological differentiation. Extensive morphological differentiation, but low genetic variation, is common for island taxa including Macaronesian plants (see Table 1 for references). Disparate morphologies arise rapidly because species undergo 'character release' but they display similar or identical DNA sequences (Givnish 1997). For instance, woodiness is surprisingly frequent in genera of different Macaronesian families (*Argyranthemum*, *Sonchus*, *Echium*, *Isoplexis*) and has been considered a relictual character (Bramwell 1976). In contrast to previous considerations, molecular evidence indicates that the woody condition is an acquired character evolving from herbaceous Mediterranean ancestors as a result of *in situ* insular evolution (Panero *et al.* 1999). In contrast, continental species of the same natural group display similar morphologies but remarkable divergent sequences (Böhle *et al.* 1996; Baldwin *et al.* 1998).

Phylogenetic analysis of molecular data allows inference of historical processes in oceanic islands. A search for patterns of ancestor-descendant relationships of critically endangered species in molecular phylogenies (Table 1) reveals that six species with distinctive morphological characters have a relatively most recent origin, as suggested by low molecular divergence and crown-based placement (*Aeonium mascaensis*, *Bencomia brachystachya*, *Echium handiense*, *Isoplexis chalcantha*, *Pericallis hadrosoma*, *Sideritis discolor*). Two more species of *Globularia* (*G. ascanii*, *G. sarcophylla*) appear to fit into this pattern of crown-based lineages (P. Comes, pers. comm.). However, the rarity of four species (*Argyranthemum sudingii*, *Cheirolophus arboreus*, *Cheirolophus junonianus*, *Bencomia sphaerocarpa*) appears to be the result of decline instead of recent species formation as inferred from a stem-based phylogenetic position. Whether a significant number of the 88 critically endangered species are the result of range contraction should be addressed in future investigations by means of evaluating phylogenetic relationships, population genetics, and historical records. Human and naturally caused catastrophies on oceanic islands lead us to interpret that limited distribution, at least for taxa forming ancestral lineages, is the result of population demise, thus resulting in the formation of geographic relicts.

Taxonomic relicts

Eighteen extinct taxa of vascular plants documented from European macrofossils (Miocene and Pliocene) have living counterparts in Macaronesia

(Bramwell 1976). Among them, at least nine were identified as extant Macaronesian species, seven had closely related siblings, and two were not properly identified (Bramwell 1976; Sundig 1979). Therefore, some living fossils support the refugium status of Macaronesia as it harbors taxonomic relicts not found on the continents. As more fossils on the mainland are uncovered, the number of relicts is expected to increase.

Lineage relicts

Relative placement of Macaronesian plant lineages in molecular phylogenies reveals that 13 lineages from 14 plant groups are unequivocally crown-based (Table 2) and 16 lineages from 18 plant groups are unequivocally stem-based (Table 3). Parsimony-based reconstructions allow recognition of the 16 stem-based lineages as lineage relicts, i.e. descended from a common ancestor with an origin that predated differentiation of closely related species on the mainland. This summary of stem-based and crown-based groups gives phylogenetic support for lineage relictualism in Macaronesia. The relative primary position of almost half of the 32 plant groups analyzed to date reveals the importance of plant dispersal and establishment in early times. Colonization has been successful since island formation (starting circa 21 mya) and the establishment of new lineages likely decreased over time (Carlquist 1965; MacArthur & Wilson 1967; Simberloff 1974; Johnson *et al.* 2000). A scenario has been envisaged in which many plant groups succeeded in colonizing immediately after island formation, but increasing competition slowed down the rate of new colonizations. Early colonization may have been dependent on subtropical habitats that were more abundant in the Pliocene and the Miocene. Therefore, a significant number of the 16 relict-inferred lineages may have been present in Macaronesia since the Tertiary, when subtropical vegetation vanished in Europe (Bramwell 1976). The extant plants considered of subtropical origin and forming part of the Macaronesian laurisilva seem to be surviving representatives of a once more widely distributed Tethyan-Tertiary flora (Bramwell 1976; Mai 1995). Both macrofossil and molecular evidence unequivocally support that Macaronesia is indeed a refugium island system.

Additionally, vegetation zones described for Macaronesia include dry habitats and offered new opportunities to Mediterranean plants for colonization during the Quaternary. A new, highly-competitive flora originated 3.2 mya as aridity increased in the Mediterranean basin (Suc 1984). The Mediterranean region has been the main floristic source for dispersal and spawning of new evolutionary lineages in Macaronesian islands. Although 13 crown-based lineages are considered relatively recent lineages, estimates of divergence times are necessary to interpret actual timing of island colonization (Figure 3).

Table 2. *List of 14 Macaronesian plant groups with only crown-based lineages, as interpreted from molecular phylogenies (see Table 1). Approximate number of taxa (Hansen & Sundig 1993), studied taxa (in brackets) if very different from total number, inferred number of introductions, particular relict status, and ancestral dispersal syndromes are also indicated.*

Plant group	Taxa number	Number of introductions	Relict status: stem-based vs. crown-based lineages	Ancestral dispersal syndrome
Aeonium alliance	c. 70	1	Crown	unassisted
Asteriscus	c.10	1	Crown	uncertain
Bencomia alliance	c. 9	1	Crown	unassisted
Cheirolophus	c. 15	1	Crown	anemochory
Echium	c. 33	1	Crown	epizoochory
Euphorbia (tabaibas)	c. 16	2	2 crown	hydrochory
Geranium	4	1 (?)	Crown	epizoochory
Gonospermum	c. 10	1	Crown (?)	unassisted
Limonium	c. 20 (3)	?	1 crown	hydrochory
Pericallis	14	1	Crown	epizoochory
Sambucus	5 (1)	?	1 crown	endozoochory
Saxifraga	2	1	Crown	unassisted
Sideritis	24	1	Crown	epizoochory (?)
Sonchus alliance	c. 40	1	Crown (?)	hydrochory

Table 3. *List of 18 Macaronesian plant groups with at least one stem-based lineage, as interpreted from molecular phylogenies (see Table 1). Approximate number of taxa (Hansen & Sundig 1993), studied taxa (in brackets) if very different from total number, inferred number of introductions, particular relict status, and ancestral dispersal syndromes are also indicated.*

Plant group	Taxa number	Number of introductions	Relict status: stem-based vs. crown-based lineages	Ancestral dispersal syndrome
Arbutus	1	1	Stem	endozoochory
Argyranthemum	c.25	1	Stem	uncertain
Armeria	1	1	Stem	anemochory
Bellis	1	1	Stem	uncertain
Chamaecytisus	2	1	Stem	endozoochory
Crambe	c. 11	1	Stem (?)	unassisted
Hedera	3	3	2 stem / 1 crown	endozoochory
Ilex	2	2	1 stem(?) / 1 crown	endozoochory
Ixanthus	1	1	Stem (?)	unassisted
Juniperus sect. *Juniperus*	2	2	1 stem / 1 crown	endozoochory
Lavatera	2	2	2 stem	hydrochory
Lolium	2 (1)	?	1 stem	endozoochory
Olea	2	2	1 stem / 1 crown	endozoochory
Pinus	1	1	Stem	uncertain
Sedum	9(2)	1	1 stem	unassisted
Sinapidendron	6(2)	1	Stem	unassisted
Solanum	c. 31 (3)	3 (native?)	1 stem / 2 crown (?)	endozoochory
Tolpis	13	1	Stem (?)	anemochory

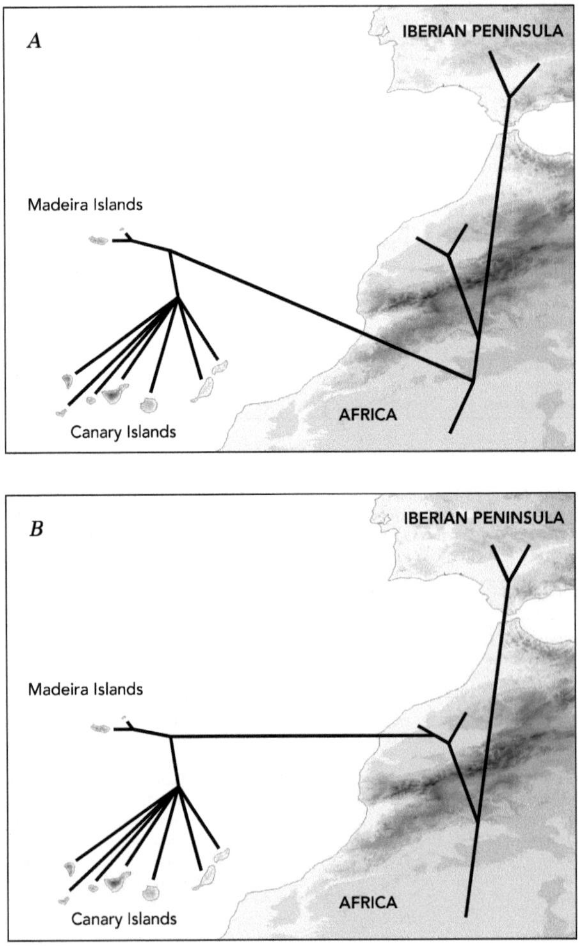

Figure 3. *Hypothetical patterns of historical biogeography (see also Cronk 1997) in the Canary Islands as inferred from* ***A)*** *stem-based lineages and* ***B)*** *crown-based lineages.*

Ancient versus stem-based lineages: ancient relicts

A higher number of old lineages is expected when applying a time scale. Recognition of a stem-based group indicates relative phylogenetic position whereas ancient refers to absolute times of origin (Wulff 1943). I consider stem-based lineages as ancient lineages that succeeded in Macaronesia in early times while continental differentiation necessarily postdated events of island colonization (Doyle & Donoghue 1993; Johnson *et al.* 2000). On the other hand, crown-based lineages may contain some ancient lineages from early times, which are not recognized because of species-poor lineages on the continent as a result of extinction or absence of speciation. When should we consider crown-based lineages as ancient? What is the time limit to include all stem-based and some crown-based groups into ancient lineages?

A criterion to designate ancient lineages in a flora generated primarily in the last 21 myr is herein reformulated (Cronk 1992). The time threshold between the Tertiary and the Quaternary (circa 1.7 mya, www.iugs.org/iugs/pubs/intstratchart.htm) is proposed because after this time limit i) a summer drought was already established (2.8 mya) in the Mediterranean climate zone (Suc 1984); ii) speciation in plants from oceanic islands can occur (Baldwin *et al.* 1998), and iii) island formation was already accomplished in Macaronesia, except for El Hierro (Figure 2). Macaronesia may have been profoundly influenced by the emerging Mediterranean flora in the late Pliocene, as adaptations to seasonal aridity have occurred on the continents since then. The ancient (Miocene) and subancient (Miocene-Pliocene) relicts defined by Cronk (1992) are herein merged into ancient (pre-Quaternary) relicts, i.e. those existing for over 1.7 million years.

Why should all stem-based lineages on islands be considered ancient lineages? Because they have, by definition, sister-relationships to groups of two or more continental species. Assuming previous speciation estimates of two and three mya in the mainland (Levin & Wilson 1976; Niklas *et al.* 1983; Vargas 2003), these lineages may have been present earlier than the Tertiary-Quaternary threshold (1.7 mya). Based on this assumption, any stem-based lineage is herein regarded as an ancient (Tertiary) lineage in the geological context of Macaronesian islands. Accordingly, the term paleoendemism (ancient), as proposed by Baldwin *et al.* (1998), includes all stem-based lineages, which are also interpreted herein as ancient lineages.

Lack of differentiation (speciation) or high extinction rates on the mainland may obscure cases of ancient lineages of Macaronesian plants identified as crown-based lineages because of their relative placement. Apart from the 16 stem-based lineages (Table 3), I suggest that some of the 13 crown-based lineages (Table2) have originated in the Tertiary and, therefore, should be also considered as ancient lineages. How can we detect ancient lineages when we find a crown placement of any plant group in a particular phylogeny? Estimates of minimum colonization times can be established by the use of molecular clocks.

Four examples of the 13 crown-based lineages (Table 2) clearly illustrate the concept of ancient versus stem-based lineages. The *Aeonium* alliance includes four genera (*Aeonium, Aichryson, Greenovia, Monanthes,*) with a crown position in the Crassulaceae phylogeny (Van Ham & Hart 1998; Mort *et al.* 2001). We should consider the *Aeonium* alliance as an ancient lineage because its ancestor's arrival, establishment, evolution, and differentiation likely occurred in ancient times, manifested by remarkable morphological differentiation (four genera). Calibration of a molecular clock in Crassulaceae is necessary to assess divergence times; however, the high molecular variation found among the four genera – in comparison to other genera of Macaronesian angiosperms – indicates that differentiation of genera may have predated the Tertiary-Quaternary limit. Two more plant groups do not have a clear stem position and their molecular clocks indicate a Tertiary origin: *Sonchus* subgenus *Dendrosonchus* (Kim *et al.* 1996) and *Crambe* (Francisco-Ortega 1999a). Further molecular-clock estimates are necessary to quantify the number of crown-based lineages that may be considered ancient lineages, i.e. lineages established in Macaronesia between the onset of island formation (circa 21 mya) and the late Pliocene (> 1.7 mya).

Success of fleshy-fruited plants and ancient lineages

Colonization of islands involves multiple factors that influence dispersal and establishment such as geography, ecological interactions, inherent biological properties of the plant, and stochastic events (Ackerley 2003). While assuming multiple factors have been crucial for colonization, I focus only on the first step for a successful establishment: dispersal. Direct observations of the arrival of present-day diaspores are difficult and inference on specific arrivals of early Macaronesian founders is a speculative exercise. Availability of molecular phylogenetic data for a wide diversity of insular plants allows scholars to evaluate whether any particular diaspore syndrome may have favored dispersal to oceanic islands through time. Dispersal syndromes are arranged into five major groups (anemochory, hydrochory, endozoochory, epizoochory, and mechanisms not associated with long-distance dispersal or unassisted). Historical reconstructions based on 31 phylogenies from studies of different plant lineages (Table 1, excluding *Pulicaria*, *Solanum*, and *Teline*) indicate 38 introductions and no syndrome shift following colonization. Bramwell (1985) estimated that 186 founders generated the extant flora of the Canary Islands, of which 63 were endozoochorous (34%), 35 epizoochorous (19%), 48 anemochorous (26%), 8 hydrochorous (4.3%), and 32 uncertain (17%). The use of molecular phylogenies and character-evolution reconstruction of diaspore types of the 38 well-documented introductions (Tables 2 & 3, excluding *Solanum*) reveals the following preliminary results of ancestral syndromes for first colonizers to Macaronesia: 13 endozoochorous (34%), 4 epizoochorous (10.5%), 3 anemochorous (8%), 6 hydrochorous (16%), 8 unassisted (21%), and 4 uncertain (10.5%). Although our dispersal syndrome classification includes one more type (unassisted) than that of Bramwell (1985), independent estimates reveal similar figures for success of endozoochory.

An additional argument for the success of endozoochory is based on a number of plant introductions. Multiple dispersals of a natural group to Macaronesia is not a common pattern in the colonization history of these archipelagos (Valcárcel *et al.* 2003). Twenty-two plant groups originated unequivocally from 22 single colonizers, whereas only eight had two or more dispersal origins (*Euphorbia*, *Hedera*, *Ilex*, *Juniperus* sect. *Juniperus*, *Lavatera*, *Olea*, *Pulicaria*, *Teline*, excluding *Solanum*)(Tables 1-3). Among them, four trees with endozoochorous fruits may have been favored not only for early dispersal (ancient lineages) but also for recurrent colonization of Macaronesia in different periods. Obviously, only a small subset (32 phylogenies) of the Macaronesian flora lends itself to this analysis. In any case, we infer that endozoochory accounts for a third of dispersal to these islands since their formation, and thus the establishment of ancient lineages. Association between the number of fleshy-fruited trees and their establishment in a particular habitat is not observed. Of those genera containing ancient lineages, three occur in the laurisilva (*Arbutus*, *Hedera*, *Ilex*) – the habitat more similar to former subtropical formations (Takhtajan 1969; Bramwell 1976) – while three others are adapted to mesic habitats (*Juniperus*, *Olea*, *Pinus*) (Table 3).

In summary, these results lead us to conclude that endozoochory has played a crucial role in successful dispersal to Macaronesia (Valcárcel *et al.* 2003), as it has been determined for the Hawaiian islands (Carlquist 1969; Vargas & Baldwin, unpublished data). Moreover, the importance of endozoochory is reflected in the proportion of diaspore syndromes favored in multiple introductions to Macaronesia. They include at least one stem-based lineage each, which indicate early colonizations and the formation of lineage relicts. It seems likely, therefore, that fleshy-fruited trees not only had higher probabilities for multiple dispersal to Macaronesia, but also a more active dissemination since ancient Tertiary times, thus fostering the origin and establishment of ancient lineages.

Acknowledgements

I am indebted to Peter Comes for his valuable comments throughout the text. Virginia Valcárcel discussed and took active interest in communicating the content of this paper in the First International Symposium of Phylogeography in Southern European Refugia (11-15 March, Vairão, Portugal). I also thank every botanist who published molecular phylogenies on Macaronesian plants (Table 1).

References

Ackerley DD (2003) Community assembly, niche conservatism, and adaptive evolution in changing environments. ***International Journal of Plant Sciences* 164** (Supplement), S165-S184.

Badr A, Martin W, Jensen U (1994) Chloroplast DNA restriction site polymorphism in *Genisteae* (Leguminosae) suggests a common origin for European and American lupines. ***Plant Systematics and Evolution* 193**, 95-106

Baldwin BG, Crawford DJ, Francisco-Ortega J, Kim, SC, Sang T, Stuessy TF (1998) Molecular phylogenetic insights on the origin and evolution of oceanic island plants. In: ***Molecular Systematics of Plants. II DNA sequencing*** (eds. Soltis D, Soltis P, Doyle JJ), pp. 410-441. Kluwer Academic Publishers, Boston.

Barber JC, Francisco-Ortega J, Santos-Guerra A, Turner KG, Jansen RK (2002) Origin of Macaronesian *Sideritis* L. (Lamioideae: Lamiaceae) inferred from nuclear and chloroplast sequence datasets. ***Molecular Phylogenetics and Evolution* 23**, 293-306.

Böhle UR, Hilger HH, Martin WF (1996) Island colonization and evolution of the insular woody habitat in *Echium* L. (Boraginaceae). ***Proceedings of the National Academy of Sciences, USA* 93**, 11740-11745.

Bohs L, Olmstead RG (1997) Phylogenetic relationships in *Solanum* (Solanaceae) based on ndhF sequences. ***Systematic Botany*, 22**, 5-17.

Bramwell D (1975) Some morphological aspects of the adaptive radiation of Canary Islands *Echium* species. ***Anales del Jardín Botánico de Madrid* 32**, 241-254.

Bramwell D (1976) The endemic flora of the Canary Islands: distribution, relationships and phytogeography. In: ***Biogeography and Ecology in the Canary Islands*** (ed. Kunkel G), pp. 207-240. Dr. W. Junk, The Hague.

Bramwell D (1985) Contribución a la biogeografía de las Islas Canarias. ***Botanica Macaronesica* 14**, 3-34.

Carlquist S (1969) The biota of the long-distance dispersal. V. Plant dispersal to Pacific Islands. ***Bulletin of the Torrey Botanical Club* 94**, 129-162.

Carlquist S (1974) ***Island Biology***. Columbia University Press, Columbia.

Carlquist S (1995) Introduction. In: ***Hawaiian Biogeography. Evolution on a hot spot archipelago*** (eds. Wagner WL, Funk VA), pp. 1-13. Smithsonian Institution Press, Washington.

Carracedo JC, Pérez FJ, Ancochea E, meco J, Hernán F, Cubas CR, Casillas R, Rodríguez E, Ahijado A (2002) Cenozoic volcanism II: the Canary Islands. In: ***The Geology of Spain*** (eds. Gibbons W, Moreno T), pp. 439-472. The Geological Society of London, London.

Charmet G, Ravel C, Balfourier F (1997) Phylogenetic analysis in the *Festuca-Lolium* complex using molecular markers and ITS rDNA. ***Theoretical and Applied Genetics* 94**, 1038-1046.

Cronk QCB (1992) Relict floras of Atlantic islands: patterns assessed. ***Biological Journal of the Linnean Society* 46**, 91-103.

Cronk QCB (1997) Islands: stability, diversity, conservation. ***Biodiversity and Conservation* 6**, 477-493.

Cuénoud P, Martínez MP, Loizeau P, Spichiger R, Andrews S, Manen J (2000). Molecular phylogeny and biogeography of the genus *Ilex* L. (Aquifoliaceae). ***Annals of Botany (Oxford)* 85**, 111-122.

Depape G (1922) Recherches sur la flore pliocêne de la Vallée du Rhône. ***Annales des Sciences Naturelles, Botanique* 10**, 73-265.

Doyle JA, Donoghue MJ (1993) Phylogenies and angiosperm diversification, ***Paleobiology* 19**, 141-167.

Emerson, BC (2002) Evolution on oceanic islands: molecular phylogenetic approaches to understanding pattern and process. ***Molecular Ecology* 11**, 951-966.

Eriksson T, Donoghue MJ (1997) Phylogenetic relationships of *Sambucus* and *Adoxa* (Adoxaceae) based on nuclear ribosomal ITS sequences and preliminar morphological data. ***Systematic Botany* 22**, 555-573.

Fernandopullé D (1976) Climatic characteristics of the Canary islands. In: ***Biogeography and Ecology in the Canary Islands*** (ed. Kunkel G.), pp. 185-205. Dr. W. Junk, The Hague.

Fiz O, Valcárcel V, Vargas P (2002) Phylogenetic position of Mediterranean Astereae and character evolution of daisies (Bellis, Asteraceae) inferred from nrDNA ITS sequences. ***Molecular Phylogenetics and Evolution* 25**, 157-171.

Francisco-Ortega J, Fuertes Aguilar J, Gómez-Campo C, Santos-Guerra A, Jansen RK (1999a) Internal transcribed spacer sequence phylogeny of *Crambe* L. (Brassicaceae): molecular data reveal two Old World disjunctions. ***Molecular Phylogenetics and Evolution* 11**, 361-380.

Francisco-Ortega J, Goertzen LR, Santos-Guerra A, Benabid A, Jansen RK (1999b) Molecular systematics of the *Asteriscus* alliance (Asteraceae: Inuleae) I: evidence from the internal transcribed spacers of nuclear ribosomal DNA. ***Systematic Botany* 24**, 249-266.

Francisco-Ortega J, Barber JC, Santos-Guerra A, Febles-Hernández R, Jansen RK (2001a) Origin and evolution of the endemic genera of Gonosperminae (Asteraceae: Anthemideae) from the Canary Islands: evidence from nucleotide sequences of the internal transcribed spacers of the nuclear ribosomal DNA. ***American Journal of Botany* 88**, 161-169.

Francisco-Ortega J, Park SJ, Santos-Guerra A, Benabid A, Jansen RK (2001b) Origin and evolution of the endemic Macaronesian Inuleae (Asteraceae): evidence from the internal transcribed spacers of nuclear ribosomal DNA. ***Biological Journal of the Linnean Society* 72**, 77-97.

Francisco-Ortega J, Fuertes-Aguilar J, Kim SC, Santos-Guerra A, Crawford DJ, Jansen RK (2002) Phylogeny of the Macaronesian endemic *Crambe* section *Dendrocrambe* (Brassicaceae) based on internal transcribed spacer sequences of nuclear ribosomal DNA. ***American Journal of Botany* 89**, 1984-1990.

Fuertes-Aguilar J, Nieto Feliner G (2003) Additive polymorphisms and reticulation in an ITS phylogeny of thrifts (*Armeria*, Plumbaginaceae). ***Molecular Phylogenetics and Evolution*** 28, 430-447.

Fuertes-Aguilar J, Ray MF, Francisco-Ortega J, Santos-Guerra A, Jansen RK (2002) Mole-cular evidence from chloroplast and nuclear markers for multiple colonizations of *Lavatera* (Malvaceae) in the Canary Islands. ***Systematic Botany* 27**, 74-83.

Givnish TJ (1997) Adaptive radiation and molecular systematics: issues and approaches. In: ***Molecular Evolution and Adaptive Radiation*** (eds. Givnish TJ, Sytsma KJ), pp. 1-54. Cambridge University Press, Cambridge.

Goertzen LR, Francisco-Ortega J, Santos-Guerra A, Mower JP, Linder CR, Jansen RK (2002) Molecular systematics of the *Asteriscus* alliance (Asteraceae: Inuleae) II: combines nuclear and chloroplast data. ***Systematic Botany* 27**, 815-823.

Hansen A, Sundig P (1993) Flora de Macaronesia. Checklist of vascular plants. 4th revised edition. ***Sommerfeltia* 17**, 1-298.

Helfgott DM, Francisco-Ortega J, Santos-Guerra A, Jansen RK, Simpson BB (2000) Biogeography and breeding system evolution of the woody *Bencomia* alliance (Rosaceae) in Macaronesia based on ITS sequence data. ***Systematic Botany*** **25**, 82-97.

Hennig W (1966) ***Phylogenetic Systematics***. University of Illinois Press, Urbana, Illinois.

Hess J, Kadereit JW, Vargas P (2000) The colonization history of *Olea europaea* L. in Macaronesia based on internal transcribed spacer 1 (ITS-1) sequences, randomly amplified polymorphic DNAs (RAPD), and intersimple sequence repeats (ISSR). ***Molecular Ecology*** **9**, 857-868.

Hewitt GM (2000) The genetic legacy of the Quaternary ice ages, ***Nature*** **405**, 907-913.

Hileman LC, Vasey MC, Parker VT (2001) Phylogeny and biogeography of the Arbutoideae (Ericaceae): implications for the Madrean-Tethyan Hypothesis. ***Systematic Botany*** **26**, 131-143.

Hobohm C (2000) Plant species diversity and endemism on islands and archipelagos, with special reference to the Macaronesian Islands. ***Flora*** **195**, 9-24.

Humphries CJ (1979) Endemism and evolution in Macaronesia. In: ***Plants and Islands*** (ed. Bramwell D), pp. 171-200. Academic Press, New York.

Johnson KP, Adler FR, Cherry JL (2000). Genetic and phylogenetic consequences of island biogeography. ***Evolution*** **54**, 387-396.

Käss E, Wink M (1995) Molecular phylogeny of the Papilionoideae (Family Leguminosae): RbcL gene sequences versus chemical taxonomy. ***Botanica Acta*** **108**, 149-162.

Kim SC, Crawford DJ, Francisco-Ortega J, Santos-Guerra A (1996) A common origin for woody *Sonchus* and five related genera in the Macaronesian islands: molecular evidence for extensive radiation. ***Proceedings of the National Academy of Sciences, USA*** **93**, 7743-7748.

Krupkin AB, Liston A, Strauss SH (1996) Phylogenetic analysis of the hard pines (*Pinus* sugenus *Pinus*, Pinaceae) from chloroplast DNA restriction site analysis. ***American Journal of Botany*** **83**, 489-498.

Levin DA, Wilson AC (1976) Rates of chromosomal evolution in seed plants: net increase in diversity of chromosome numbers and species numbers through time. ***Proceedings of the National Academy of Sciences, USA*** **73**, 2086-2090.

Lledó MD, Crespo MB, Cameron KM, Fay MF, Chase M (1998) Systematics of Plumbagineaceae based upon cladistic analysis of *rbc*L sequence data. ***Systematic Botany*** **23**, 21-29.

MacArthur RH, Wilson EO (1967) ***The Theory of Island Biogeography***. Princeton University Press, New Jersey.

Mai DH (1995) ***Tertiäre Vegetationsgeschichte Europas***. Gustav Fischer, Jena, Stuttgart, New York.

Martínez J, Vargas P (2002) Chloroplast differentiation and active dispersal of *Juniperus oxycedrus* in the Mediterranean basin based on trnL-trnF sequences. Paper held at the I International Symposium of Phylogeography in Southern European Refugia. 11–15 March, Vairão, Portugal.

Médail F, Quézel P (1999) The phytogeographical significance of S. W. Morocco compared to the Canary Islands. ***Plant Ecology*** **140**, 221-244.

Mes THM, van Brederode J, Hart H. (1996) Origin of the woody Macaronesian *Sempervivoideae* and the phylogenetic position of the East African species of *Aeonium*. ***Botanica Acta*** **109**, 477-491.

Molero J, Garnatje T, Rovira A, Garcia-Jacas N, Susanna A (2002) Karyological evolution and molecular phylogeny in Macaronesian dendroid spurges (Euphorbia subsect. Pachycladae). ***Plant Systematics and Evolution* 231**, 109-132.

Moore MJ, Francisco-Ortega J, Santos-Guerra A, Jansen RK (2002) Chloroplast DNA evidence for the roles of island colonization and extinction in *Tolpis* (Asteraceae: Lactuceae). ***American Journal of Botany* 89**, 518-526.

Mort ME, Soltis DE, Soltis PS, Francisco-Ortega J, Santos-Guerra A (2001) Phylogenetic relationships and evolution of Crassulaceae inferred from matK sequence data. ***American Journal of Botany* 88**, 76-91.

Niklas KJ, Tiffney BH, Knoll AJ (1983) Patterns in vascular land plant differentiation. ***Nature* 303**, 614-616.

Panero JL, Francisco-Ortega J, Jansen RK, Santos-Guerra A (1999) Molecular evidence for multiple origins of woodiness and a New World biogeographic connection of the Macaronesian island endemic *Pericallis* (Asteraceae: Senecioneae). ***Proceedings of the National Academy of Sciences, USA* 96**, 13886-13891.

Park S, Korompai EJ, Francisco-Ortega J, Santos-Guerra A, Jansen RK (2001) Phylogenetic relationships of *Tolpis* (Asteraceae: Lactuceae) based on *ndh*F sequence data. ***Plant Systematics and Evolution* 226**, 23-33.

Percy DM, Cronk QCB (2002) Different fates of island brooms: contrasting evolution in *Adenocarpus*, *Genista*, and *Teline* (Genisteae, Fabaceae) in the Canary Islands and Madeira. ***American Journal of Botany* 89**, 854-864.

Ray MF (1995) Systematics of *Lavatera* and *Malva* (Malvaceae, Malveae) - a new perspective. ***Plant Systematics and Evolution* 198**, 29-53.

Ridley, HN (1930) ***The Dispersal of Plants throughout the World.*** L. Reeve & Co., Kent.

Rivas-Martínez S., Wildpret W., Arco-Aguilar M., Rodríguez O., Perez-de-Paz PL, García-Gallo A, Acebes JR, Díaz TE, Fernandez-Gonzalez F (1993). Las comunidades vegetales de la Isla de Tenerife (Islas Canarias). ***Itinera Geobotanica* 7**, 169-374.

Rothe P (1996) ***Kanarische Inseln.*** 2nd edition, Sammlung geologischer Führer, Band 81, Berlin, Stuttgart.

Saporta MG (1889) Dernières adjonctions a la Flore fossile d´Aix-en-Provence. Description des espèces, 2ª partie. ***Annales des Sciences Naturelles, Botanique* 7**, 1-192.

Simberloff DS (1974) Equilibrium theory of islands biogeography and ecology. ***Annual Review of Ecology and Systematics* 5**, 161-182.

Suc JP (1984) Origin and evolution of the Mediterranean vegetation and climate in Europe. ***Nature* 307**, 429-432.

Sundig P (1979) Origins of the Macaronesian Flora. In: ***Plants and Islands*** (ed. Bramwell D), pp. 13-40. Academic Press, New York.

Susanna A, Garnatje T, García-Jacas N (1999) Molecular phylogeny of *Cheirolophus* (Asteraceae: Cardueae-Centaureinae) based on ITS sequences of nuclear ribosomal DNA. ***Plant Systematics and Evolution* 214**, 147-160.

Takhtajan A (1969) ***Flowering Plants: Origin and Dispersal.*** Smithsonian Institute, Washington.

Thiv M, Struwe L, Kadereit JW (1999) The phylogenetic relationships and evolution of the Canarian laurel forest endemic *Ixanthus viscosus* (Aiton) Griseb. (Gentianaceae): evidence from matK and ITS sequences, and floral morphology and anatomy. ***Plant Systematics and Evolution* 218**, 299-317.

Valcárcel V, Fiz O, Vargas P (2003) Chloroplast and nuclear evidence for multiple origins of polyploids and diploids of *Hedera* (Araliaceae) in the Mediterranean basin. ***Molecular Phylogenetic and Evolution* 26**, 1-20.

Van Ham RCH, Hart H (1998) Phylogenetic relationships in the Crassulaceae inferred from chloroplast DNA restriction-site variation. ***American Journal of Botany* 85**, 123-134.

Van der Pijl, L (1979) ***Principles of Dispersal in Higher Plants***. Springer Verlag, New York.

Vargas P (2003) Molecular evidence for multiple diversification patterns of alpine plants in Mediterranean Europe. ***Taxon* 52**, 463-476.

Vargas P, McAllister HA, Morton CM, Jury SL, Wilkinson MJ (1999a) Polyploid speciation in *Hedera* L.: phylogenetic and biogeographic insights based on chromosome counts and ITS sequences. ***Plant Systematics and Evolution* 219**, 165-179.

Vargas P, Morton CM, Jury SL (1999b) Biogeographic patterns in Mediterranean and Macaronesian species of *Saxifraga* (Saxifragaceae) inferred from phylogenetic analyses of ITS sequences. ***American Journal of Botany* 86**, 724-734.

VV. AA. (2000) Lista roja de flora vascular Española (valoración según categorías UICN). ***Conservación Vegetal* 6** (extra), 11-38.

Warwick SI, Black LD (1993) Molecular relationships in subtribe Brassicinae (Cruciferae, tribe Brassiceae). ***Canadian Journal of Botany* 71**, 906-918.

Wulff EV (1943) ***An Introduction to Historical Plant Geography***. Chronica Botanica, Waltham, Massachusetts.

Chapter 12

Nested clade phylogeographical analysis of barbel (*Barbus barbus*) mitochondrial DNA variation

Petr Kotlík* and Patrick Berrebi
*E-mail: kotlik@iapg.cas.cz

Abstract
We applied nested clade phylogeographical analysis (NCPA) to the mitochondrial DNA phylogeographical data of the barbel *Barbus barbus* to assess the historical biogeography scenario suggested for this species by a traditional phylogeographical approach. Major previously inferred historical events received support from the NCPA: i) twofold range fragmentation, an ancient one between the central European and Balkan/Anatolian populations and a more recent one ascribed to the survival of the central European lineage in two refugia during the latest glacial, and ii) contiguous range expansion from the Danubian refuge into the more north-western river basins. Likely due to insufficient genetic variation, the hypothesized dispersion from the more western central European refuge was not detected by the NCPA as was not the hypothesized expansion throughout the Balkans and Anatolia. The NCPA interpretation of the significant pattern within the Danube river basin as reflecting a recurrent gene flow restricted through isolation by distance should be taken with caution. Similar patterns can reflect non-equilibrium conditions, such as population growth, which seems a plausible alternative interpretation given the star-like genealogy of the Danubian population, and its presumably short period of demographic stability.

Keywords: isolation by distance, mtDNA, phylogeography, Pleistocene, range fragmentation, range expansion

Introduction

Phylogeography is a standard tool for interpreting geographical patterns of genetic variation from a genealogical perspective. For a decade, phylogeographical studies relied on the visual examination of how estimated gene trees overlay upon the mapped geographical sources of genetic data, and the deduction of evolutionary processes compatible with the observed patterns (Avise *et al.* 1987; Avise 2000). Several caveats related to this intuitive approach have been appreciated. The population genetic structure can be the result of a combination of contemporary processes as well as historical events, and the traditional phylogeographical approach may not fully allow for the estima-

S. Weiss and N. Ferrand (eds.), Phylogeography of Southern European Refugia, 315–325.

tion of the dynamic structure and temporal juxtaposition of different evolutionary factors (Templeton *et al.* 1995; Templeton 1998). In particular, an advantage has been advocated of an approach that would provide assessment of the statistical significance of association between the phylogeny and geography and allow less subjective interpretation (Templeton *et al.* 1995). The recently developed nested clade phylogeographical analysis (NCPA; Templeton *et al.* 1995; Templeton 1998, 2004) represents a potentially useful method in these respects (e.g. Durand *et al.* 2000; Bernatchez 2001; Seddon *et al.* 2001; Branco *et al.* 2002; Pfenninger *et al.* 2003; reviewed in Templeton 2004). The NCPA is an analytical framework devised to measure and statistically assess the strength of the association between haplotype positions in a gene tree with their geographical origin, and to biologically interpret the significant patterns through an inference key incorporating prediction about the outcomes of various evolutionary processes (Templeton *et al.* 1995; Templeton 1998, 2002, 2004). The concern logically brought about by the development of this formal method is what new conclusions could be obtained through the assessment of phylogeographical data with the NCPA compared to the traditional approach (Avise 2000; Turner *et al.* 2000; Paulo *et al.* 2002). It has been suggested that, under particular circumstances, NCPA may result in false inference or biological misinterpretation (Paulo *et al.* 2002). This is likely to occur because the likelihood of the significant phylogeographical pattern being the result of a specific process or event is not assessed by the NCPA (Knowles & Maddison 2002; Knowles 2004). A way to address these issues is to use actual data sets previously analyzed with the traditional approach (Templeton 1998; Branco *et al.* 2002; Paulo *et al.* 2002). Application of the NCPA to such data sets with prior expectations about the processes underlying the present phylogeographical patterns can be then used to evaluate the performance of this method (Templeton 1998, 2004).

Based on a phylogeographical analysis of the variation at the mitochondrial DNA (mtDNA) cytochrome *b* gene we have recently proposed a historical biogeography scenario for the barbel *Barbus barbus*, a widely distributed European cyprinid (Kotlík & Berrebi 2001). The barbel has likely been sundered into two refugia for several later glacial cycles. The Danubian refuge served as the major source of postglacial colonization of non-Mediterranean Europe, although the most recent dispersion throughout western Europe may have taken place from an additional more western refuge where the species survived the last glacial. While likely reduced in their effective sizes, the Balkan and Anatolian populations did not expand further north postglacially (Kotlík & Berrebi 2001).

In this paper we apply the NCPA to the barbel phylogeographical data with the aims to assess this historical biogeography hypothesis and to evaluate concordance between the traditional and nested clade phylogeographical approaches.

Material and methods

The barbel mtDNA data set

We used the data obtained by the DNA sequence analysis of 594 bp of the barbel mtDNA cytochrome *b* gene (Kotlík & Berrebi 2001). The data set consisted of 87 individual sequences sampled from 33 geographical locations throughout Europe and Anatolia (Figure 1, Table 1). Among these sequences 15 nucleotide positions were variable, and they defined altogether 11 haplotypes (h1 to h11) differing from each other by one to nine nucleotides with pairwise sequence divergence ranging from 0.002 to 0.015.

The overall data fitted the infinite-site model of sequence evolution, which resulted in the absence of evidence for homoplasy. The method of Templeton *et al.* (1992) was used to estimate the maximum number of pairwise site differences in the data, for which none of the observed differences is due to more than one mutation with the probability greater then 0.95. Haplotypes differing by up to 10 nucleotide differences showed this probability. Because the number of differences between any pair of haplotypes did not exceed this limit, we applied the maximum parsimony criterion to construct a minimum-spanning tree (MST) as a portrayal of the genealogical relationship among the haplotypes. The TCS software, version 1.06 (Clement *et al.* 2000) was used to estimate the probability of parsimony and to construct the MST.

Nested clade phylogeographical analysis

We used the NCPA (Templeton *et al.* 1995; Templeton 1998) to detect and biologically interpret statistically significant phylogeographical patterns in the barbel mtDNA data set. A hierarchical set of nested clades has been defined from the haplotype MST using the algorithm of Templeton *et al.* (1987) and Templeton & Sing (1993). For each clade at each hierarchical level, statistical significance of deviation from the expectation under null hypothesis of no geographical association was determined for the clade geographical range (clade distance, Dc), it's average geographical distance from the geographical center of the entire nesting clade (nested clade distance, Dn), and for the contrast of each measure between the interior and tip clades (I-T). Great circle geographical distances were assumed (Bernatchez 2001). Calculations were performed using the GeoDis software, version 2.0, with the null distribution generated from 1000 random permutations of the haplotypes or clades against the sampling locations (Posada *et al.* 2000). The processes compatible with the significant patterns within nested clades were identified using the inference key provided in Templeton (1998) and at http://zoology.byu.edu/crandall_lab/geodis.htm.

A

B

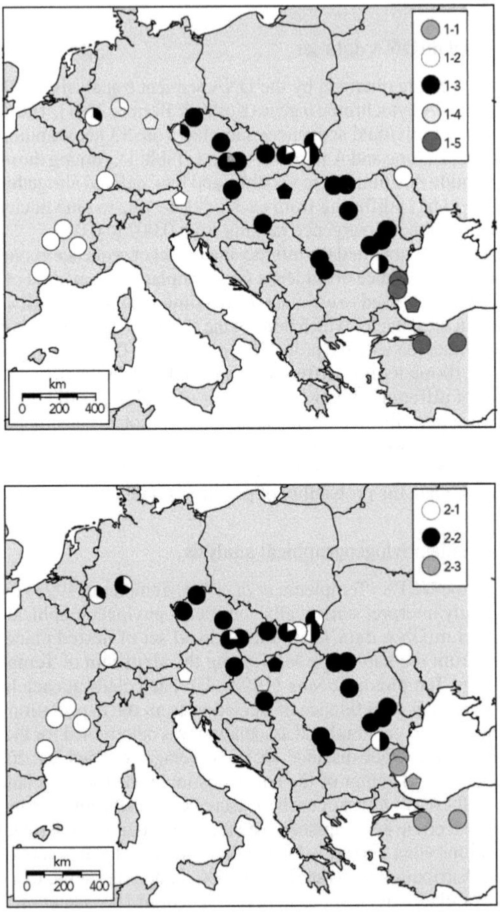

Figure 1. *Geographical distribution of* ***A****) one-step and* ***B****) two-step clades. Polygons indicate the geographical centers of clades with geographical variation and of the entire tree (empty symbol in* ***B****). A detailed description of single haplotype distributions is provided in Table 1.*

Table 1 *. Geographical origin of the barbel samples and distribution and absolute sample frequencies of the 11 mitochondrial DNA haplotypes.*

					Coordinates		
No.	River	Country*	Drainage	*n*	Latitude	Longitude	Haplotype counts
1	Kamchyia	BG	Kamchyia	3	43° 13′ N	27° 55′ E	h4(3)
2	Dvojnica	BG	Dvojnica	7	42° 49′ N	27° 53′ E	h4(7)
3	Kirmir†	TY	Sakarya	3	40° 10′ N	31° 57′ E	h4(3)
4	Karasu†	TY	Sakarya	1	40° 05′ N	29° 31′ E	h4(1)
5	Dniester	MO	Dniester	5	48° 09′ N	28° 17′ E	h1(4) h2(1)
6	Danube	BG	Danube	2	44° 03′ N	26° 37′ E	h1(1) h5(1)
7	Ialomiţa	RO	Danube	1	44° 56′ N	26° 02′ E	h5(1)
8	Putna	RO	Danube	1	45° 41′ N	27° 11′ E	h5(1)
9	Buzău	RO	Danube	1	45° 09′ N	26° 49′ E	h5(1)
10	Lăpuş	RO	Danube	3	47° 40′ N	23° 35′ E	h5(3)
11	Iza	RO	Danube	1	47° 44′ N	24° 22′ E	h9(1)
12	Mureş	RO	Danube	3	46° 33′ N	24° 33′ E	h5(3)
13	Archar	BG	Danube	3	43° 49′ N	22° 57′ E	h5(3)
14	Danube	YU	Danube	2	44° 20′ N	22° 36′ E	h5(1) h6(1)
15	Danube	H	Danube	5	45° 59′ N	18° 42′ E	h5(4) h8(1)
16	Laborec	SK	Danube	2	49° 01′ N	21° 58′ E	h1(1) h5(1)
17	Torysa	SK	Danube	3	49° 06′ N	21° 06′ E	h1(3)
18	Váh	SK	Danube	2	49° 05′ N	18° 56′ E	h5(2)
19	Dyje	CZ	Danube	2	48° 45′ N	17° 01′ E	h5(1) h7(1)
20	Dyje	CZ	Danube	4	48° 45′ N	16° 16′ E	h3(1) h5(3)
21	Rabnitz	A	Danube	2	47° 29′ N	16° 38′ E	h5(2)
22	Poprad	SK	Vistula	5	49° 03′ N	20° 18′ E	h1(1) h5(4)
23	San	PL	Vistula	5	49° 34′ N	22° 13′ E	h1(2) h5(3)
24	Sázava	CZ	Elbe	1	49° 34′ N	15° 53′ E	h5(1)
25	Elbe	CZ	Elbe	2	50° 40′ N	14° 02′ E	h5(2)
26	Berounka	CZ	Elbe	3	50° 00′ N	13° 06′ E	h5(1) h10(2)
27	Eder	D	Weser	3	51° 09′ N	08° 55′ E	h1(1) h10(1) h11(1)
28	Rhine	D	Rhine	3	50° 44′ N	07° 05′ E	h1(2) h5(1)
29	Rhine	D	Rhine	3	47° 59′ N	07° 51′ E	h1(3)
30	Rhône	F	Rhône	5	45° 45′ N	04° 51′ E	h1(5)
31	Drôme†	F	Rhône	4	44° 56′ N	05° 54′ E	h1(4)
32	Loire	F	Loire	3	45° 12′ N	04° 02′ E	h1(3)
33	Lergue	F	Hérault	2	43° 44′ N	03° 20′ E	h1(2)

* BG, Bulgaria; TY, Turkey; MO, Republic of Moldova; RO, Romania; YU, Yugoslavia; H, Hungary; SK, Slovakia; CZ, Czech Republic; A, Austria; PL, Poland; D, Germany; F, France.
† Data from Tsigenopoulos & Berrebi (2000).

Results

The nested clade design of the barbel data consisted of three levels (Figure 2). The null hypothesis of no association between the position of haplotypes or clades in the tree with the geographical location was rejected for at least one nesting clade at each level (Table 2). Application of the inference key indicated that gene flow restricted by isolation by distance best explained the geographical distribution of haplotypes nested within clade1-3 and found in the Danube River basin. A pattern implying contiguous population range expansion was observed among clades nested within clade 2-2. The large Dn value involved in the significant I-T Dn contrast and characteristic of a range expansion (Templeton *et al.* 1995) was observed for the nested clade 1-4 confined to the Elbe and Weser river basins. A signal of past fragmentation best explained the distribution of the three clades at the highest nesting level (Table 2).

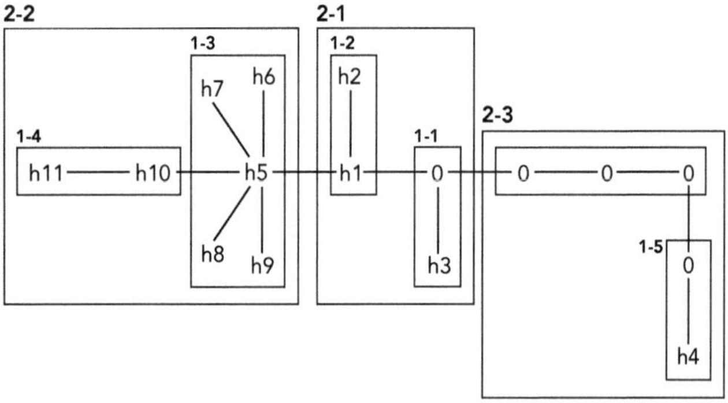

Figure 2. *Nested clade design in the minimum spanning tree for the barbel mitochondrial DNA haplotypes. Rectangles indicate one-step (1-1 to 1-5) and two-step (2-1 to 2-3) clades. Zeros indicate hypothetical internal haplotypes that were not observed.*

Discussion

Application of the NCPA to barbel phylogeographical data yielded inference of three different evolutionary processes, with a clear temporal juxtaposition, as the major determinants of population structure and divergence in this species.

Isolation by distance

At the lowest nesting level, tip haplotypes within the Danubian clade 1-3 had restricted geographical ranges (quantified by significantly large I-T Dc), which were nested within the significantly large range of the haplotype immediately interior to them. The NCPA inferred a process of gene flow restricted through isolation by distance from this pattern (Table 2). This may seem somehow

Table 2. *Nested clade analysis of barbel mitochondrial DNA phylogeographical data. Clade (Dc) and nesting clade (Dn) distances are shown for each nesting level following the haplotype or clade number (No.); the superscripts refer to significantly (at the 0.05 level) small (S) or large (L) values. The average difference between interior (bold) and tip nested clades is shown for both distance measures (I-T). Inference chain that refers to the key of Templeton (2004), followed by the biological inference, is given at the bottom of a nested set of clades among which one or more significant distance measures were detected. Inferences are shown from most recent (left) to oldest (right).*

Haplotypes			One-step clades			Two-step clades		
No.	Dc	Dn	No.	Dc	Dn	No.	Dc	Dn
h3	-	-	**1-1**	0	327^S			
h1	694	699						
h2	0	1184						
I-T	694	−484	1-2	709	710			
			I-T	−709	-383^S			
			1-2-11-17 No: IN		**2-1**	701	831^L	
h5	393^L	393^L						
h6	0	358						
h7	0	328						
h8	0	223						
h9	0	266						
I-T	393^L	100						
1-2-3-4 No: IBD			**1-3**	381^S	384^S			
h10	143	161						
h11	0	160						
I-T	143	1	1-4	161	747^L			
			I-T	221	-363^S			
			1-2-11-12 No: RE	2-2	410^S	435^S		
h4	-	-	1-5	-	-	2-3	209^S	1000^L
					I-T	337^L	266^L	
					1-2-3-5-15 No: RF			

IN, inconclusive outcome; IBD, gene flow restricted through isolation by distance; RE, range expansion; RF, past range fragmentation.

counterintuitive because of the star-like structure of the clade 1-3 (Figure 2) (Paulo *et al.* 2002), which is an expected consequence of a population growth (Hudson 1990; Harpending *et al.* 1998; Galtier *et al.* 2000). Although the Danube River basin could support the barbel population through the last glaciation (Bănărescu 1991), it seems unlikely that the effective population size has been maintained stable (Dynesius & Jansson 2000; Hewitt 2000). The NCPA interpretation of the pattern within this population as reflecting isolation by distance, which entails contemporary gene flow in an equilibrium population (Wright 1943), should be therefore viewed with caution. A recently expanded, non-equilibrium population may show localized distribution of new haplotypes, and can converge to the NCPA inference of isolation by distance. Because the recent history determined population structure in many temperate species (Dynesius & Jansson 2000; Hewitt 2000), causing them not to have had sufficient time to reach equilibrium (Whitlock & McCauley 1999; Schaal & Olsen 2000), the isolation by distance may actually be less conceivable in natural populations than suggested by the NCPA.

Range expansion

The contiguous range expansion evidenced by the significant signal within the nesting clade 2-2 supported the dispersion of the barbel from the Danubian refuge throughout the Elbe and Weser river basins, which was suggested on the basis of the phylogeographical analysis to have followed the retreat of ice and warming at the end of the last (Würm) glacial (Kotlík & Berrebi 2001).

Other two previously suggested range expansions have not been detected by the NCPA, however. All barbels sampled from France carried the same haplotype (h1) nested within the clade 1-2, and all fish from the Balkans and Anatolia had the single haplotype (h4) of the clade 1-5 (and 2-3) (Figure 1; Table 1). We explained this phylogeographical pattern by temporary reductions in effective population sizes within these two regions, followed by post-Pleistocene range expansions (Kotlík & Berrebi 2001). As an apparent consequence of the insufficient genetic variation (Templeton 1998), the NCPA failed to detect any significant pattern within the clades 1-2 and 1-5 (Table 2). This limitation of inferences to patterns based upon sufficient sampling and genetic resolution to have a significant phylogeographical signal is indeed considered the major advantage of the NCPA over the traditional phylogeography (Templeton 1998, 2002).

Paulo *et al.* (2002) concluded that the NCPA is unable to detect range expansion once an ancestral haplotype became fixed due to an accompanying founder effect. However, a characteristic pattern is expected to emerge as new tip clades are created in the new geographical areas through the mutation process. Such tip clades would typically show restricted geographical ranges located distantly from the geographical centers of their ancestral haplotypes, causing high Dn values for the tip clades (Templeton 1998, 2002). This is what apparently happened in the case of the barbel populations from the Elbe and Weser river basins. The Dn value for the tip clade 1-4 confined to these rivers (Figure 1) was significantly large, as was the I-T Dn contrast for the entire nesting clade 2-2, which provided the inference key with a signal for range expansion (Table 2).

Thus, although a range expansion may eradicate a genetic diversity and reduce the phylogeographical signal, the NCPA makes a clear prediction about the pattern expected to arise in the new geographical area (Templeton 1998, 2002). This makes it feasible to detect range expansion even after a severe founder effect. It seems reasonable to assume that an increased individual and/or nucleotide sampling could reveal new tip clades within many newly colonized areas, and provide the signal used by the NCPA to infer range expansion.

Past range fragmentation

The NCPA recovered the process of past fragmentation into the Balkan/Anatolian (nesting clade 2-3) and central European (nesting clades 2-1 and 2-2) phylogeographical groupings, inferred by the phylogeographical analysis as lineages I and II (Kotlík & Berrebi 2001). This separation coincides with a long branch (Figure 2) suggesting a relatively ancient fragmentation event (Templeton 1998), which supports the interpretation of refugial isolation of the lineages through several later glacial cycles (Kotlík & Berrebi 2001). In addition to the clade 2-3, a significantly small Dc value characteristic of a geographical isolate (Templeton *et al.* 1995) was observed for the other tip clade at the highest nesting level, the central European clade 2-2 (Table 2). This pattern, along with the largely non-overlapping geographical ranges of the clades 2-1 and 2-2 (Figure 1), supported the range fragmentation within the lineage II suggested by Kotlík & Berrebi (2001). The partly exclusive geographical distributions of the two clades and the short branch connecting them imply only a recent fragmentation (Templeton 1998), probably during the last glacial, followed by incomplete lineage sorting within the descendant populations and/or a secondary contact (Kotlík & Berrebi 2001).

Conclusions

Application of the NCPA to the barbel phylogeographical data did not unambiguously detect new historical events compared to what has been inferred thought the visual examination of the overlay of the gene tree upon geography. It did, however, provide support to the major events and processes (though not all) integrated in the historical biogeography hypothesis previously proposed for this species, by which this scenario gained significant confidence.

Acknowledgements

P. K. was an invited scientist at the Paris Muséum National d'Histoire Naturelle, and was supported by the French 'Reseau formation recherche du Ministère de la Recherche' managed by R. Billard, and by the Academy of Sciences of the Czech Republic (grant numbers KJB600450507 and IRP IAPG No. AVOZ50450515).

References

Avise JC (2000) ***Phylogeography: The History and Formation of Species.*** Harvard University Press, Cambridge.

Avise JC, Arnold J, Ball RM, Birmingham E, Lamb T, Neigel JE, Reeb CA, Saunders NC (1987) Intraspecific phylogeography: the mitochondrial DNA bridge between population genetics and systematics. ***Annual Review of Ecology and Systematics*** **18**, 489-522.

Bănărescu P (1991) ***Zoogeography of Fresh Waters: Distribution and Dispersal of Freshwater Animals in North America and Eurasia.*** vol. 2. AULA-Verlag, Wiesbaden.

Bernatchez L (2001) The evolutionary history of brown trout (*Salmo trutta*) inferred from phylogeographic, nested clade, and mismatch analyses of mitochondrial DNA variation. ***Evolution*** **55**, 351-379.

Branco M, Monnerot M, Ferrand N, Templeton AR (2002) Postglacial dispersal of the European rabbit (*Oryctolagus cuniculus*) on the Iberian Peninsula reconstructed from nested clade and mismatch analyses of mitochondrial genetic variation. ***Evolution*** **56**, 792-803.

Clement M, Posada D, Crandall KA (2000) TCS: a computer program to estimate gene genealogies. ***Molecular Ecology*** **9**, 1657-1660.

Durand JD, Ünlü E, Doadrio I, Pipoyan S, Templeton AR (2000) Origin, radiation, dispersion and allopatric hybridization in the chub *Leuciscus cephalus*. ***Proceedings of the Royal Society of London, B*** **267**, 1687-1697.

Dynesius M, Jansson R (2000) Evolutionary consequences of changes in species' geographical distributions driven by Milankovitch climate oscillations. ***Proceedings of the National Academy of Sciences, USA*** **97**, 9115-9120.

Galtier N, Depaulis F, Barton NH (2000) Detecting bottlenecks and selective sweeps from DNA sequence polymorphism. ***Genetics*** **155**, 981-987.

Harpending HC, Barzer MA, Gurven M, Jorde LB, Rogers AR, Sherry ST (1998) Genetic traces of ancient demography. ***Proceedings of the National Academy of Sciences, USA*** **95**, 1961-1967.

Hewitt GM (2000) The genetic legacy of the Quaternary ice ages. ***Nature*** **405**, 907-913.

Hudson RR (1990) Gene genealogies and the coalescent process. ***Oxford Surveys in Evolutionary Biology*** **7**, 1-44.

Knowles LL (2004) The burgeoning field of statistical phylogeography. ***Journal of Evolutionary Biology*** **17**, 1-10.

Knowles LL, Maddison WP (2002) Statistical phylogeography. ***Molecular Ecology*** **11**, 2623-2635.

Kotlík P, Berrebi P (2001) Phylogeography of the barbel (*Barbus barbus*) assessed by mitochondrial DNA variation. ***Molecular Ecology*** **10**, 2177-2186.

Paulo OS, Jordan WC, Bruford MW, Nichols RA (2002) Using nested clade analysis to assess the history of colonization and the persistence of populations of an Iberian Lizard. ***Molecular Ecology*** **11**, 809-819.

Pfenninger M, Posada D, Magnin F (2003) Evidence for survival of Pleistocene climatic changes in northern refugia by the land snail *Trochoidea geyeri* (Soos 1926) (Helicellinae, Stylommatophora). ***BMC Evolutionary Biology*** **3**, art. no.-8.

Posada D, Crandal KA, Templeton AR (2000) GeoDis: a program for the cladistic nested analysis of the geographical distribution of genetic haplotypes. ***Molecular Ecology*** **9**, 487-488.

Schaal BA, Olsen KM (2000) Gene genealogies and population variation in plants. ***Proceedings of the National Academy of Sciences, USA*** **97**, 7024-7029.

Seddon JM, Santucci F, Reeve NJ, Hewitt GM (2001) Dna footprints of European hedgehogs, *Erinaceus europaeus* and *E. concolor*: Pleistocene refugia, postglacial expansion and colonization routes. ***Molecular Ecology*** **10**, 2187-2198.

Templeton AR (1998) Nested clade analyses of phylogeographic data: testing hypotheses about gene flow and population history. ***Molecular Ecology*** **7**, 381-397.

Templeton AR (2002) Out of Africa again and again. ***Nature*** **416**, 45-51.

Templeton AR (2004) Statistical phylogeography: methods of evaluating and minimizing inference errors. ***Molecular Ecology*** **13**, 789-809.

Templeton AR, Sing CF (1993) A cladistic analysis of phenotypic association with haplotypes inferred from restriction endonuclease mapping. IV. Nested analyses with cladogram uncertainty and recombination. ***Genetics*** **134**, 659-669.

Templeton AR, Boerwinkle E, Sing CF (1987) A cladistic analysis of phenotypic association with haplotypes inferred from restriction endonuclease mapping. I. Basic theory and an analysis of alcohol dehydrogenase activity in *Drosophila*. ***Genetics*** **117**, 343-351.

Templeton AR, Crandall KA, Sing CF (1992) A cladistic analysis of phenotypic association with haplotypes inferred from restriction endonuclease mapping and DNA sequence data. III. Cladogram estimation. ***Genetics*** **132**, 619-633.

Templeton AR, Routman E, Phillips CA (1995) Separating population structure from population history: a cladistic analysis of the geographic distribution of mitochondrial DNA haplotypes in the tiger salamander, *Ambystoma tigrinum*. ***Genetics*** **140**, 767-782.

Tsigenopoulos CS, Berrebi P (2000) Molecular phylogeny of North Mediterranean freshwater barbs (genus *Barbus*: Cyprinidae) inferred from cytochrome *b* sequences: biogeographic and systematic implications. ***Molecular Phylogenetics and Evolution*** **14**, 165-179.

Turner TF, Trexler JC, Harris JL, Haynes JL (2000) Nested cladistic analysis indicates population fragmentation shapes genetic diversity in a freshwater mussel. ***Genetics*** **154**, 777-785.

Whitlock MC, McCauley DE (1999) Indirect measures of gene flow and migration: $F_{ST} \neq 1/(4Nm+1)$. ***Heredity*** **82**, 117-125.

Wright S (1943) Isolation by distance. ***Genetics*** **28**, 114-138.

Chapter 13

Cross-section of a refugium: genetic diversity of amphibian and reptile populations in the Balkans

Jelka Crnobrnja-Isailovic
E-mail: jelka@ibiss.bg.ac.yu

Abstract
The Balkan Peninsula has been a 'transitional area' for the historical migrations of amphibian and reptile species from east to west, north to south and *vice versa*. As a result of dynamic paleogeographic and climatic history, central parts of the Balkans today harbor representatives of Mediterranean, middle European, Boreal and Steppe herpetofauna. According to current species lists, a high level of responsibility for the conservation of European herpetofauna is proposed for all Balkan countries (Croatia, Bosnia, Serbia, Montenegro, Albania, Macedonia, Romania, Bulgaria, Greece and Turkey-European part). This study evaluates the genetic diversity (the core of biodiversity evaluation) of the Balkan herpetofauna, based on literature review as well as the author's own results of various molecular studies made on representative European herpetofauna.

Keywords: herpetofauna, Balkan Peninsula, refugium, genetic diversity

Introduction

The conservation of biodiversity has become a crisis issue. The recent rate of taxa loss is almost 100-fold greater than during historical mass extinctions. Attempts to conserve and repair the world's biodiversity follow two main approaches: actions focused toward restoring populations of seriously endangered taxa, and attempts to diminish human impact primarily in those parts of the globe recognized as 'centers of biodiversity'. The most popular and least expensive approach to recognizing areas of conservation priority is to estimate species richness ('alpha diversity' by Whittaker 1972). This is not a bad 'first step', as the biodiversity of a given area represents the diversity of species living there, the genetic diversity harbored in their populations, and the diversity of communities and ecosystems these species inhabit. The number of species, however, is not a very powerful measure if we agree that "*biodiversity is genetic diversity*" (Avise & Hamrick 1995). The degree of population genetic structuring within species together with the di-

S. Weiss and N. Ferrand (eds.), Phylogeography of Southern European Refugia, 327–337.

versity of habitats could be a more comprehensive measure of local biodiversity and a more appropriate measure of the conservation priority of a particular region.

Why is genetic diversity so important? Survival of any population is related, among the other factors, to its ability to produce as many genetic combinations as possible. Species with homogeneous population genetic structure (low genetic diversity) are, at least theoretically, more susceptible to extinction than species with apparent population genetic structuring. Conservation priorities below the species level focus on genetically divergent populations, often characterized by different alleles, as well as allelic frequencies. The important question is, do genetically diverse populations – those characterized with high genetic potential (high heterozygosity, high allelic richness and many polymorphic loci) – also have priority in conservation actions?

Phylogeographic studies confirm that peninsulas of southern Europe (Iberian, Apennine and Balkan) were refugia of genetic diversity during the Pleistocene (Hewitt 1996, 1999). Recently Kryçtufek & Reed (2004) provided a review of the biological diversity in the Balkan Peninsula. To what extent is the Balkan Peninsula unique and important from a conservation perspective? This chapter attempts to answer the question by reviewing the available molecular genetic data for amphibians and reptiles.

1. Herpetofauna species richness in southern Europe

What can the distribution of amphibian and reptile species tell us about the biological richness and complexity of a particular region? Reptiles and amphibians are poor dispersers, and many express homing behavior. They are assigned as indicators of ecosystem disturbance – directly because of their sensitivity to environmental changes and certain pollutants, and indirectly, as intermediate constituents of food chains in natural communities. According to Gasc *et al.* (1997), the percent of European amphibian (90%) and reptile (70%) species (not including island endemics and sea turtles) inhabiting southern peninsulas (islands omitted from calculation) is extraordinarily high (Table 1). The Iberian, Apennine and Balkan peninsulas of Europe provide suitable habitats for 36% of the amphibian and 28% of the reptile species distributed throughout the continent. The very low number of species shared between them (Table 1) represents the uniqueness and importance of each of the three peninsular refugia.

Some concordance exists in the composition of vascular plant taxa in southern Europe: Junikka & Uotila (2002) point to the abundance of Iberian vascular plant taxa in western Europe, which shows strong affinity with the Atlantic Province and weaker affinity with the central European and Mediterranean Provinces. By contrast, the Apennine Peninsula has a clear floristic affinity with the central European Province, but less so with the Atlantic Province. The Iberian and Apennine Peninsulas are very similar in species number, but the Iberian Peninsula has much more of its 'own taxa' than the Apennine Peninsula. Floristically the Balkan and the Apennine peninsulas have a closer affinity with each other than either has with the Iberian Peninsula.

Table 1. *Degree of endemism in European amphibians and reptiles. Endemic island species and sea turtles were omitted from counting. Data used from Gasc* et al. *1997.*

Area	Asia Minor	Balkans
Iberian Peninsula only	8	12
Apennine Peninsula only	8	1
Balkan Peninsula only	4	15
Continental Europe only	5	29
Iberian - Apennine	0	0
Iberian - Balkan	0	1
Iberian - continental Europe only	7	7
Apennine - Balkan	3	2
Apennine - continental Europe	3	1
Balkan - continental Europe	7	13
Iberian - Apennine - Balkan	0	0
Iberian - Apennine - continental Europe	4	4
Iberian - Balkan - continental Europe	0	0
Apennine - Balkan - continental Europe	7	8
Iberian - Apennine - Balkan - continental Europe	8	16
Sum	64	109
Total (including islands and excluding sea turtles)	79	125

2. Balkan Peninsula

The Balkan Peninsula in southeastern Europe is bordered on the east with the Black, Marmara and Aegean seas, on the south with the Mediterranean Sea, and on the west with the Adriatic and Ionian seas. The northern boundary is not easily recognized. It may be defined with a line drawn arbitrarily from the Adriatic near Rijeka, in Croatia to the upper Sava River, and then along the Sava River to its confluence with the Danube River, at Belgrade in Serbia, and along the Danube until it reaches the Black Sea.

Most of the Balkan Peninsula is mountainous. Lowlands are localized along the lower reaches of rivers. Rivers are grouped into three major catchments, draining into the Black, Adriatic and Aegean seas. The mountain ranges include so-called Outer and Inner Dinaric Alps distributed parallel to the Adriatic coast, the old Rhodope chain between Macedonia and the Maritsa valley, the Sara-Pindus range in northern Greece, eastern Albania and western Macedonia, the Balkan Mountains distributed mostly in Bulgaria, part of the Carpathian Mountain range in eastern Serbia and isolated summits, including Mounts Olympus, Pelion, and Óssa in Greece.

The richness of physiogeographic features together with its specific topographic position in Europe influences the biogeographic diversity of the Balkan Peninsula. According to Matvejev & Puncer (1989), most of the Balkans (territory of former Yugoslavia) comprise seven landscape types (biomes). These are: 1) evergreen Mediterranean maritime woodlands and maquis; 2) sub-Mediterranean Adriatic mostly oak woodlands; 3) South European, mostly deciduous woodlands; 4) European, mostly coniferous boreal type woodlands; 5) Alpine and high nordic rock-grounds, pastures and snow patches; 6) biomes of steppes and woodland steppe and 7) rocky substrates, pastures and woodlands on rocky grounds of Mediterranean mountains. For comparison – there are only three such landscape types in northern and central Europe. East-southernmost parts of former Yugoslavia also contain elements of Irano-Turanian semideserts and elements of the tropical and subtropical biomes.

Table 2. *Amphibian and reptile species inhabiting territory of the Balkan Peninsula (endemic island species and sea turtles are omitted from counting). Data from Gasc* et al. *(1997).*

Countries	Amphibian species	Reptile species
Croatia (south of the Sava River)	20	32
Bosnia & Herzegovina	20	33
Serbia (south of the Sava and Danube rivers)	21	22
Romania (Dobrogean region)	12	20
Bulgaria	17	33
Montenegro	15	32
Albania	19	35
Former Yugoslav Republic of Macedonia	14	32
Greece (without islands)	17	46
Turkey (European part)	15	32

3. Herpetofauna of the Balkans and its genetic diversity

According to Gasc *et al.* (1997), 29 amphibian and 54 reptilian species are distributed throughout the Balkan Peninsula. Their degree of endemism is roughly presented in Table 2. All Balkan countries (most of Croatia, Bosnia and Herzegovina, Macedonia, most of Serbia, Montenegro, Albania, Greece, Dobrogean part of Romania in the Danube delta, Bulgaria and European Turkey) have a high level of responsibility in the conservation of European herpetofauna, as seen from the apparent species diversity (Astudillo & Arano 1995).

Table 3. *Haplotype diversity of amphibian and reptile species present in the Balkans.*

Taxon	Number of haplotypes/ population			Number of unique haplotypes		
	The Balkans	WE	EE	The Balkans	The Balkans & WE	The Balkans & EE
Salamandra atra (Riberon *et al.* 2001)	1	1	/	1	0	/
Triturus cristatus (Wallis & Arntzen 1989; Arntzen & Wallis 1999)	1	1	1.2	1	0	0
Triturus carnifex (Wallis & Arntzen 1989; Arntzen & Wallis 1999)	1	1.5	/	1	0	/
Triturus dobrogicus (Wallis & Arntzen 1989; Arntzen & Wallis 1999)	1	1	1	0	0	0
Bombina bombina (Szymura *et al.* 2000)	1*	1	1	No data	1	0
Bombina variegata (Szymura *et al.* 2000)	1.4	1.3	1	1	1	0
Rana temporaria (Pidancier *et al.* 2001)	Mixed WE & EE haplotypes	No data	No data	No data	No data	No data
Lacerta vivipara-viviparous (Surget-Groba *et al.* 2001)	1.2	1.1	1.1	5	0	1
Natrix tessellata (Guicking *et al.* in print)	Mixed haplotypes	No data	No data	No data	No data	No data

WE = Western Europe; EE = Eastern Europe.
* Population from Adapazari (Turkey) counted as Balkan.

a) Haplotype diversity

Phylogeographic research on European herpetofauna can be categorized into those studies presenting haplotype diversity in species absent from the Balkan Peninsula and those analyzing species whose distribution area comprise the Balkans (see Table 3 & 4). However, data from this area are still scarce and the present evaluation of genetic diversity must be considered preliminary. The most fundamental result obtained from these studies is that the Balkan Peninsula undoubtedly was an important refugium for European herpetofauna during the Pleistocene Ice Ages. Presently, for most of the species analyzed, Balkan populations seem to be genetically closer to western rather than eastern European conspecifics, suggesting the primary direction of recolonization routes (Szymura *et al.* 2000; Pidancier *et al.* 2001; Surget-Groba *et al.* 2001; Ursenbacher *et al.* 2001; see Table 4). Concerning haplotype diversity, Balkan populations in general are characterized by both unique haplotypes and a mixture of major 'western' and 'eastern' lineages (Table 3). It would not be surprising if future studies confirm similar haplotype diversity patterns among species belonging to the same biogeographical entity. In the Balkans, for example, *Salamandra atra*, *Rana temporaria*, *Lacerta vivipara* and *Vipera berus* inhabit biomes of boreal woodlands and/or high Alpine rocky substrates.

Apart from differences in the size of their distribution areas, their populations probably followed the same migration routes during Pleistocene events. Nagy *et al.* (2004) offered the same explanation for similar spatial patterns of genetic differentiation in Mediterranean species *Coluber* (*Hierophis*) *viridiflavus* (Nagy *et al.* 2004) and *Emys orbicularis* (Lenk *et al.* 1999) on the Appenine Peninsula. As they note: "*...common zoogeographical patterns can be traced back to the glacial zoogeographic history in all likelihood.*"

Table 4*. Genetic differentiation of amphibian and reptile species present in the Balkans.*

Taxon	Molecular markers	Number of unique haplotypes		
		The Balkans	The Balkans *vs.* WE	The Balkans *vs.* EE
Salamandra atra (Riberon *et al.* 2001)	mt cyt b 1050bp	0.6*	1.1-3	/
Triturus cristatus (Wallis & Arntzen 1989; Arntzen & Wallis 1999)	mt DNA various fragments	No data	0.15 (0.01)	No data
Triturus dobrogicus (Wallis & Arntzen 1989; Arntzen & Wallis 1999)	mt DNA various fragments	0.0	0.0	0.0
Bombina bombina (Szymura *et al.* 2000)	mt DNA various fragments	No data	0.6 (0.5)**	0.0-0.5 (0.5)**
Bombina variegata (Szymura *et al.* 2000) various fragments	mt DNA	0.3(0.4)	0.3(0.4)	4.7(1.2)-5.2(1.3)
Rana temporaria (Pidancier & Miaud, pers. comm.)	mt cyt b 375 bp	0.27	1.2-1.5	2.5-3
Lacerta vivipara -viviparous (Surget-Groba *et al.* 2001)	mt cyt b 406bp mt Glu-tRNA 23bp	No data	No data	No data
Natrix tessellata (Guicking *et al.* in print)	mt cyt b complete seq.	0.27-8.1	0.8-8.1	0.8-8.1
Vipera berus (Ursenbacher, pers. comm.)	mt cyt b 1083bp mt CR 970 bp	0.19-4.6	No data	No data

WE = Western Europe; EE = Eastern Europe.
Where possible, standard deviations are presented in parentheses.
* Population from Slovenia counted as Balkan.
** Population from Adapazari (Turkey) counted as Balkan.

b) Allozyme diversity

Comparable allozyme studies, according to Widmer & Lexer's (2001) criteria (heterozygosity, gene diversity, proportion of polymorphic loci, allelic richness, etc.) refer to the genus *Triturus* (Table 5), and to a lesser extent, some European lacertids (Mayer 1981; Crnobrnja-Isailovic *et al.* 1995). Widmer & Lexer (2001) reported measures of genetic diversity in a manner similar to

Leberg (1992). They proposed that the average number of alleles per locus (n_a) is highly dependent on effective population size and is more useful for identifying historical processes such as bottlenecks and population admixture than average heterozygosity. The data presented in Table5 point to greater allelic richness in Balkan populations of the European genus *Triturus*. If this comparison is reliable (electrophoretic results from different laboratories may not be directly comparable; for details see Widmer & Lexer 2001) one can assign them priority status as natural areas of gene pool conservation.

***Table 5**. Allozymic diversity in species of genus* Triturus *present in Balkan Peninsula.*

Species	Region	Sub region	N	N_l	N_a	H	P
Triturus alpestris	WE	Great Britain, Italy, Spain (Frelow *et al.*, unpublished)	3	22	1.00-1.27	/	0.00-0.23
	EE	Poland (Rafinski & Arntzen 1987)	1	19	1.53	0.18	0.47
	BP	Central Balkans (Kalezic & Hedgecock 1979)	2	18	1.54-1.70	0.10-0.17	0.46-0.52
		Greece (Sotiropoulos *et al.* 2001)	11	18	1.60(avg)	0.11-0.32	0.53(avg)
Triturus vulgaris	WE	Great Britain, Italy (Frelow *et al.*, unpublished)	3	22	1.41-1.50	/	0.36
	EE	Poland (Rafinski & Arntzen 1987)	1	19	1.32	0.08	0.21
	BP	Central Balkans (Kalezic & Hedgecock 1979)	2	16.5	1.43-1.53	0.07-0.08	0.32-0.34
		Central Balkans (Kalezic 1983, 1984)	9	20.6	1.29-1.59	0.06-0.13	0.22-0.46
			5	21.5	1.32-1.54	0.04-0.10	0.23-0.36
			4	22	1.14-1.23	0.02-0.04	0.09-0.18
			3	22	1.14-1.32	0.03-0.06	0.09-0.23
Triturus cristatus superspecies	WE	Great Britain, Germany, Italy (Frelow *et al.*, unpublished)	6	22	1.18-1.64	/	0.18-0.50
		France (Rafinski & Arntzen 1987)	1	19	1.11	0.01	0.11
	EE	Russia, Romania, Ukraine (Litvinchuk *et al.* 1999)	4	13	1.08-1.62	0.01-0.17	0.08-0.62
	BP	Central Balkans (Kalezic & Hedgecock 1979)	1	21	1.24	0.04	0.24
		Central Balkans (Crnobrnja & Kalezic 1990)	7	18.4	1.25-1.72	0.05-0.11	0.20-0.58
		Central Balkans (Litvinchuk *et al.* 1999)	4	13	1.23-1.69	0.07-0.12	0.23-0.46

WE = Western Europe; EE = Eastern Europe; BP = Balkan Peninsula; N = number of samples;
N_l = number of analyzed loci; N_a = average number of alleles per locus. H = average heterozigosity/locus;
P = proportion of polymorphic loci/population; avg = average value.

4. The Balkans as a refugium of genetic diversity

The central Balkans today are inhabited by a mixed herpetofauna with Mediterranean, middle European, Boreal and Steppe representatives (Radovanovic 1951; Crnobrnja-Isailovic & Dzukic 1995; Dzukic 1995; Crnobrnja-Isailovic & Aleksic 1999; Ajtic & Tomovic 2001), supporting the existence of a transitional area involving historical migrations of amphibian and reptile species from east to west, north to south and *vice versa*. The paleogeographic events in the Balkans were dynamic and complex even before the Pleistocene: changes of relief, loss of ancient corridors and establishment of new ones were coupled with climatic and vegetational changes. The specific geographical position of the Balkans in Europe (vicinity of Caspian sea region; existence of land-bridge connections to Asia Minor during the past; closeness to the Apennine Peninsula, etc.) compared to the other two southern peninsular refugia greatly influences the diversity of taxa present today. Again, there is concordance with the diversity patterns of vascular plants: of the three South European peninsulas the Balkan Peninsula is the richest in vascular plant taxa, but it also covers the most extensive area. The Balkan Peninsula is floristically the most diverse in this sample and the distribution of taxa over Europe is extensive not only in central European and Boreal Provinces, but also in the Apennine Peninsula (Junikka & Uotila 2002).

The genetic potential of Balkan populations of animal and plant taxa has not yet been appropriately evaluated. Concerning herpetofauna, some authors argue that the central part of the Balkans apparently harbors a number of subspecies (Dzukic 1995). However, for many of them genetic confirmation is needed in order to avoid misinterpretation based on phenotypic plasticity (Futuyma 1998). An alpha-diversity approach modified to count the number of subspecies is a somewhat naive attempt to measure genetic diversity of a certain area, if subspecies recognition is done traditionally (i.e. without genetic analyses). For example, an electrophoretic study of *Triturus alpestris* populations by Arano & Arntzen 1987 reinforced the hypothesis of a Balkan origin of this species, but annulled previous subspecific status of several neotenic Balkan populations. Phenotypically very different populations of *Podarcis muralis* from the Lake Skadar region (Montenegro) showed only high allelic richness together with a high observed level of heterozygosity and no genetic divergence from a population near Belgrade, some 500 km to the north (Crnobrnja-Isailovic *et al.* 1995). Low nucleotide divergence between isolated Balkan populations of boreal and high-Alpine species in comparison to some other European regions shows recent origin of their insular population structure, established after Pleistocene glaciations. Future phylogeographical studies of European amphibian and reptile species should include more samples from different parts of the Balkans. It is important to evaluate how much genetic divergence really exists between subsets of Balkan populations for every species. Careful and more detailed sampling is undoubtedly needed for appropriate evaluation of extant biodiversity in the Balkans as for any region characterized by complex physiogeographic features.

Acknowledgements

I am grateful to N. Pidancier, C. Miaud and S. Ursenbacher for sharing their unpublished results with me, also to D. Guicking for providing me with an *in print* version of her manuscript and last but not the least to M. Carretero, S. Weiss, ICETA/CECA and WUS Austria for encouragement and logistic support.

References

Ajtic R, Tomovic LJ (2001) First record of Kotschy's gecko *Cyrtodactylus kotschyi* (Steindachner 1870) (Gekkonidae, Lacertilia) in FR Yugoslavia. ***Archives of Biological Sciences Belgrade* 53**, 23P-24P.

Arano B, Arntzen JW (1987) Genetic differentiation in the alpine newt, *Triturus alpestris*. In: ***Proceedings of the Fourth OGM SEH Nijmegen 1987*** (Faculty of Sciences) (eds. van Gelder JJ, Strijbosh H, Bergers PJM), pp. 21-24. SEH, Nijmegen.

Arntzen JW, Wallis GP (1999) Geographic variation and taxonomy of crested newts (*Triturus cristatus* superspecies): morphological and mitochondrial DNA data. ***Contributions to Zoology* 68**, 181-203.

Astudillo G, Arano B (1995) Europa y su herpetofauna: responsabilidades de cada país en lo referente a su conservación. ***Boletín Asociación Herpetológica Española* 6**, 14-45.

Avise JC, Hamrick JL (eds.) (1995) ***Conservation Genetics: Case Histories from Nature.*** Chapman & Hall, New York.

Crnobrnja J, Kalezic ML (1990) The organization of the genetic variation in the crested newt (*Triturus cristatus* complex) from Yugoslavia. ***Genetika* 22**, 133-146.

Crnobrnja-Isailovic J, Aleksic I (1999) First record of Coluber *najadum* in Serbia. ***Archives of Biological Sciences Belgrade* 51**, 41P-42P.

Crnobrnja-Isailovic J, Dzukic G (1995) First report about conservation status of herpetofauna in the Skadar Lake region (Montenegro): current situation and perspectives. In: ***Scientia Herpetologica*** (eds. Llorente G, Montori A, Santos X, Carretero MA), pp. 373-380. SEH, Barcelona.

Crnobrnja-Isailovic J, Dzukic G, Aleksic I, Vujicic L, Avramov S (1995) *Podarcis muralis* and *Lacerta oxycephala* (Reptilia, Lacertidae) on the islands of Skadar Lake: distribution and genetic relationships of populations. In: ***Scientia Herpetologica*** (eds. Llorente G, Montori A, Santos X, Carretero MA), pp. 71-80. SEH, Barcelona.

Dzukic G (1995) Diverzitet vodozemaca (Amphibia) I gmizavaca (Reptilia) Jugoslavije, sa pregledom vrsta od medjunarodnog znacaja. In: ***Biodiverzitet Jugoslavije sa pregledom vrsta od medjunarodnog znacaja*** (eds. Stevanovic V, Vasic V), pp. 447-470. Bioloski fakultet & Ecolibri, Beograd.

Frelow MM, Macgregor H, Wake DB *Genetic differentiation among newts of the genus Triturus in Western Europe*. Unpublished.

Futuyma D (1998) *Evolutionary Biology*. Sinnauer Associates, Inc. Publishers, Sunderland, Massachusetts.

Gasc J-P, Cabela A, Crnobrnja-Isailovic J, Dolmen D, Grossenbacher K, Haffner P, Lescure J, Martens H, Martinez Rica JP, Maurin H, Oliveira ME, Sofianidou TS, Veith M, Zuiderwijk A (eds.) (1997) *Atlas of Amphibians and Reptiles in Europe*. SEH&MNHN (IEGB/SPN), Paris.

Guicking D, Joger U, Wink M (in press) Molecular phylogeography of the viperine snake (*Natrix maura*) and the dice snake (*Natrix tessellata*): First results. In: *Proceedings of the 11th Ordinary General Meeting of Societas Europaea Herpetologica* (ed. Vogrin M). SEH.

Hewitt GM (1996) Some genetic consequences of ice ages, and their role in divergence and speciation. *Biological Journal of the Linnean Society* **58**, 247-276.

Hewitt GM (1999) Post-glacial re-colonization of European biota. *Biological Journal of the Linnean Society* **68**, 87-112.

Junikka L, Uotila P (2002) *Comparison of the vascular flora of Mediterranean peninsulas on the basis of distribution mapping*. OPTIMA Meeting Palermo, Sicily, September 12-23, 2002, Abstracts.

Kalezic ML (1983) Geographical aspects of genetic variability in the smooth newt (*Triturus vulgaris*). *Genetika* **15**, 93-103.

Kalezic ML (1984) Evolutionary divergences in the smooth newt, *Triturus vulgaris* (Urodela, Salamandridae): Electrophoretic evidence. *Amphibia-Reptilia* **5**, 221-230.

Kalezic ML, Hedgecock D (1979) Genetic variation and differentiation of three common european newts (*Triturus*) in Yugoslavia. *British Journal of Herpetology* **6**, 49-57.

Kryçtufek B, Reed M (2004) *Pattern and Process in Balkan Biodiversity – an overview*. In : *Balkan Biodiversity: Pattern and Process in the European Hotspot* (eds. Griffiths HI, Kryçtufek B, Reed JM), pp.203-217. Kluwer Academic Publishers, Dordrecht.

Leberg PL (1992) Effects of population bottlenecks on genetic diversity as measured by allozyme electrophoresis. *Evolution* **46**, 477-494.

Lenk P, Fritz U, Joger U, Wink M (1999) Mitochondrial phylogeography of the European pond turtle, *Emys orbicularis* (Linnaeus 1758). *Molecular Ecology* **8**, 1911-1922.

Litvinchuk SN, Borkin LJ, Dzukic G, Kalezic ML, Khalturin MD, Rosanov JM (1999) Taxonomic status of *Triturus karelinii* on the Balkans, with some comments about other crested newt taxa. *Russian Journal of Herpetology* **6**, 153-163.

Matvejev S, Puncer I (1989) *Map of Biomes-landscapes of Yugoslavia*. Special Issues 36. Natural History Museum, Belgrade.

Mayer W (1981) Elektrophoretische Untersuchungen an europäischen Arten der Gattungen *Lacerta* und *Podarcis*. *Zoologischer Anzeiger Jena* **207**, 151-157.

Nagy ZT, Schmidtler JF, Joger U, Wink M (2004) Systematik der Zwergnattern (Reptilia: Colubridae: Eirenis) und verwandter Gruppen anhand von DNA-Sequenzen und morphologischen Daten. *Salamandra* **39**, 149-168.

Pidancier N, Miaud C, Taberlet P (2001) Phylogeography of the common frog *Rana temporaria* inferred from mitochondrial DNA sequences. 11th OGM SEH, Zalec, Slovenia, July 13-17, 2001, *Biota* **2**, Supplement 101.

Radovanovic M (1951) *Vodozemci i gmizavci nase zemlje*. Naucna knjiga, Beograd.

Rafinski J, Arntzen JW (1987) Biochemical systematics of the old world newts, genus *Triturus*: allozyme data. ***Herpetologica* 43**, 446-457.

Riberon A, Miaud C, Grossenbacher K, Taberlet P (2001) Phylogeography of the Alpine salamander, *Salamandra atra* (Salamandridae) and the influence of the Pleistocene climatic oscillations on population divergence. ***Molecular Ecology* 10**, 2555-2560.

Sotiropoulos K, Polymeni RM, Legakis A (2001) *Triturus alpestris* in Greece: extensive genetic variability and population differentiation. 11th OGM SEH, Zalec, Slovenia, July 13-17, 2001, ***Biota* 2**, Supplement 112.

Surget-Groba Y, Ueulin B, Guillaume C-P, Thorpe RS, Kupriyanova L, Vogrin N, Maslak R, Mazzoti S, Venczel M, Ghira J, Odierna G, Leontyeva O, Monney J-C, Smith N (2001) Intraspecific phylogeography of *Lacerta vivipara* and the evolution of viviparity. ***Molecular Phylogenetics and Evolution* 18**, 449-459.

Szymura JM, Uzzell T, Spolsky B (2000) Mitochondrial DNA variation in the hybridizing fire-bellied toads, *Bombina bombina* and *B. variegata*. ***Molecular Ecology* 9**, 891-899.

Ursenbacher S, Helfer V, Carlsson M, Fumagalli L (2001) Molecular phylogeography of the adder (*Vipera berus*): preliminary results. 11th OGM SEH, Zalec, Slovenia, July 13-17, 2001, summary of poster presentation.

Wallis GP, Arntzen JW (1989) Mitochondrial DNA variation in the crested newt superspecies: limited cytoplasmic gene flow among species. ***Evolution* 43**, 88-104.

Whittaker RH (1972) Evolution and measurement of species diversity. ***Taxon* 21**, 213-251.

Widmer A, Lexer C (2001) Glacial refugia: sanctuaries for allelic richness, but not for gene diversity. ***Trends in Ecology & Evolution* 16**, 267-269.

Part V

Conclusions

Chapter 14

Current perspectives in phylogeography and the significance of South European refugia in the creation and maintenance of European biodiversity

Steven Weiss* and Nuno Ferrand**
E-mail: *steven.weiss@uni-graz.at; **nferrand@mail.icav.up.pt

Abstract
This contribution briefly summarizes the current state of phylogeography and its significance in evolutionary research, particularly involving organisms existing in southern European refugia. These refugia, characterized by a highly heterogeneous landscape, harbor a large percentage of Europe's organismal diversity. They are also shown to offer model systems where the long-term dynamics of particular evolutionary phenomena can be fruitfully explored within well-defined phylogeographic contexts. It is suggested, for example, that the dynamics of gene-flow and natural selection in hybrid zones within long-term glacial refugia are much more complex than those in previously glaciated regions. Emphasis in this review is also given to the breadth of available analytical approaches to phylogeographic analysis, including current controversies and ongoing developments such as the aim to better integrate coalescent theory and both large(geographic) and small scale (landscape) environmental variables. Despite different schools of thought on how to approach phylogeographic analysis, a plea is made to maintain pluralism when dealing with such complex, multi-disciplinary and stochastically influenced data sets. In summary, phylogeography is shown to be a highly successful and popular field of inquiry with a high potential for growth, especially as cutting edge analytical and genomic-oriented techniques, often developed within the field of human genetics, are applied across broader taxonomic scales.

Keywords: conservation, coalescence, nested clade phylogeographic analysis, hybrid zones, statistical phylogeography, landscape genetics

Introduction

The phylogeographic approach has proven its utility in evolutionary research across a broad array of taxa and areas of inquiry. Its greatest contributions thus far have been its emphasis on non-equilibrium aspects of population genetics, the relationship between demography and historical genealogy, and the interfacing of formerly distinct fields of population genetics and phylo-

S. Weiss and N. Ferrand (eds.), Phylogeography of Southern European Refugia, 341–357.

genetics (Avise 2000). The phylogeographic structure of an organism provides the biologist with a broad-scale evolutionary framework, which in itself can answer some straightforward questions, but moreover provides a historical foundation upon which more precise hypotheses can be efficiently formulated and tested. The universality of this perspective is underscored by the observation that nearly all species in nature exhibit some level of genetic structuring associated with geography. This structure can be highly complex, especially for species whose ranges encompass regions subject to high-amplitude paleoclimatic fluctuations. When the 'genealogical' dimension is added to geographic patterns, phylogeographic structure arises. Phylogeographic structure reflects the interaction between both demographic and genealogical processes and landscape level dynamics. The enlightening appeal of phylogeography stems in part from the puzzle-like challenge of uncovering the historical role of geographical processes in shaping current genetic structure. Sea- or lake-level fluctuations, land-bridge formations, tectonic shifts, glacial advance and retreat and river course dynamics all represent biologically independent phenomena that can affect an organism's demography. The coupling of historical events with geographic breaks in an organism's genealogy underscores the success of the phylogeographic perspective.

European phylogeography developed rather quickly with continent-wide studies, which were soon related to a variable model of how southern peninsular refugia contributed to the postglacial colonization of northern Europe (Hewitt 1996, 1999, 2000; Chapter 3). Implicit in this model is that southern refugia harbor higher levels of biodiversity, which serve both as a source for future demographic expansions as well as evolutionary radiations. While the southern peninsulas obviously experienced more climatic stability than regions frequented by glaciers or tundra-like conditions, they are also highly fragmented by mountain ranges and bordered by seas, creating highly heterogeneous landscapes that foster organismal diversification. In recent years, many researchers have built a strong argument for the role of additional refugia in central or unglaciated portions of both northern and eastern Europe for some groups of organisms (Stewart & Lister 2001). This has been particularly apparent for cold-tolerant fishes, such as *Cottus gobio*, brown trout *Salmo trutta* and European grayling *Thymallus thymallus*, for which more northern refugia may have been solely responsible for the recolonization of formerly glaciated regions in northern Europe after the Last Glacial Maximum (Volckaert *et al.* 2002; García-Marín *et al.* 1999; Weiss *et al.* 2000, 2002). This perspective has also been addressed for boreal-temperate plants such as in *Betula* (Palmé *et al.* 2003) and *Calluna* (Rendell & Ennos 2002). This growing appreciation for northern refugia calls into question the role of peninsular refugia in the recolonization of previously glaciated regions, implying that for some organisms, these southern populations may be better characterized as long-term relicts, rather than sources for future expansions or radiations. Nonetheless, the three peninsular refugia of southern Europe are centers of biodiversity and due to both natural fragmentation as well as anthropogenic influences, contain numerous threatened or endangered taxa and populations. In fact, the Mediterranean region as a whole is considered one of the 25 biodiversity hotspots on earth (Myers *et al.* 2000). While the species richness of southern Europe is well documented, research assimilating phylogeographic approaches demonstrate an underappreciated genealogical complexity of organisms within large refugial habitats. This is exemplified herein by the 'refugia-within-refugia' model (Chapter 5) demonstrating multiple highly divergent lineages for a number of species within one peninsula, or, for example in the wall lizard *Podarcis* sp., numerous undescribed taxa or cryptic species diversity (Harris & Sá-Sousa 2002; Pinho *et al.* in press).

Simple interpretations of high genetic diversity within southern refugia, however, have been challenged. In particular, the results have been shown to depend on how genetic diversity is quantified (Comps *et al.* 2001; Petit *et al.* 2003). This stems from the fact that different measures of genetic diversity, such as allelic richness, heterozygosity and population divergence are differentially influenced by demographic factors. Heterozygosity, for example, can be very high in hybrid zones formed by expanding lineages north of refuge areas, as expected when two formerly divergent lineages come into secondary contact. However, secondary contact zones within southern refugia have received much less attention, and may be characterized by extremely high pairwise population divergences and much more complex evolutionary dynamics. For example, the hybrid zone of the European rabbit *Oryctolagus cuniculus* in Iberia contains two lineages thought to have arisen in allopatry approximately two million years ago. This hybrid zone has been used as a model system to demonstrate highly differential modes of introgression across markers (Chapters 7 & 8), saturation of mutation spectra in microsatellites (Queney *et al.* 2001), and to suggest balancing selection in several protein loci (Ferrand *et al.*, unpublished data). Within the same geographic region, a hybrid zone of the endemic Schreiber's green lizard *Lacerta schreiberi* has revealed not only differential introgression across markers, but also recombinant haplotypes across relatively short genomic regions (Godinho *et al.* 2006). Such marker discordance in areas of secondary contact clearly reflects more complex and long-term interactions of fluctuating demography and selection than expected in postglacial hybrid zones. While the studies referenced above all derive from the Iberian Peninsula, it is expected that hybrid zones within refugia exist on other peninsulas, though they await future study.

Caveats, criticisms and challenges of phylogeographic approaches

Phylogeographic research continues to be dominated by studies utilizing single-locus uni-parentally inherited markers, namely mtDNA or cpDNA genes. There are numerous caveats to exclusively relying on organelle markers (reviewed in Ballard & Whitlock 2004), but their utility in many systems is unquestionable, and it is perhaps remarkable to note the number of initial inferences drawn from single marker phylogeographic studies that withstand extended analysis including more genetic markers as well as sampled individuals. In contrast, where sex-biased dispersal is strong, or interspecific hybridization and capture of an organelle genome has occurred, the single organelle gene may be completely misleading (Bernatchez *et al.* 1995; Melo-Ferreira *et al.* 2005). Regardless of the frequency of such events, the problem is that there is no way to predict the reliability of such single-locus data sets. If theory is followed, many gene genealogies must be sampled in order to have some degree of confidence that the history of an organismal lineage is being reasonably recovered. A counter-intuitive illustration of this point, especially for those trained in classical population genetics or ecological studies, is that the number of individuals needed to be sampled in a population in order to uncover the genealogical diversity is quite small, with a sample size of 10 sufficing for an approximate 90% probability of revealing the most ancestral haplotype (Hein *et al.* 2005). In contrast, the variance in coalescence trees reveals that the most efficient way of increasing the accuracy of inferences drawn from gene genealogies is to increase the number of independent (i.e. unlinked) loci sampled. An additional aspect, however, is the number of distinct demes existing across a species range. The best phylogeographic studies sample the entire range of the organism, and using some prior knowledge of an organism's genetic or phenotypic diversity, or predictions of how landscape fragmentation may have molded extant genetic structure, an attempt is made to provide sample coverage of potential major demes. More suspect or less convincing are phylogeographic studies that cover only small portions of a species range, or highly fragmented or opportunistic sampling of predicted or known demes, and then attempt to draw large-scale inferences. Such efforts suffer both from poor sampling and the uncertainties of a single-locus data set, and thus should be highly discouraged.

Another aspect of phylogeography that needs improvement is the approach often taken to estimate divergence times among populations or sister taxa, although this problem is certainly not specific to phylogeography. While the theoretical understanding that the coalescence of two lineages as measured with DNA sequence data will always predate the splitting of the actual populations has been with us for some time, many researchers ignore this difference and overestimate divergence, by as much as 50% or even more in highly structured populations (Edwards & Beerli 2000; Arbogast *et al.* 2002). For study-specific calibrations based on geological evidence this caveat may not be too important (but see Arbogast *et al.* 2002), but numerous studies still rely on universal clock rates or apply a calibration in one study across many taxa or wholly different time scales. Again, this problem is exacerbated when only a single locus is screened.

A central area of much debate is the application of statistical tools for making inferences in an explicit phylogeographic context. Phylogeographic analysis began with a fusion of monumental technical advances in the laboratory combined with hand-drawn intuitively satisfying genealogical networks and purely *ad hoc* inferences (Avise *et al.* 1979). Subsequently, various theoretical developments and statistical tools were applied to the analysis of genes in a phylogeographic context, such as pairwise mismatch analysis (Slatkin & Hudson 1991; Rogers & Harpending 1992); analysis of molecular variance, AMOVA (Excoffier *et al.* 1992), and several approaches to reconstructing haplotype networks, recently reviewed in Posada & Crandall (2001). However, the most comprehensive attempt to develop a systematic approach to phylogeographic analysis has been described in a series of publications, now best summarized as nested clade phylogeographic analysis (NCPA) (Templeton 1998, 2004). Some studies have pointed out deficiencies in the inferences drawn from NCPA involving specific historical scenarios or demographic events that lead to false inferences (Alexandrino *et al.* 2002; Paulo *et al.* 2002; Masta *et al.* 2003). These criticisms have led to either changes in the methodology, or explicit caveats of when the approach is more likely to fail (Templeton 2004). For example, range expansion is often not inferred even in systems where common sense or other information dictates that range expansion has occurred. However, no analysis can support range expansion when a genetic marker lacks the resolution to do so, as is often the case for mtDNA and recent (i.e. postglacial) range expansion. A much more sweeping criticism is that NCPA is *ad hoc*, non-statistical and not amenable to falsification (Knowles & Maddison 2002; Knowles 2004). This emerging debate has resulted in the coining of the term '*Statistical Phylogeography*' for alternative model-based approaches.

NCPA, however, does involve several explicitly statistical steps, the first of which is the reconstruction of a 95% parsimony network, and the second the construction of a nested design for testing genetic/geographic associations based on this network. This design is three-dimensional, incorporating nested spatial distances, genetic variation, and a qualitative temporal component based on the genealogy (the level of nesting). The most common alternative to such a design is a simple Mantel test on the global data set, an approach that produces only a single statistic, but moreover lacks a temporal component. The clear weakness of the nested statistical design in NCPA is that it rests wholly on the reliability of the haplotype network. Petit & Grivet (2002) further question the appropriateness of the permutational procedure that determines statistical significance, arguing that populations rather than individuals should be permutated (but see Templeton 2002a). There have also been other attempts to capture the temporal component of allelic data without using trees or networks (Pons & Petit 1996; Grivet & Petit 2002).

The most unorthodox component of NCPA is the use of a key to draw inferences from the correlation statistics. Such an approach can be considered *a posteriori* as opposed to *a priori* hypothesis testing, the common mode of natural scientists. However, the use of an inference key is not *ad hoc* and combined with the frequent conclusion that a data set lacks genetic resolution or adequate geographic sampling to draw an inference, NCPA inhibits rather than promotes *ad hoc* inferences. Templeton (2004) discusses in detail fundamental considerations, limitations and complementarity of '*a priori* and *a posteriori*' approaches to phylogeographic analysis. More broadly, any gene-tree based methodology, including NCPA, can be viewed as a graphical approach as opposed to a more traditional mathematical approach utilizing summary statistics, whereby both approaches clearly have their limitations (Hey & Machado 2003). Smouse (1998) concludes that both trees (bi-furcating) and networks (multi-furcating) have serious drawbacks, and researchers should rather spend more time analyzing the haplotype data with more straightforward statistical approaches.

'Statistical Phylogeography', as presented by Knowles (2004), proposes to steer the discipline into a model-driven, hypothesis-testing framework, which nonetheless considers some of the additional, at times non-biological information that may be important for recovering or drawing additional meaning to a phylogeographic scenario. A major concern of Knowles (2004) is an underappreciation of the stochastic variance inherent in gene genealogies (not reflected in a single haplotype network), which will lead to inaccurate or misleading interpretations and thus must be incorporated into the analytical approach. The challenge is to define *a priori* model-based predictions, which can be falsified in a statistical framework that incorporates the stochasticity of gene coalescence. Whereas recognition of coalescent theory has been implicitly integrated into phylogeographic interpretation since its onset (Avise *et al.* 1979) and used to justify inference frameworks in NCPA (e.g. Castelloe & Templeton 1994), it is not often explicitly or computationally integrated into hypothesis testing and the generation of statistical confidence. Thus, the advocacy for so-called 'Statistical Phylogeography' reflects several of the discipline's most common caveats and further parallels the prognosis that better integration of coalescent theory into phylogeographic analysis will not only improve our confidence in phylogeographic inferences, but also aid in the advancement of population genetics as a whole (Wakeley 2003).

Current trends and state-of-the-art (2001-present)

Despite the theoretical advantages of multi-locus approaches, the majority of current manuscripts in phylogeography is still based on single-locus uni-parentally inherited markers. Viewing the last 12 issues of *Molecular Ecology* (Oct. 2005-Nov. 2004), there were 42 manuscripts with the term 'phylogeography' in the title, among which 34 (81%) were based exclusively on organelle genes (mtDNA or cpDNA). Nonetheless, these single-locus studies improve upon past efforts in that more nucleotides are screened and sample coverage often includes continent-wide or circumpolar coverage (Alsos *et al.* 2005; Van Houdt *et al.* 2005). Additionally, due to concerns about heterogeneous substitution rates, many studies routinely use more than one organelle gene (11 of the 34 noted above). Coalescent simulations and the so-called 'gene-tree/population tree' approach are also applied to answer more discrete phylogeographic scenarios (e.g. one, two or several refugia) still using mtDNA data sets (Knowles 2001; Carstens *et al.* 2005) with particular subroutines within the modular software package Mesquite (Maddison & Maddison 2004).

The more important advance to true multi-locus phylogeographic studies, however, has been rather slow despite the optimistic prospects outlined in Hare (2001). Relevant theoretical issues include an underappreciation of the stochasticity of the coalescent process (Hudson & Turelli 2003), the statistical treatment of recombinants (Templeton 2004), and both the computational complexity and lack of empirical data on the among locus variance in substitution rates (Yang 1997; Arbogast *et al.* 2002). Practical hurdles involve locating nDNA regions with adequate variation, and resolving haplotypes from diploid genotypes. This latter problem can be approached with a variety of either statistical or experimental methods (reviewed in Zhang & Hewitt 2003). For a typical phylogeographic study reasonable throughput can be achieved through the combination of PCR-SSCP and subsample sequencing or initial PCR and allele-specific PCR (ARMS, Newton *et al.* 1989) to resolve the potential ambiguities of heterozygotes, which can be extreme when intron indels are present. For large data sets, probable haplotypes at a level of confidence adjustable to the study's demands can be estimated using maximum likelihood or Bayesian approaches (Excoffier and Slatkin 1995; Stephens *et al.* 2001)

Templeton (2002b) provides an example of a multi-locus phylogeographic study using 10 genes to argue for a more complex out of Africa colonization by humans, based on both NCPA and some coalescent simulations, though his conclusions are somewhat controversial. We must emphasize that the use of a single nuclear gene fragment as a token consideration of the nuclear genome for a broad-scale phylogeographic study may achieve very little. Furthermore, the application of multi-locus microsatellite surveys to phylogeographic studies may serve as a poor surrogate to nDNA haplotype data. The combined use of microsatellites and mtDNA in an animal study can control for extreme sex-biased dispersal or interspecific introgression, events that can make inferences based solely on mtDNA data extremely misleading. However, it is difficult to analyze microsatellite allele variation in a strictly phylogeographic context. This is because the available analytical techniques do not explicitly integrate a temporal perspective, and more importantly the high and variable substitution rate of repeat loci results in an unpredictable and non-linear relation between genetic divergence and time. This problem is magnified across deeper phylogeographic breaks, where homoplasy will lead to vastly underestimated divergence estimates among some, but not all pairwise population comparisons. One of the best examples of this is the demonstration of stationary distributions of microsatellite allele frequencies in a phylogeographic context with the European rabbit (Queney *et al.* 2001), an observation that has been further substantiated when it was shown that sex-biased dispersal is not prevalent in the system (Geraldes & Ferrand in press). Thus, while theorists bemoan the lack of true multi-locus phylogeographic studies, it is apparent that both theoretical and logistic constraints continue to limit such efforts.

One relatively recent approach, however, combines nDNA single-nucleotide polymorphisms (SNPs) with a linked microsatellite. Thus, one can take advantage of the high mutation rate of a microsatellite incorporated into haplotype data to age alleles in a phylogeographic context (Tishkoff *et al.* 2001). Such data can also be used with new software developed for more flexible coalescent simulations allowing incorporation of differing mutation rates and effective population sizes in estimating the divergence of populations (Hey & Nielsen 2004). This approach was recently used to evaluate the number of founders in the colonization of America by humans (Hey 2005) as well as to investigate the divergence times among Lake Malawi cichlids (Won *et al.* 2005). There are numerous opportunities to use such approaches to answer explicit questions within a phylogeographic framework.

An additional avenue of inquiry that is directly applicable to phylogeographic analysis involves attempts to incorporate geography into coalescent simulations. The development of the software DANCING TREES will give birth to an inference tool that explicitly incorporates the spatial component in the coalescent (S. Baird, personal communication). The incorporation of climatic and other landscape variables into simulated phylogeographic scenarios across, for example, postglacial time scales, has also been accomplished with the development of the software SPLATCHE (Currat *et al.* 2004), which has been applied to questions concerning the geographic origin of early modern humans (Ray *et al.* 2005). At finer scales, the link between environmental factors and population genetics is developing into a field of its own – landscape genetics – whereby the newest software allowing for such analysis is GENELAND (Guillot *et al.* 2005a,b).

Yet another theme that has been until most recently ignored in phylogeographic research is the effects of natural selection on the markers we choose to use, as well as its large-scale influence on genetic architecture and the adaptive landscape. One view is that estimates of ancient demography are best done with selectively neutral markers, and thus effort is expended to detect markers under selection in order that they be removed from particular analyses (Vitalis *et al.* 2001). More problematic is mtDNA, so often used as a putatively neutral marker in phylogeographic studies, but difficult to replace with a nuclear gene region with the same level of genetic structure. Recently, natural selection has been hypothesized to have played a major role in shaping the present-day diversity of mtDNA in humans (Elson *et al.* 2004). Thus, the concern that selection on the mtDNA molecule may have affected the phylogeographic structure in other organisms is increasing (Ballard & Whitlock 2004). For nuclear gene loci, methods to detect markers under selection have been developing (see reviews of Beaumont 2005 & Storz 2005) and under particular population models the detection of F_{ST} outlier loci (Beaumont & Nichols 1996) is currently gaining use. In contrast to viewing natural selection as problematic, phylogeographers can also use markers under selection in order to gain a better understanding of the evolutionary mechanisms that have accompanied or perhaps even directly influenced demographic declines and expansions at a large scale.

Consideration of a broader array of evolutionary mechanisms, and incorporation of selection, recombination and linkage disequilibrium into phylogeographic studies is easily feasible only for model organisms. In humans, the use of compound haplotypes such as SNPSTRs (Mountain *et al.* 2002) or HapSTRs (Hey *et al.* 2004) has been used to describe the signature of natural selection across broad geographic scales in the glucose-6-phosphate dehydrogenase and lactase loci (Tishkoff *et al.* 2001; Coelho *et al.* 2005). In *Drosophila melanogaster* screening of whole genomic regions across large geographic scales not only suggests that selection has played a major role in global demographic expansion, but also underscores the effect that both selection and genomically heterogeneous recombination rates have on the behavior of nuclear gene markers across phylogeographically relevant scales (Kauer *et al.* 2002, 2003; Catania *et al.* 2004). As the genetic knowledge base of nonmodel organisms continues to increase, approaches that incorporate genomewide information, and/or markers under selection will be more broadly applied to questions carried out within a phylogeographic context. It is at this point that a synergy between population genetics and phylogeography can be recognized that can pave the way for many future evolutionary research programs.

In fact, population geneticists who appreciate phylogeography recognize its utility in providing a framework within which more focused, evolutionary genetic questions can be pursued. Thus, the study system of the European rabbit in the Iberian Peninsula presents two distinct phylogroups that meet in a contact zone within a long-term refugial environment, providing a setting to study differential rates of introgression, selection, drift and migration across multiple genetic markers over significant periods of time (Chapter 8; Queney *et al.* 2001; Geraldes & Ferrand in press). Research on the evolutionary history of the golden-striped salamander *Chioglossa lusitanica* began with the study of allozyme and mtDNA variation across the entire species range but has more recently focused on the contact zone of two major phylogroups, revealing seemingly discordant patterns among loci, fine-scaled gene flow and vicariance (Chapter 6; Sequeira *et al.* 2005). These data have been coupled to climatic modeling of suitable habitat across relevant phylogeographic times scales, supporting the hypothesized vicariance scenario as well as allowing future predictions (Teixeira *et al.* 2001; Teixeira & Arntzen 2002). While using only a single nDNA locus, Godinho *et al.* (2006) demonstrated that a nuclear genealogy involving recombinant haplotypes can aid in the investigation of ancient demography and admixture dynamics within the framework of an mtDNA-defined phylogeographic structure. On a larger scale, the combination of comparative phylogeography and paleoclimatic modeling was used to verify the location of refugia and predict species responses to future environmental changes (Hugall *et al.* 2002). Drawing on one of the largest sample collections of a wild organism, Petit *et al.* (2002a,b) integrate fossil tree pollen data to verify the location of glacial refugia and increase both the resolution and confidence of prior phylogeographic inferences that were made based on organelle genetics alone. This perspective of considering how biota respond to climatic change, embedded in a broad-scale phylogeographic framework has recently led to the consideration of the conservation relevance of the rear (i.e. southern) edge of species distributions, in contrast to the frequent focus on the postglacial expanding edge (Hampe & Petit 2005), further illuminating the importance of understanding the role of southern peninsulas for the conservation of European biodiversity (Taberlet & Cheddadi 2002). These broad-scale environmental perspectives invariably fuel our attention to the interaction of genetics, landscape dynamics and habitat requirements at the organismic level, forming the basis of landscape genetic approaches that integrate GIS-based modeling and the estimation of traditional population genetic parameters (Manel *et al.* 2003; Spear *et al.* 2005). Thus, extending on the simplistic dichotomous characterization of postglacial expansion stemming from dispersal from refugia into formerly uninhabited environments, landscape genetics seeks to understand both dispersal and population differentiation in terms of the landscape variables that promote or inhibit individual movement and colonization (Geffen *et al.* 2004; Arnaud 2003).

In summary, current trends directly or indirectly related to phylogeography seek to expand upon the integration of an array of data external to genetic architecture, in order to understand the true nature and importance of spatial genetic structure to both the evolutionary legacy and future of an organism. At the genomic level, developments are almost all directed towards improving upon the legitimacy of the genetic signal in reflecting the true history of populations, by increasing genomic coverage and incorporating coalescent theory and statistical inference.

Prospects and predictions for European phylogeography

Phylogeography is not a discipline without detractors. Recently, the entire paradigm of the importance of phylogeographic structure in Europe was challenged. Based on a study of ancient DNA, which revealed a lack of phylogeographic structure in mammals of the last interglacial, Hofreiter *et al.* (2004) suggest that the current structure of extant animals is in a transitory phase. Despite the continued reliance on initial patterns seen at organelle genes, for some organisms, like forest trees, these genes may reveal a very limited view on the geographic structure of the genome due to limited dispersal, especially for seeds, as well as insufficient mutation rates. Considering the consistent discordance between organelle and nDNA markers in animal studies in contact zones (particularly older hybrid zones) this contention may be relevant for other taxa as well. Nonetheless, the caveats are known, and the phylogeographer's tool chest has grown significantly so there is little doubt that our understanding of the pattern and importance of genealogical distributions in time and space within species will also expand. And, if current literature is any indication, the phylogeographic perspective will remain an important baseline component of evolutionary research.

While obvious to some, young researchers engaging in phylogeographic research should be aware that a large percentage of the approaches applied in the field were first developed, and continue to be developed in the highly competitive field of human genetics. If this field is any indication of what will become of our understanding of phylogeographic structure and inferences on paleodemography for non-human organisms, then much controversy lies ahead. Increasingly sophisticated models of the out-of-Africa expansion of modern humans continue to fuel debate on the level of ancient population size, substructure, selection and migration and the relevance of these factors to the observed genomic architecture (Harding & McVean 2004). Nonetheless, human studies are fueling the development of coalescent based approaches, integration of more complex demographic scenarios such as metapopulation dynamics, and examination of gene expression and selection regimes in time and space for phenotypic traits of interest, approaches that are paving the way for the evolutionary biologist interested in applying cutting edge technology to the evolutionary study of non-human species.

As noted by John Avise at the first European conference on Phylogeography in Vairão, Portugal (2002), European phylogeographers have done a remarkable job in putting together large-scale data sets encompassing whole species ranges and most notably, have often resolved questions that would have been difficult or impossible to answer without a phylogeographic perspective. The description of the phylogeographic structure of an organismal lineage is often just the starting point before delving into more specific questions or seeking statistical support for a particular inference or historical scenario. Nevertheless, obtaining an evolutionary framework of a species, in space and time, is an irreplaceable milestone in understanding that can affect both the direction and interpretation of genetic data at any scale of inquiry.

Acknowledgements

Nuno Ferrand was supported by a sabbatical grant (SFRH/BSAB/318/2002) from Fundação para a Ciência e a Tecnologia (FCT). Much of the results described in this chapter were supported, in part, by FCT research projects, including POCTI/BIA/11174/98, POCTI/BSE/40987/2001 and POCTI/BSE/34151/2000. Nuno Ferrand and Steven Weiss want to acknowledge all the members of CIBIO for their enthusiasm and dedication, as well as for providing an excellent and inspiring atmosphere. We want to thank Rémy Petit for his comments on an earlier draft of this paper.

References

Alexandrino J, Arntzen JW, Ferrand N (2002) Nested clade analysis and the genetic evidence for population expansion in the phylogeography of the golden-striped salamander, *Chioglossa lusitanica* (Amphibia : Urodela). ***Heredity* 88**, 66-74.

Alsos IG, Engelskjon T, Gielly L, Taberlet P, Brochman C (2005) Impact of ice ages on circumpolar molecular diversity: insights from an ecological key species. ***Molecular Ecology* 14**, 2739-2754.

Arbogast BS, Edwards SV, Wakeley J, Beerli P, Slowinski JB (2002) Estimating divergence times from molecular data on phylogenetic and population genetic timescales. ***Annual Review in Ecology and Systematics* 33**, 707-740.

Arnaud JF (2003) Metapopulation genetic structure and migration pathways in the land snail Helix aspersa: influence of landscape heterogeneity. ***Landscape Ecology* 18**, 333-346.

Avise JC (2000) ***Phylogeography: The History and Formation of Species***. Harvard University Press, Cambridge.

Avise JC, Lansman RA, Shade RO (1979) The use of restriction endonucleases to measure mitochondrial DNA sequence relatedness in natural populations. I. Population structure and evolution of the genus Peromyscus. ***Genetics* 92**, 279-295.

Ballard JWO, Whitlock M (2004) The incomplete natural history of mitochondria. ***Molecular Ecology* 13**, 729-744.

Beaumont M (2005) Adaptation and speciation: what can Fst tell us? ***Trends in Ecology and Evolution* 20**, 435-440

Beaumont M, Nichols RA (1996) Evaluating loci for use in the genetic analysis of population structure. ***Proceedings of the Royal Society London, B* 263**, 1619-1626.

Bernatchez L, Glémet H, Wilson C, Danzman RG (1995) Fixation of introgressed mitochondrial genome of arctic charr (*Salvelinus alpinus* L.) in an allopatric population of brook charr (*Salvelinus fontinalis* Mitchill). ***Canadian Journal of Fisheries and Aquatic Sciences* 52**, 179-185.

Carstens BC, Stevenson AL, Degenhardt JD, Sullivan J (2005) Testing nested phylogenetic and phylogeographic hypotheses in the *Plethodon vandykei* species group. ***Systematic Biology* 53**, 781-792.

Castelloe J, Templeton AR (1994) Root probabilities for intraspecfic gene trees under neutral coalescent theory. ***Molecular Evolution and Phylogenetics* 3**, 102-113.

Catania F, Kauer MO, Daborn PJ, Yen JL, Ffrench-Constant RH, Schlötterer C (2004) World-wide survey of an Accord insertion and its association with DDT resistance in *Drosophila melanogaster*. ***Molecular Ecology* 13**, 2491-2504.

Coelho M, Luiselli, D, Bertorelle G, Lopes AI, Seixas S, Destro-Bisol G, Rocha J (2005). Microsatellite variation and evolution of human lactase persistence. ***Human Genetics* 117**, 329-339.

Comps B, Gömöry D, Letouzey J, Thiebaut B, Petit RJ (2001) Diverging trends between heterozygosity and allelic richness during postglacial colonization in the European beech. ***Genetics* 157**, 389-397.

Currat M, Ray N, Excoffier L (2004) SPLATCHE: a program to simulate genetic diversity taking into account environmental heterogeneity. ***Molecular Ecology Notes* 4**, 139-142.

Edwards SV, Beerli P (2000) Gene divergence, population divergence, and the variance in coalescence time in phylogeographic studies. ***Evolution* 54**, 1839-1854.

Elson JL, Turnbull DM, Howell N (2004) Comparative genomics and the evolution of human mitochondrial DNA: assessing the effects of selection. ***American Journal of Human Genetics* 74**, 229-238.

Excoffier L, Slatkin M (1995) Maximum-likelihood estimation of molecular haplotype frequencies in a diploid population. ***Molecular Biology and Evolution* 12**, 921-927.

Excoffier L, Smouse PE, Quattro JM (1992) Analysis of molecular variance inferred from metric distances among DNA haplotypes: application to human mitochondrial DNA restriction data. ***Genetics* 131**, 479-491.

Garcia-Marin JL, Utter FM, Pla C (1999) Postglacial colonization of brown trout in Europe based on distribution of allozyme variants. ***Heredity* 82**, 46-56.

Geffen E, Anderson MJ, Wayne RK (2004) Climate and habitat barriers to dispersal in the highly mobile grey wolf. ***Molecular Ecology* 13**, 2481-2490.

Geraldes A, Ferrand N (in press) A 7-bp insertion 3' to the HGM box suggests the duplication and concerted evolution of the rabbit SRY gene. ***Genetics Selection Evolution***.

Godinho R, Mendonça B, Crespo EG, Ferrand N (2006) Genealogy of the nuclear β-fibrinogen locus in a highly structured lizard species: comparison with mtDNA and evidence for intragenic recombination in the hybrid zone. ***Heredity***, doi:10.1038/sj.hdy.6800823.

Grivet D, Petit RJ (2002) Phylogeography of the common ivy (*Hedera* sp.) in Europe: genetic differentiation through space and time. ***Molecular Ecology* 11**, 1351-1362.

Guillot G, Mortier F, Estoup A (2005a) Geneland: a program for landscape genetics. ***Molecular Ecology Notes* 5**, 703-711.

Guillot G, Estoup A, Mortier F, Cosson JF (2005b) A spatial statistical model for landscape genetics. ***Genetics* 170**, 1261-1280.

Hampe A, Petit RJ (2005) Conserving biodiversity under climate change: the rear edge matters. ***Ecology Letters* 8**, 461-467.

Harding RM, McVean G (2004) A structured ancestral population for the evolution of modern humans. ***Current Opinion in Genetics and Development* 14**, 667-674.

Hare MP (2001) Prospects for nuclear gene phylogeography. ***Trends in Ecology and Evolution*** **16**, 700-706.

Harris DJ, Sá-Sousa P (2002) Molecular Phylogenetics of Iberian Wall lizards (*Podarcis*): is *Podarcis hispanica* a species complex? ***Molecular Phylogenetics and Evolution*** **23**, 75-81.

Hein J, Schierup MH, Wiuf C (2005) ***Gene Genealogies, Variation and Evolution: A Primer in Coalescent Theory***. Oxford University Press, Oxford.

Hewitt, GM (1996) Some genetic consequences of ice ages, and their role in divergence and speciation. ***Biological Journal of the Linnean Society 58***, 247-276.

Hewitt, GM (1999) Post-glacial recolonization of European biota. ***Biological Journal of the Linnean Society*** **68**, 87-112.

Hewitt, GM (2000) The genetic legacy of the Quaternary ice ages. ***Nature*** **405**, 907-913.

Hey J (2005) On the number of New world founders: a population genetic portrait of the peopling of the Americas. ***PloS Biology*** **3**, 1-11.

Hey J, Machado CA (2003) The study of structured populations--new hope for a difficult and divided science. ***Nature Reviews Genetics*** **4**, 535-543.

Hey J, Nielsen R (2004) Multilocus methods for estimating population sizes, migration rates and divergence time, with applications to the divergence time of *Drosophila pseudoobscura* and *D. persimilis*. ***Genetics*** **167**, 747-760.

Hey J, Won Y-J, Sivasundar A, Nielson R, Markert JA (2004) Using nuclear haplotypes with microsatellites to study gene flow between recently separated populations. ***Molecular Ecology*** **13**, 909-919.

Hofreiter M, Serre D, Rohland N, Rabeder G, Nagel D, Conrad N, Münzel S, Pääbo S. (2004) Lack of phylogeography in European mammals before the last glaciation. ***Proceedings of the National Academy of Sciences, USA*** **101**, 12963-12968.

Hudson RR, Turelli M (2003) Stochasticity overrules the 'Three-Times Rule': genetic drift, genetic draft, and coalescence times for nuclear loci versus mitochondrial DNA. ***Evolution*** **57**, 182-190.

Hugall A, Moritz C, Moussalli A, Stanisic J (2002). Reconciling paleodistribution models and comparative phylogeography in the Wet Tropic rainforest land snail *Gnarosophia bellendenkerensis* (Brazier 1875). ***Proceedings of the National Academy of Sciences, USA*** **99**, 6112-6117.

Kauer M, Zangerl B, Dieringer D, Schlötterer C (2002) Background selection and hitchhiking - an explanation for the contrasting chromosomal patterns of microsatellite variability in African and non-African populations of *D. melanogaster*. ***Genetics*** **160**, 247-256.

Kauer MO, Dieringer D, Schlötterer C (2003) A microsatellite variability screen for positive selection associated with the 'out of Africa' habitat expansion of *Drosophila melanogaster*. ***Genetics*** **165**, 1137-1148.

Knowles LL (2001) Genealogical portraits of speciation in montane grasshoppers (genus *Melanoplus*) from the sky islands of the Rocky Mountains. ***Proceedings of the Royal Society London, B*** **268**, 1-6.

Knowles LL (2004) The burgeoning field of statistical phylogeography. ***Journal of Evolutionary Biology*** **17**, 1-10.

Knowles LL, Maddison WP (2002) Statistical phylogeography. ***Molecular Ecology*** **11**, 2623-2635.

Maddison WP, Maddison DR (2004) ***Mesquite: a Modular System for Evolutionary Analysis, version 1.01.*** available at http://mesquiteproject.org.

Manel S, Schwartz MK, Luikart G, Taberlet P (2003) Landscape genetics: combining landscape ecology and population genetics. ***Trends in Ecology & Evolution*** **18**, 189-197.

Masta SE, Laurent NM, Routman EJ (2003) Population genetic structure of the toad *Bufo woodhousii*: an empirical assessment of the effects of haplotype extinction on nested cladistic analysis. ***Molecular Ecology*** **12**,1541-1454.

Melo-Ferreira J, Boursot P, Suchentrunk F, Ferrand N, Alves PC (2005) Invasion from the cold past: extensive introgression of mountain hare (*Lepus timidus*) mitochondrial DNA into three other hare species in northern Iberia. ***Molecular Ecology*** **14**, 2459-2464.

Mountain JL, Knight A, Jobin M, Gignoux C, Miller A, Lin AA, Underhill PA (2002) SNPSTRs: empirically derived, rapidly typed, autosomal haplotypes for inference of population history and mutational processes. ***Genome Research*** **12**, 1766-1772.

Myers N, Mittermeier RA, Mittermeier CG, da Fonseca GAB, Jennifer K (2000) Biodiversity hotspots for conservation priorities. ***Nature*** **403**, 853-858.

Newton CR, Graham A, Heptinsall LE, Powell SJ, Summers C, Kalsheker N, Smith JC, Markham AF (1989) Analysis of any point mutation in DNA: the amplification refractory mutation system (ARMS). ***Nucleic Acids Research*** **17**, 2503-2516.

Palmé AE, Su Q, Rautenberg A, Manni F, Lascoux M (2003) Postglacial recolonization and cpDNA variation of silver birch *Betula pendula*. ***Molecular Ecology*** **12**, 201-212

Paulo OS, Jordan WC, Bruford MW, Nichols RA (2002) Using nested clade analysis to assess the history of colonization and the persistence of populations of an Iberian Lizard. ***Molecular Ecology*** **11**, 809-820.

Petit RJ, Grivet D (2002) Optimal randomization strategies when testing the existence of a phylogeographic structure. ***Genetics*** **161**, 469-471.

Petit RJ, Brewer S, Bordács S, Burg K, Cheddadi R, Coart E, Cottrell J, Czaikl UM, van Dam B, Deans JD, Espinel S, Fineschi S, Finkeldey R, Glaz I, Goicoechea PG, Jensen JS, König AO, Lowe AJ, Madsen SF, Mátyás G, Munro RC, Popescu F, Slade D, Tabbener H, deVries SGM, Ziegenhagen B, de Beaulieu J-L, Kremer A (2002a) Identification of refugia and post-glacial colonisation routes of European white oaks based on chloroplast DNA and fossil pollen evidence. ***Forest Ecology and Management*** **156**, 49-74.

Petit RJ, Csaikl UM, Bordács S, Burg K, Coart E, Cottrell J, van Dam B, Deans JD, Dumolin-Lapègue S, Fineschi S, Finkeldey R, Gillies A, Glaz I, Goicoechea PG, Jensen JS, König AO, Lowe AJ, Madsen SF, Mátyás G, Munro RC, Olalde M, Pemonge M, Popescu F, Slade D, Tabbener H, Tauchini D, Vries SGM, Ziegenhagen B, Kremer A (2002b) Chloroplast DNA variation in European white oaks: phylogeography and patterns of diversity based on data from over 2600 populations. ***Forest Ecology and Management*** **156**, 5-26.

Petit JR, Aguinagalde I, Beaulieu J-L, Bittkau C, Brewer S, Cheddadi R, Ennos R, Fineschi S, Grivet D, Lascoux M, Mohanty A, Müller-Starck G, Demesure-Musch B, Palme A, Martin JP, Rendell S, Vendramin GG (2003) Glacial refugia: hotspots but not melting pots of genetic diversity. ***Science*** **300**, 1563-1565.

Pinho C, Ferrand N, Harris DJ (in press) Reexamination of the Iberian and North African Podarcis (Squamata: Lacertidae) phylogeny based on increased mitochondrial DNA sequencing. ***Molecular Phylogenetics and Evolution***.

Pons O, Petit RJ (1996) Measuring and testing genetic differentiation with ordered versus unordered alleles. ***Genetics* 144**, 1237-1245.

Posada D and Crandall KA (2001) Intraspecific phylogenetics: trees grafting into networks. ***Trends in Ecology and Evolution* 16**, 37-45.

Queney G, Ferrand N, Weiss S, Mougel F, Monnerot M (2001) Stationary distributions of microsatellite loci between divergent population groups of the European rabbit (*Oryctolagus cuniculus*). ***Molecular Biology and Evolution* 18**, 2169-2178.

Ray N, Currat M, Berthier P, Excoffier L (2005) Recovering the geographic origin of early modern humans by realistic and spatially explicit simulations. ***Genome Research* 15**, 1161-1167.

Rendell S, Ennos RA (2002) Chloroplast DNA diversity in *Calluna vulgaris* (heather) populations in Europe. ***Molecular Ecology* 11**, 69-78.

Rogers J, Harpending H (1992) Population growth makes waves in the distribution of pairwise genetic differences. ***Molecular Biology and Evolution* 9**, 552-569.

Sequeira F, Alexandrino J, Rocha R, Arntzen JW, Ferrand N (2005) Genetic exchange across a hybrid zone within the Iberian endemic golden-striped salamander, *Chloglossa lusitanica*. ***Molecular Ecology* 14**, 245-254.

Slatkin M, Hudson RR (1991) Pairwise comparison for mitochondrial DNA sequences in stable and exponentially growing populations. ***Genetics* 129**, 555-562.

Smouse PE (1998) To tree or not to tree. ***Molecular Ecology* 7**, 399-412.

Spear SF, Peterson CR, Matocq MD, Storfer A (2005) Landscape genetics of the blotched tiger salamander (*Ambystoma tigrinum melanostrictum*). ***Molecular Ecology* 14**, 2553-2564.

Stephens M, Smith NJ, Donnelly P (2001) A new statistical method for haplotype reconstruction from population data. ***American Journal Human Genetics* 68**, 978-989.

Stewart JR, Lister AM (2001) Cryptic northern refugia and the origins of the modern biota. ***Trends in Ecology and Evolution* 16**, 608–613.

Storz (2005) Using genome scans of DNA polymorphism to infer adaptive population divergence. ***Molecular Ecology* 14**, 671-688.

Taberlet P, Cheddadi R (2002) Quaternary refugia and persistence of biodiversity. ***Science* 297**, 2009-2010.

Teixeira J, Arntzen JW (2002) Potential impact of climate warming on the distribution of the Golden-striped salamander, *Chioglossa lusitanica*, on the Iberian Peninsula. ***Biodiversity and Conservation* 11**, 2167-2176.

Teixeira J, Ferrand N, Arntzen JW (2001) Biogeography of the golden-striped salamander, *Chioglossa lusitanica*: a field survey and spatial modelling approach. ***Ecography* 24**, 618-624.

Templeton AR (1998) Nested clade analyses of phylogeographic data: testing hypotheses about gene flow and population history. ***Molecular Ecology* 7**, 381-397.

Templeton AR (2002a) 'Optimal' randomization strategies when testing the existence of a phylogeographic structure: a reply to Petit and Grivet. ***Genetics* 161**, 473-475

Templeton AR (2002b) Out of Africa again and again. ***Nature* 416**, 45–51.

Templeton AR (2004) Statistical phylogeography: methods of evaluating and minimizing inference errors. ***Molecular Ecology* 13**, 789-809.

Tishkoff SA, Varkonyi R, Cahinhian N, Abbes S, Argyropoulos G, Destro-Bisol G, Drousioutou A, Danbgerfield B, Lefranc G, Loiselet J, Piro A, Stoneking M, Tagarelli A, Tagarelli G, Touma EH, Williams SM, Clark AG (2001) Haplotype diversity and linkage disequilibrium at human G6PD: recent origin of alleles that confer malarial resistance. ***Science*** **293**, 455-462.

Van Houdt JKJ, Cleyn DE, Perretti A, Volckaert AM (2005) A mitogenic view on the evolutionary history fo the Holarctic freshwater gadoid, burbot (*Lota lota*). ***Molecular Ecology*** **14**, 2445-2457.

Vitalis R, Dawson K, Boursot P (2001) Interpretation of variation across marker loci as evidence of selection. ***Genetics*** **158**, 1811-1823.

Volckaert FAM, Hänfling B, Hellmans B, Carvalho GR (2002) Timing of the population dynamics of bullhead *Cottus gobio* (Teleostei: Cottidae) during the Pleistocene. ***Journal of Evolutionary Biology*** **15**, 930-944.

Wakeley J (2004) Inferences about the structure and history of populations: coalescents and intraspecific phylogeography. In: ***The Evolution of Population Biology*** (eds. Singh RS, Uyenoyama MK), pp. 193-215. Cambridge University Press, Cambridge.

Weiss S, Antunes A, Schlötterer C, Alexandrino P (2000) Mitochondrial haplotype diversity among Portuguese brown trout *Salmo trutta L.* populations: relevance to the post-Pleistocene recolonization of northern Europe. ***Molecular Ecology*** **9**, 691-698.

Weiss S, Persat H, Eppe R, Schlötterer C, Ublein F (2002) Complex patterns of colonization and refugia revealed for European grayling *Thymallus thymallus*, based on complete sequencing of the mtDNA control region. ***Molecular Ecology*** **11**, 1393-1407.

Won Y-J, Sivasundar A, Wang Y, Hey J (2005) On the origin of Lake Malawi cichlids: a population genetic analysis of divergence. ***Proceedings of the National Academy of Sciences, USA*** **102**, 6581-6586.

Yang Z (1997) On the estimation of ancestral population sizes of modern humans. ***Genetical Research*** **69**, 111-116.

Zhang DX, Hewitt GM (2003) Nuclear DNA analysis in genetic studies of populations: practice, problems and prospects. ***Molecular Ecology*** **12**, 563-584.

Index

MIX
Papier aus verantwortungsvollen Quellen
Paper from responsible sources
FSC® C105338

If you have any concerns about our products,
you can contact us on
ProductSafety@springernature.com

In case Publisher is established outside the EU,
the EU authorized representative is:
Springer Nature Customer Service Center GmbH
Europaplatz 3, 69115 Heidelberg, Germany

Printed by Libri Plureos GmbH
in Hamburg, Germany